Springer Series in Statistics

Springer Series in Statistics

Alho/Spencer: Statistical Demography and Forecasting.
Andersen/Borgan/Gill/Keiding: Statistical Models Based on Counting Processes.
Atkinson/Riani: Robust Diagnostic Regression Analysis.
Atkinson/Riani/Ceriloi: Exploring Multivariate Data with the Forward Search.
Berger: Statistical Decision Theory and Bayesian Analysis, 2nd edition.
Borg/Groenen: Modern Multidimensional Scaling: Theory and Applications, 2nd edition.
Brockwell/Davis: Time Series: Theory and Methods, 2nd edition.
Bucklew: Introduction to Rare Event Simulation.
Cappé/Moulines/Rydén: Inference in Hidden Markov Models.
Chan/Tong: Chaos: A Statistical Perspective.
Chen/Shao/Ibrahim: Monte Carlo Methods in Bayesian Computation.
Coles: An Introduction to Statistical Modeling of Extreme Values.
Devroye/Lugosi: Combinatorial Methods in Density Estimation.
Diggle/Ribeiro: Model-based Geostatistics.
Dudoit/Van der Laan: Multiple Testing Procedures with Applications to Genomics.
Efromovich: Nonparametric Curve Estimation: Methods, Theory, and Applications.
Eggermont/LaRiccia: Maximum Penalized Likelihood Estimation, Volume I: Density Estimation.
Fahrmeir/Tutz: Multivariate Statistical Modeling Based on Generalized Linear Models, 2nd edition.
Fan/Yao: Nonlinear Time Series: Nonparametric and Parametric Methods.
Ferraty/Vieu: Nonparametric Functional Data Analysis: Theory and Practice.
Ferreira/Lee: Multiscale Modeling: A Bayesian Perspective.
Fienberg/Hoaglin: Selected Papers of Frederick Mosteller.
Frühwirth-Schnatter: Finite Mixture and Markov Switching Models.
Ghosh/Ramamoorthi: Bayesian Nonparametrics.
Glaz/Naus/Wallenstein: Scan Statistics.
Good: Permutation Tests: Parametric and Bootstrap Tests of Hypotheses, 3rd edition.
Gouriéroux: ARCH Models and Financial Applications.
Gu: Smoothing Spline ANOVA Models.
Györfi/Kohler/Krzyżak/Walk: A Distribution-Free Theory of Nonparametric Regression.
Haberman: Advanced Statistics, Volume I: Description of Populations.
Hall: The Bootstrap and Edgeworth Expansion.
Härdle: Smoothing Techniques: With Implementation in S.
Harrell: Regression Modeling Strategies: With Applications to Linear Models, Logistic Regression, and Survival Analysis.
Hart: Nonparametric Smoothing and Lack-of-Fit Tests.
Hastie/Tibshirani/Friedman: The Elements of Statistical Learning: Data Mining, Inference, and Prediction.
Hedayat/Sloane/Stufken: Orthogonal Arrays: Theory and Applications.
Heyde: Quasi-Likelihood and its Application: A General Approach to Optimal Parameter Estimation.
Huet/Bouvier/Poursat/Jolivet: Statistical Tools for Nonlinear Regression: A Practical Guide with S-PLUS and R Examples, 2nd edition.
Ibrahim/Chen/Sinha: Bayesian Survival Analysis.
Jiang: Linear and Generalized Linear Mixed Models and Their Applications.
Jolliffe: Principal Component Analysis, 2nd edition.
Knottnerus: Sample Survey Theory: Some Pythagorean Perspectives.
Konishi/Kitagawa: Information Criteria and Statistical Modeling.

(continued after index)

Sadanori Konishi
Genshiro Kitagawa

Information Criteria and Statistical Modeling

Sadanori Konishi
Faculty of Mathematics
Kyushu University
6-10-1 Hakozaki, Higashi-ku
Fukuoka 812-8581
Japan
konishi@math.kyushu-u.ac.jp

Genshiro Kitagawa
The Institute of Statistical Mathematics
4-6-7 Minami-Azabu, Minato-ku
Tokyo 106-8569
Japan
kitagawa@ism.ac.jp

ISBN: 978-1-4419-2456-8 e-ISBN: 978-0-387-71887-3

Printed on acid-free paper

9 8 7 6 5 4 3 2 1

springer.com

Preface

Statistical modeling is a critical tool in scientific research. Statistical models are used to understand phenomena with uncertainty, to determine the structure of complex systems, and to control such systems as well as to make reliable predictions in various natural and social science fields. The objective of statistical analysis is to express the information contained in the data of the phenomenon and system under consideration. This information can be expressed in an understandable form using a statistical model. A model also allows inferences to be made about unknown aspects of stochastic phenomena and to help reveal causal relationships. In practice, model selection and evaluation are central issues, and a crucial aspect is selecting the most appropriate model from a set of candidate models.

In the information-theoretic approach advocated by Akaike (1973, 1974), the Kullback–Leibler (1951) information discrepancy is considered as the basic criterion for evaluating the goodness of a model as an approximation to the true distribution that generates the data. The Akaike information criterion (AIC) was derived as an asymptotic approximate estimate of the Kullback–Leibler information discrepancy and provides a useful tool for evaluating models estimated by the maximum likelihood method. Numerous successful applications of the AIC in statistical sciences have been reported [see, e.g., Akaike and Kitagawa (1998) and Bozdogan (1994)]. In practice, the Bayesian information criterion (BIC) proposed by Schwarz (1978) is also widely used as a model selection criterion. The BIC is based on Bayesian probability and can be applied to models estimated by the maximum likelihood method.

The wide availability of fast and inexpensive computers enables the construction of various types of nonlinear models for analyzing data with complex structure. Nonlinear statistical modeling has received considerable attention in various fields of research, such as statistical science, information science, computer science, engineering, and artificial intelligence. Considerable effort has been made in establishing practical methods of modeling complex structures of stochastic phenomena. Realistic models for complex nonlinear phenomena are generally characterized by a large number of parameters. Since the maximum likelihood method yields meaningless or unstable parameter estimates and leads to overfitting, such models are usually estimated by such methods as the maximum penalized likelihood method [Good and Gaskins (1971), Green and Silverman (1994)] or the Bayes approach. With the development of these flexible modeling techniques, it has become necessary to develop model selection and evaluation criteria for models estimated by methods other than the maximum likelihood method, relaxing the assumptions imposed on the AIC and BIC.

One of the main objectives of this book is to provide comprehensive explanations of the concepts and derivations of the AIC, BIC, and related criteria, together with a wide range of practical examples of model selection and evaluation criteria. A secondary objective is to provide a theoretical basis for the analysis and extension of information criteria via a statistical functional approach. A generalized information criterion (GIC) and a bootstrap information criterion are presented, which provide unified tools for modeling and model evaluation for a diverse range of models, including various types of nonlinear models and model estimation procedures such as robust estimation, the maximum penalized likelihood method and a Bayesian approach. A general framework for constructing the BIC is also described.

In Chapter 1, the basic concepts of statistical modeling are discussed. In Chapter 2, models are presented that express the mechanism of the occurrence of stochastic phenomena. Chapter 3, the central part of this book, explains the basic ideas of model evaluation and presents the definition and derivation of the AIC, in both its theoretical and practical aspects, together with a wide range of practical applications. Chapter 4 presents various examples of statistical modeling based on the AIC. Chapter 5 presents a unified information-theoretic approach to statistical model selection and evaluation problems in terms of a statistical functional and introduces the GIC [Konishi and Kitagawa (1996)] for the evaluation of a broad class of models, including models estimated by robust procedures, maximum penalized likelihood methods, and the Bayes approach. In Chapter 6, the GIC is illustrated through nonlinear statistical modeling in regression and discriminant analyses. Chapter 7 presents the derivation of the GIC and investigates its asymptotic properties, along with some theoretical and numerical improvements. Chapter 8 is devoted to the bootstrap version of information criteria, including the variance reduction technique that substantially reduces the variance associated with a Monte Carlo simulation. In Chapter 9, the Bayesian approach to model evaluation, such as the BIC, ABIC [Akaike (1980b)] and the predictive information criterion [Kitagawa (1997)] are discussed. The BIC is also extended such that it can be applied to the evaluation of models estimated by the method of regularization. Finally, in Chapter 10, several model selection and evaluation criteria such as cross-validation, generalized cross-validation, final prediction error (FPE), Mallows' C_p, the Hannan–Quinn criterion, and ICOMP are introduced as related topics.

We would like to acknowledge the many people who contributed to the preparation and completion of this book. In particular, we would like to acknowledge with our sincere thanks Hirotugu Akaike, from whom we have learned so much about the seminal ideas of statistical modeling.

We have been greatly influenced through discussions with Z. D. Bai, H. Bozdogan, D. F. Findley, Y. Fujikoshi, W. Gersch, A. K. Gupta, T. Higuchi, M. Ichikawa, S. Imoto, M. Ishiguro, N. Matsumoto, Y. Maesono, N. Nakamura, R. Nishii, Y. Ogata, K. Ohtsu, C. R. Rao, Y. Sakamoto, R. Shibata, M. S. Srivastava, T. Takanami, K. Tanabe, M. Uchida, N. Yoshida, T. Yanagawa, and Y. Wu.

We are grateful to three anonymous reviewers for comments and suggestions that allowed us to improve the original manuscript. Y. Araki, T. Fujii, S. Kawano, M. Kayano, H. Masuda, H. Matsui, Y. Ninomiya, Y. Nonaka, and Y. Tanokura read parts of the manuscript and offered helpful suggestions. We would especially like to express our gratitude to D. F. Findley for his previous reading of this manuscript and his constructive comments. We are also deeply thankful to S. Ono for her help in preparing the manuscript by LaTeX. John Kimmel patiently encouraged and supported us throughout the final preparation of this book. We express our sincere thanks to all of these people.

Sadanori Konishi
Genshiro Kitagawa

Fukuoka and Tokyo, Japan
February 2007

Contents

1 Concept of Statistical Modeling 1
- 1.1 Role of Statistical Models 1
  - 1.1.1 Description of Stochastic Structures by Statistical Models 1
  - 1.1.2 Predictions by Statistical Models 2
  - 1.1.3 Extraction of Information by Statistical Models 3
- 1.2 Constructing Statistical Models 4
  - 1.2.1 Evaluation of Statistical Models–Road to the Information Criterion 4
  - 1.2.2 Modeling Methodology 5
- 1.3 Organization of This Book 7

2 Statistical Models 9
- 2.1 Modeling of Probabilistic Events and Statistical Models 9
- 2.2 Probability Distribution Models 10
- 2.3 Conditional Distribution Models 17
  - 2.3.1 Regression Models 17
  - 2.3.2 Time Series Model 24
  - 2.3.3 Spatial Models 27

3 Information Criterion 29
- 3.1 Kullback–Leibler Information 29
  - 3.1.1 Definition and Properties 29
  - 3.1.2 Examples of K-L Information 32
  - 3.1.3 Topics on K-L Information 33
- 3.2 Expected Log-Likelihood and Corresponding Estimator 35
- 3.3 Maximum Likelihood Method and Maximum Likelihood Estimators 37
  - 3.3.1 Log-Likelihood Function and Maximum Likelihood Estimators 37

3.3.2 Implementation of the Maximum Likelihood Method by Means of Likelihood Equations ... 38
3.3.3 Implementation of the Maximum Likelihood Method by Numerical Optimization ... 40
3.3.4 Fluctuations of the Maximum Likelihood Estimators ... 44
3.3.5 Asymptotic Properties of the Maximum Likelihood Estimators ... 47
3.4 Information Criterion AIC ... 51
3.4.1 Log-Likelihood and Expected Log-Likelihood ... 51
3.4.2 Necessity of Bias Correction for the Log-Likelihood ... 52
3.4.3 Derivation of Bias of the Log-Likelihood ... 55
3.4.4 Akaike Information Criterion (AIC) ... 60
3.5 Properties of MAICE ... 69
3.5.1 Finite Correction of the Information Criterion ... 69
3.5.2 Distribution of Orders Selected by AIC ... 71
3.5.3 Discussion ... 73

4 Statistical Modeling by AIC ... 75
4.1 Checking the Equality of Two Discrete Distributions ... 75
4.2 Determining the Bin Size of a Histogram ... 77
4.3 Equality of the Means and/or the Variances of Normal Distributions ... 79
4.4 Variable Selection for Regression Model ... 84
4.5 Generalized Linear Models ... 88
4.6 Selection of Order of Autoregressive Model ... 92
4.7 Detection of Structural Changes ... 96
4.7.1 Detection of Level Shift ... 96
4.7.2 Arrival Time of a Signal ... 99
4.8 Comparison of Shapes of Distributions ... 101
4.9 Selection of Box–Cox Transformations ... 104

5 Generalized Information Criterion (GIC) ... 107
5.1 Approach Based on Statistical Functionals ... 107
5.1.1 Estimators Defined in Terms of Statistical Functionals ... 107
5.1.2 Derivatives of the Functional and the Influence Function ... 111
5.1.3 Extension of the Information Criteria AIC and TIC ... 115
5.2 Generalized Information Criterion (GIC) ... 118
5.2.1 Definition of the GIC ... 119
5.2.2 Maximum Likelihood Method: Relationship Among AIC, TIC, and GIC ... 124
5.2.3 Robust Estimation ... 128
5.2.4 Maximum Penalized Likelihood Methods ... 134

6 **Statistical Modeling by GIC** 139
6.1 Nonlinear Regression Modeling via Basis Expansions 139
6.2 Basis Functions 143
6.2.1 *B*-Splines 143
6.2.2 Radial Basis Functions 146
6.3 Logistic Regression Models for Discrete Data 149
6.3.1 Linear Logistic Regression Model 149
6.3.2 Nonlinear Logistic Regression Models 152
6.4 Logistic Discriminant Analysis 156
6.4.1 Linear Logistic Discrimination 157
6.4.2 Nonlinear Logistic Discrimination 159
6.5 Penalized Least Squares Methods 160
6.6 Effective Number of Parameters 162

7 **Theoretical Development and Asymptotic Properties of the GIC** 167
7.1 Derivation of the GIC 167
7.1.1 Introduction 167
7.1.2 Stochastic Expansion of an Estimator 170
7.1.3 Derivation of the GIC 171
7.2 Asymptotic Properties and Higher-Order Bias Correction 176
7.2.1 Asymptotic Properties of Information Criteria 176
7.2.2 Higher-Order Bias Correction 178

8 **Bootstrap Information Criterion** 187
8.1 Bootstrap Method 187
8.2 Bootstrap Information Criterion 192
8.2.1 Bootstrap Estimation of Bias 192
8.2.2 Bootstrap Information Criterion, EIC 195
8.3 Variance Reduction Method 195
8.3.1 Sampling Fluctuation by the Bootstrap Method 195
8.3.2 Efficient Bootstrap Simulation 196
8.3.3 Accuracy of Bias Correction 202
8.3.4 Relation Between Bootstrap Bias Correction Terms 205
8.4 Applications of Bootstrap Information Criterion 206
8.4.1 Change Point Model 206
8.4.2 Subset Selection in a Regression Model 208

9 **Bayesian Information Criteria** 211
9.1 Bayesian Model Evaluation Criterion (BIC) 211
9.1.1 Definition of BIC 211
9.1.2 Laplace Approximation for Integrals 213
9.1.3 Derivation of the BIC 215
9.1.4 Extension of the BIC 218
9.2 Akaike's Bayesian Information Criterion (ABIC) 222

9.3 Bayesian Predictive Distributions 224
9.3.1 Predictive Distributions and Predictive Likelihood 224
9.3.2 Information Criterion for Bayesian Normal Linear Models 226
9.3.3 Derivation of the PIC 227
9.3.4 Numerical Example 230
9.4 Bayesian Predictive Distributions by Laplace Approximation 231
9.5 Deviance Information Criterion (DIC) 236

10 Various Model Evaluation Criteria 239
10.1 Cross-Validation 239
10.1.1 Prediction and Cross-Validation 239
10.1.2 Selecting a Smoothing Parameter by Cross-Validation 242
10.1.3 Generalized Cross-Validation 243
10.1.4 Asymptotic Equivalence Between AIC-Type Criteria and Cross-Validation 245
10.2 Final Prediction Error (FPE) 247
10.2.1 FPE 247
10.2.2 Relationship Between the AIC and FPE 249
10.3 Mallows' C_p 251
10.4 Hannan–Quinn's Criterion 253
10.5 ICOMP 254

References 255

Index 269

1

Concept of Statistical Modeling

Statistical modeling is a crucial issue in scientific data analysis. Models are used to represent stochastic structures, predict future behavior, and extract useful information from data. In this chapter, we discuss statistical models and modeling methodologies such as parameter estimation, model selection, regularization method, and hierarchical Bayesian modeling. Finally, the organization of this book is described.

1.1 Role of Statistical Models

Models play a critical role in statistical data analysis. Once a model has been identified, various forms of inferences such as prediction, control, information extraction, knowledge discovery, validation, risk evaluation, and decision making can be done within the framework of deductive argument. Thus, the key to solving complex real-world problems lies in the development and construction of a suitable model. In this section, we consider the fundamental problem of statistical modeling, namely, our basic standpoint in statistical modeling, particularly model evaluation.

1.1.1 Description of Stochastic Structures by Statistical Models

A statistical model is a probability distribution that uses observed data to approximate the true distribution of probabilistic events. As such, the purpose of statistical modeling is to construct a model that approximates the true structure as accurately as possible through the use of available data (Figure 1.1). This is a natural requirement for practitioners who are engaged in data analysis. For example, in fitting a regression model, this assumption involves detecting the "true set of explanatory variables." In fitting polynomial regression models or autoregressive models, it entails selecting the true order. This appears to be a natural requirement, and in conventional mathematical

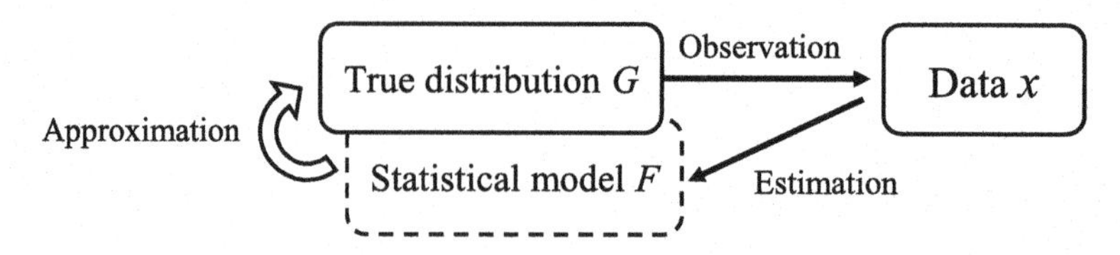

Fig. 1.1. Estimation of a true structure based on statistical modeling.

statistics it is only considered as background for problem settings. In practice, however, it is rare that linear regression models with a finite number of explanatory variables or AR models with a finite order can represent the true structure. Therefore, these models must be considered as an approximation that represents only one aspect of complex phenomena.

The important issue here is whether we should pursue a structure that is as close as possible to the true model. In other words, the critical question is whether the evaluation of a model should be performed under the requirement that models should be unbiased.

1.1.2 Predictions by Statistical Models

Based on the previous discussion, the question arises as to whether the objective of selecting a correct order or a correct model is fraught with problems. To answer this question, we need to consider the following questions: "What is the purpose of modeling?" and "What is the model to be used for?" As a critical point of view for statistical models, Akaike singled out the problem of prediction [Akaike (1974, 1985)]. Akaike considered that the purpose of statistical modeling is not to accurately describe current data or to infer the "true distribution." Rather, he thought that the purpose of statistical modeling is to predict future data as accurately as possible. In this book, we refer to this viewpoint as the *predictive point of view.*

There may be no significant difference between the point of view of inferring the true structure and that of making a prediction if an infinitely large quantity of data is available or if the data are noiseless. However, in modeling based on a finite quantity of real data, there is a significant gap between these two points of view, because an optimal model for prediction purposes may differ from one obtained by estimating the "true model."

In fact, as indicated by the information criteria given in this book for evaluating models intended for making predictions, simple models, even those containing biases, are often capable of giving better predictive distributions than models obtained by estimating the true structure.

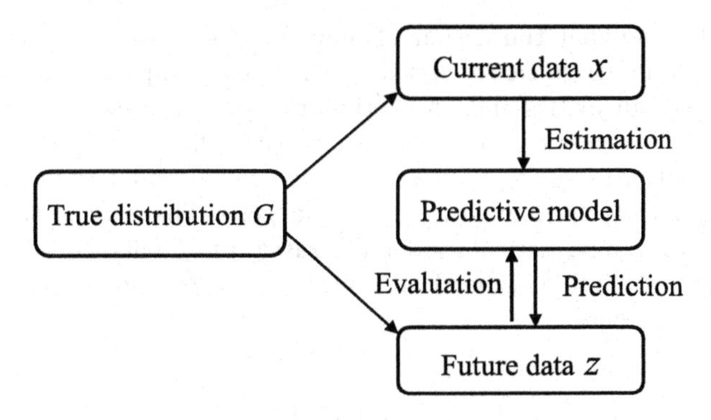

Fig. 1.2. Statistical modeling and the predictive point of view.

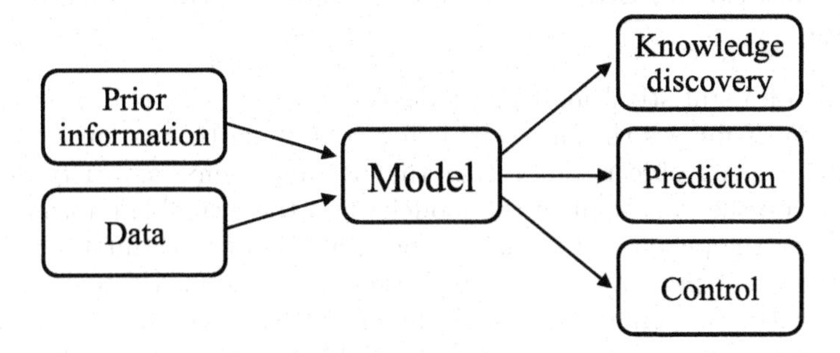

Fig. 1.3. Statistical modeling for extracting information.

1.1.3 Extraction of Information by Statistical Models

Another important point of view is the extraction of information. Many conventional statistical inferences assume that the "true" model that governs the object of modeling is a known entity, or at least that a "true" model exists. Also, conventional statistical inferences have adopted the approach of defining a problem as that of estimating a small number of unknown parameters based on data, given that the "true" model exists and that these parameters are contained in the model. However, a recent trend that has been gaining popularity is the idea that models are tools of convenience that are used for extracting information and discovering knowledge.

In this viewpoint, a statistical model is not something that exists in the objective world; rather, it is something that is constructed based on the prior knowledge and expectations of the analyst concerning the modeling objective, e.g., his knowledge based on past experience and data and based on the purpose of the analysis, such as the specific type of information to be extracted from the data and what is to be accomplished by the analysis. Therefore, if a specific model is obtained as a result of statistical modeling, we do not necessarily believe that the actual phenomenon behaves in accordance with the model in the strict sense. Actual events are complex, containing various kinds of nonlinearities and nonstationarities. Furthermore, in many cases they should be considered to be subject to the influence of other variables. Even in such situations, however, a relatively simple model often proves to be more appropriate for achieving a specific purpose. The crux of the matter is not whether a given statistical model accurately represents the true structure of a phenomenon, but whether it is suitable as a tool for extracting useful information from data.

1.2 Constructing Statistical Models

1.2.1 Evaluation of Statistical Models–Road to the Information Criterion

If the role of a statistical model is understood as being a tool for extracting information, it follows that a model is not something that is uniquely determined for a given object, but rather that it can assume a variety of forms depending on the viewpoint of the modeler and the available information. In other words, the purpose of statistical modeling is not to estimate or identify the "unique" or "perfect" model, but rather to construct a "good" model as a tool for extracting information according to the characteristics of the object and the purpose of the modeling [Akaike and Kitagawa (1998), Chapter 23].

This means that, as a general rule, the results of inference and evaluation will vary according to the specific model. A good model will generally yield good results; however, one cannot expect to obtain good results when using an inappropriate model. Herein lies the importance of model evaluation criteria for assessing the "goodness" of a subjective model.

How shall we set about evaluating the goodness of a model? In considering the circumstances under which statistical models are actually used, Akaike considered that a model should be evaluated in terms of the goodness of the results when the model is used for prediction. Furthermore, for the general evaluation of the goodness of a statistical model, he thought that it is important to assess the closeness between the predictive distribution $f(x)$ defined by the model and the true distribution $g(x)$, rather than simply minimizing the prediction error. Based on this concept, he proposed evaluating statistical models in terms of Kullback–Leibler information (divergence) [Akaike (1973)]. In this book, we refer to the model evaluation criterion derived from this fundamental model evaluation concept based on Kullback–Leibler information as the *information criterion.* This information criterion is derived from three fundamental concepts: (1) a prediction-based viewpoint of modeling; (2) evaluation of prediction accuracy in terms of distributions; and (3) evaluation of the closeness of distributions in terms of Kullback–Leibler information.

1.2.2 Modeling Methodology

The information criterion suggests several concrete methods for developing good models based on a limited quantity of data. First, it is obvious that the larger its log-likelihood, the better the model. The information criterion indicates, however, that given a finite quantity of data available for modeling, a model having an excessively high degrees of freedom will lead to an increase in the instability of the estimated model, and this will result in a reduced prediction ability. In other words, it is not beneficial to needlessly increase the number of free parameters without any restriction. Under these considerations, several methods are appropriate for assessing a good model based on a given set of data.

(1) Point estimation and model selection The first such method involves applying the information criterion directly to determine the number of unknown parameters to be estimated and to select the specific model to use. In this method, many alternative models $M_1, \ldots, M_k$ are considered, and the unknown parameters $\boldsymbol{\theta}_1, \ldots, \boldsymbol{\theta}_k$ associated with these models are estimated using the maximum likelihood method or another estimation method such as the robust estimation method. In this case, since the corresponding information criterion represents the goodness (or badness) of each model, the best model in terms of the information criterion can be obtained by selecting the model that minimizes the information criterion.

A simple and popular model selection method is order selection. If we assume a model with parameters $(\theta_1, \ldots, \theta_p)$, and if we denote the restricted model assuming that $\theta_{k+1} = \cdots = \theta_p = 0$ by M_k, then hierarchical models satisfying the relationship $M_0 \subset M_1 \subset \cdots \subset M_p$ can be obtained. In this case, a good model that strikes an acceptable balance between increasing the log-likelihood attained by increasing the number of parameters and increasing the number of the penalty terms can be obtained by selecting the order that minimizes the information criterion.

(2) Regularization and Bayesian modeling Another method for obtaining a good model involves imposing appropriate restrictions on parameters using a large number of parameters, without restricting the number of parameters. This strategy requires the integration of various types of information, such as the information from data $\boldsymbol{x}_n$, the modeling objective, empirical knowledge and tentative models based on past data, the theory related to the subject, and the purpose of the analysis. This information can be integrated using methods such as a regularization method or a maximum penalized likelihood method that maximizes the quantity

$$\log f(\boldsymbol{x}_n|\boldsymbol{\theta}) - Q(\boldsymbol{\theta}), \tag{1.1}$$

including the addition to the log-likelihood function of regularization terms or penalty terms, which is equivalent to imposing restrictions on the number of parameters.

It has been suggested that in many cases these model construction methods can be implemented in terms of a Bayesian model that combines information from prior distribution and data [Akaike (1980)]. In Bayesian modeling, a model can be constructed by obtaining the posterior distribution

$$\pi(\boldsymbol{\theta}|\boldsymbol{x}_n) = \frac{f(\boldsymbol{x}_n|\boldsymbol{\theta})\pi(\boldsymbol{\theta})}{\displaystyle\int f(\boldsymbol{x}_n|\boldsymbol{\theta})\pi(\boldsymbol{\theta})d\boldsymbol{\theta}}, \tag{1.2}$$

by introducing an appropriate prior distribution $\pi(\boldsymbol{\theta})$ for an unknown parameter vector $\boldsymbol{\theta}$ that defines the data distribution $f(\boldsymbol{x}|\boldsymbol{\theta})$.

(3) Hierarchical Bayesian modeling By generalizing Bayesian modeling, we can consider the situation in which multiple Bayesian models $M_1, \ldots, M_k$ exist. If $P(M_j)$ denotes the prior probability of model M_j, $f(x|\boldsymbol{\theta}_j, M_j)$ denotes a data distribution, and $\pi(\boldsymbol{\theta}_j|M_j)$ denotes the prior distribution of parameters, then the posterior probability of the models can be defined as

$$P(M_j|\boldsymbol{x}_n) \propto P(M_j)p(\boldsymbol{x}_n|M_j), \tag{1.3}$$

where $p(\boldsymbol{x}_n|M_j)$ is the likelihood of model M_j defined as

$$p(\boldsymbol{x}_n|M_j) = \prod_{\alpha=1}^{n} \int f(x_\alpha|\boldsymbol{\theta}_j, M_j)\pi(\boldsymbol{\theta}_j|M_j)d\boldsymbol{\theta}_j\,. \tag{1.4}$$

Suppose that the posterior predictive distribution of model M_j is defined by

$$p(z|\boldsymbol{x}_n, M_j) = \int f(z|\boldsymbol{\theta}_j, M_j)\pi(\boldsymbol{\theta}_j|\boldsymbol{x}_n, M_j)d\boldsymbol{\theta}_j\,, \tag{1.5}$$

where $\pi(\boldsymbol{\theta}_j|\boldsymbol{x}_n, M_j)$ is the posterior distribution of $\boldsymbol{\theta}_j$ defined by (1.2). Then the predictive distribution based on all of the models is given by

$$p(z|\boldsymbol{x}_n) = \sum_{j=1}^{k} P(M_j|\boldsymbol{x}_n)p(z|\boldsymbol{x}_n, M_j)\,. \tag{1.6}$$

In constructing a hierarchical Bayesian model, if the prior distribution of parameters is improper, it is not possible to determine the likelihood of the Bayesian model, $p(x_n|M_j)$, based on its definition. However, if $\mathrm{IC}(M_j)$ denotes an appropriately defined information criterion, the likelihood of the model can be defined as [Akaike (1978, 1980a)]

$$\exp\left\{-\frac{1}{2}\mathrm{IC}(M_j)\right\}. \tag{1.7}$$

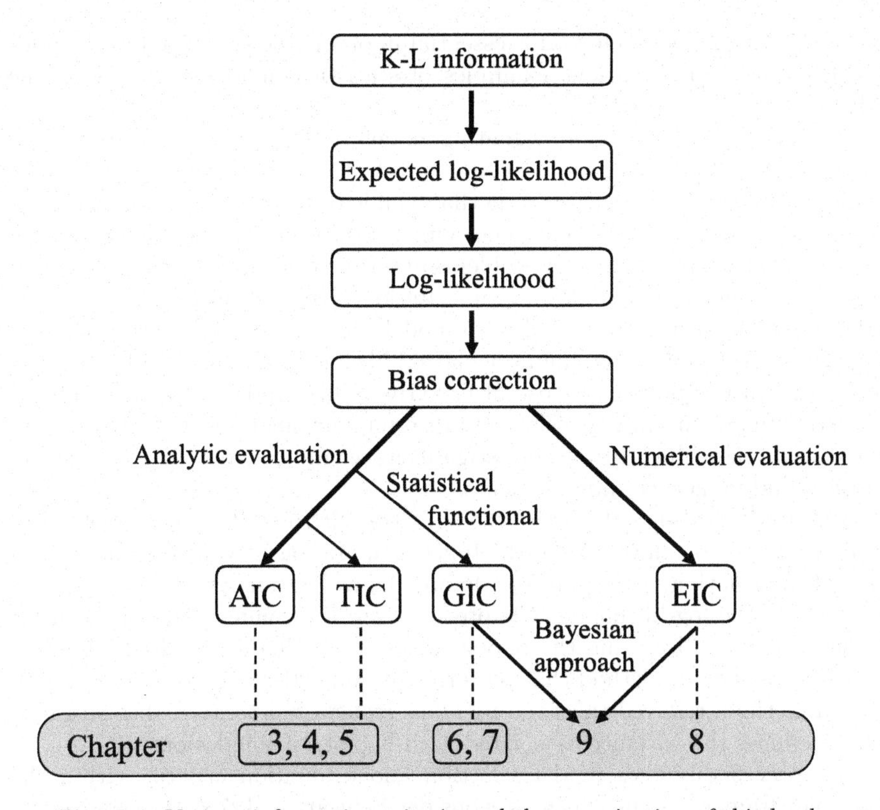

Fig. 1.4. Various information criteria and the organization of this book.

1.3 Organization of This Book

The main aim of this book is to explain the information criteria that play a critical role in statistical modeling as has been described in the previous subsections. Chapter 2 discusses the main subject of this book, namely the question "What is a statistical model?" and introduces probability distribution models employed as the base for statistical models. In addition, Chapter 2 also shows that using conditional distributions is essential for utilizing various forms of information in real-world modeling and describes linear and nonlinear regression, time series, and spatial models as specific forms of conditional distributions.

Chapter 3 provides the basis of this book. First, Kullback–Leibler information is used as a criterion for evaluating the goodness of a statistical model that approximates the true distribution, which generates the data, and in consequence the log-likelihood and the maximum likelihood estimates are demonstrated to derive naturally from this criterion. Second, the AIC is derived by showing that when estimating the Kullback–Leibler information, bias correction of the log-likelihood is essential in order to compare multiple models. Chapter 4 gives various examples of statistical modeling based on the AIC.

The AIC is a criterion for evaluating models estimated using the maximum likelihood method. With the development of modeling techniques, it has become necessary to construct criteria that enable us to evaluate various types of statistical models. Chapter 5 presents a unified information-theoretic approach to statistical model evaluation problems in terms of statistical functionals and introduces a generalized information criterion (GIC) [Konishi and Kitagawa (1996)] for evaluating a broad class of models, including models estimated using robust procedures, maximum penalized likelihood methods and the Bayes approach. In Chapter 6, the use of the GIC is illustrated through nonlinear statistical modeling in regression and discriminant analyses. Chapter 7 gives the derivation of the GIC and investigates its asymptotic properties with theoretical and numerical improvements.

Chapter 8 discusses the use of the bootstrap [Efron (1979)] in model evaluation problems by emphasizing the functional approach. Whereas the derivation of information criteria up to Chapter 7 involves analytical evaluation of the bias of the log-likelihood, Chapter 8 describes a numerical approach for evaluating biases by using the bootstrap method. Chapter 8 also presents a modified bootstrap method that performs second-order bias corrections along with a method, referred to as the *variance reduction procedure*, that substantially reduces the variance associated with bootstrap simulations.

Chapter 9 discusses model selection and evaluation criteria within the Bayesian framework, in which we consider Schwarz's (1978) Bayesian information criterion, Akaike's (1980b) Bayesian information criterion (ABIC), a predictive information criterion (PIC) [Kitagawa (1997)] as a criterion for evaluating the prediction likelihood of the Bayesian model, and a deviance information criterion (DIC) [Spiegelhalter et al. (2002)]. Furthermore, the BIC is extended in such a way that it can be used to evaluate models estimated by the maximum penalized likelihood method. Chapter 10 introduces various model selection and evaluation criteria as related topics. Specifically, we briefly touch upon cross-validation [Stone (1974)], final prediction error [Akaike (1969)], Mallows' (1973) C_p, Hannan–Quinn's (1979) criterion, and the information measure of model complexity (ICOMP) [Bozdogan (1988)].

2

Statistical Models

In this chapter, we describe probability distributions, which provide fundamental tools for statistical models, and show that conditional distributions are used to acquire various types of information in the model-building process. By using regression and time series models as specific examples, we also discuss why evaluation of statistical models is necessary.

2.1 Modeling of Probabilistic Events and Statistical Models

Before considering statistical models, let us first discuss how to represent events that we know occur in a deterministic way. In the simple case in which an event is fixed and invariable, the state of the event can be expressed in the form $x = a$. In general, however, x varies depending on some factor. If x is dependent on an external factor u, then it can be expressed as a function of u, e.g., $x = h(u)$. In some cases, x is determined according to past events or based on the present state, in which case x can be expressed as some function of the factor.

Most real-life events, however, contain uncertainty, and in many cases our information about external factors is incomplete. In such cases, the value of x cannot be specified as a fixed value or a deterministic function of factors, and in such cases we use a probability distribution.

Given a random variable X defined on the sample space Ω, for any real value $x(\in \mathrm{R})$, the probability $\Pr(\{\omega \in \Omega\,; X(\omega) \leq x\})$ of an event such that $X(\omega) \leq x$ can be determined. If we regard such a probability as a function of x and express it as

$$\begin{aligned} G(x) &= \Pr(\{\omega \in \Omega\,; X(\omega) \leq x\}) \\ &= \Pr(X \leq x)\,, \end{aligned} \tag{2.1}$$

then the function $G(x)$ is referred to as the *distribution function* of X. By determining the distribution function $G(x)$, we can characterize the random variable X. In particular, if there exists a nonnegative function $g(t) \geq 0$ that satisfies

$$G(x) = \int_{-\infty}^{x} g(t)dt, \tag{2.2}$$

then X is said to be *continuous*, and the function $g(t)$ is called a *probability density function.* A continuous probability distribution can be defined by determining the density function $g(t)$.

On the other hand, if the random variable X takes either a finite or a countably infinite number of discrete values $x_1, x_2, \ldots$, then the variable X is said to be *discrete*. The probability of taking a discrete point $X = x_i$ is determined by

$$\begin{aligned} g_i = g(x_i) &= \Pr(\{\omega \in \Omega\,; X(\omega) = x_i\}) \\ &= \Pr(X = x_i), \qquad\qquad i = 1, 2, \ldots, \end{aligned} \tag{2.3}$$

where $g(x)$ is called a *probability function*, for which the distribution function is given by $G(x) = \sum_{\{i; x_i \leq x\}} g(x_i)$, where $\sum_{\{i; x_i \leq x\}}$ represents the sum of the discrete values such that $x_i \leq x$.

If we assume that the observations $\boldsymbol{x}_n = \{x_1, x_2, \ldots, x_n\}$ are generated from the distribution function $G(x)$, then $G(x)$ is referred to as the *true distribution*, or the *true model.* On the other hand, the distribution function $F(x)$ used to approximate the true distribution is referred to as a *model* and is assumed to have either a density function or a probability function $f(x)$. If a model is specified by p-dimensional parameters $\boldsymbol{\theta} = (\theta_1, \theta_2, \ldots, \theta_p)^T$, then the model can be written as $f(x|\boldsymbol{\theta})$. If the parameters are represented as a point in the set $\Theta \subset \mathrm{R}^p$, then $\{f(x|\boldsymbol{\theta}); \boldsymbol{\theta} \in \Theta\}$ is called a *parametric family of probability distributions or models.*

An estimated model $f(x|\hat{\boldsymbol{\theta}})$ obtained by replacing an unknown parameter $\boldsymbol{\theta}$ with an estimator $\hat{\boldsymbol{\theta}}$ is referred to as a *statistical model.* The process of constructing a model that appropriately represents some phenomenon is referred to as *modeling.* In statistical modeling, it is necessary to estimate unknown parameters. However, settng up an appropriate family of probability models prior to estimating the parameters is of greater importance.

We first describe some probability distributions as fundamental models. After that, we will show that the mechanism of incorporating information from other variables can be represented in the form of a conditional distribution model.

2.2 Probability Distribution Models

The most fundamental form of a model is the probability distribution model or the probability model. More sophisticated models, such as conditional distribution models described in the next section, are also constructed using the probability distribution model.

Example 1 (Normal distribution model) The most widely used continuous probability distribution model is the normal distribution model, or Gaussian distribution model. The probability density function for the normal distribution is given by

$$f(x|\mu,\sigma^2) = \frac{1}{\sqrt{2\pi\sigma^2}} \exp\left\{-\frac{(x-\mu)^2}{2\sigma^2}\right\}, \quad -\infty < x < \infty. \tag{2.4}$$

This distribution is completely specified by the two parameters μ and σ^2, which are the mean and the variance, respectively. A probability distribution model, such as the normal distribution model, that can be expressed in a specific functional form containing a finite number of parameters $\boldsymbol{\theta} = (\mu, \sigma^2)^T$ is called a *parametric probability distribution model.*

In addition to the normal distribution model, the following parametric probability distribution models are well known:

Example 2 (Cauchy distribution model) If the probability density function is given by

$$f(x|\mu,\tau^2) = \frac{1}{\pi} \frac{\tau}{(x-\mu)^2 + \tau^2}, \quad -\infty < x < \infty, \tag{2.5}$$

then the distribution is called a *Cauchy distribution.* The parameters μ and τ^2 define the center of the distribution and the spread of the distribution, respectively. While the Cauchy distribution is symmetric with respect to the mode at μ, its mean and variance are not well-defined.

Example 3 (Laplace distribution model) A random variable X is said to have a Laplace distribution if its probability density function is

$$f(x|\mu,\tau) = \frac{1}{2\tau} \exp\left(-\frac{|x-\mu|}{\tau}\right), \quad -\infty < x < \infty, \tag{2.6}$$

where $-\infty < \mu < \infty$ and $\tau > 0$. The mean and variance are respectively given by $E[X] = \mu$ and $V(X) = 2\tau^2$. The distribution function of the Laplace random variable is

$$F(x|\mu,\tau) = \begin{cases} \dfrac{1}{2} \exp\left(\dfrac{x-\mu}{\tau}\right), & x \le \mu, \\ 1 - \dfrac{1}{2} \exp\left(-\dfrac{x-\mu}{\tau}\right), & x > \mu. \end{cases} \tag{2.7}$$

Example 4 (Pearson's family of distributions model) If the probability density function is given by

$$f(x|\mu,\tau^2,b) = \frac{\Gamma(b)\tau^{2b-1}}{\Gamma(b-\frac{1}{2})\Gamma(\frac{1}{2})}\frac{1}{\{(x-\mu)^2+\tau^2\}^b}, \quad -\infty < x < \infty, \tag{2.8}$$

then the distribution is known as a *Pearson's family of distributions*, in which the quantities μ and τ^2 are referred to as the *center* and *dispersion parameters*, as in the case of the Cauchy distribution. The quantity b is a parameter that specifies the shape of the distribution. By varying the value of b, it is possible to represent a variety of distributions. When $b = 1$, the distribution is Cauchy, and when $b = (k+1)/2$ where k is an integer, the distribution is a t-distribution with k degrees of freedom. Also, the distribution becomes a normal distribution when $b \to \infty$.

Example 5 (Mixture of normal distributions model) If the density function can be represented by

$$f(x|m,\boldsymbol{\theta}) = \sum_{j=1}^{m} \alpha_j \frac{1}{\sqrt{2\pi\sigma_j^2}} \exp\left\{-\frac{(x-\mu_j)^2}{2\sigma_j^2}\right\}, \quad -\infty < x < \infty, \tag{2.9}$$

then the distribution is called a *mixture of normal distributions*, where $\boldsymbol{\theta} = (\mu_1,\ldots,\mu_m,\sigma_1^2,\ldots,\sigma_m^2,\alpha_1,\ldots,\alpha_{m-1})^T$ and $\sum_{j=1}^m \alpha_j = 1$. A mixture of normal distributions is constructed by combining m normal distributions with weights α_j, in which case m is referred to as the number of components. A wide range of probability distribution models can be expressed by appropriate selection of the parameters m, α_j, μ_j, and σ_j^2.

Figure 2.1 shows various examples of probability distribution models. The model in the upper left panel is the standard normal distribution model with mean 0 and variance 1. The model in the upper right panel is a Cauchy distribution model with $\mu = 0$ and $\tau^2 = 1$. One feature of this model is that it has fatter left and right tails. By using a Cauchy distribution rather than a normal distribution, it is possible to model a phenomenon in which large absolute values have small but nonnegligible probabilities. This property can be used to detect outliers, perform a robust estimation, or detect jumps in a trend. The lower left panel shows Pearson distributions with $b = 0.6, 0.75, 1, 1.5$, and 3. By varying the value of b, it is possible to continuously represent various distributions, ranging from distributions that have even fatter tails than the Cauchy distribution to the normal distribution. The lower right panel shows an example of a mixture of normal distributions, which is capable of representing complex distributions even in the simplest case when $m = 2$.

Example 6 (Binomial distribution model) Let X be a binary random variable taking the values of either 0 or 1, and let the probability of an event's occurring be given by

$$\Pr(X=1) = p, \qquad \Pr(X=0) = 1-p, \qquad (0 < p < 1). \tag{2.10}$$

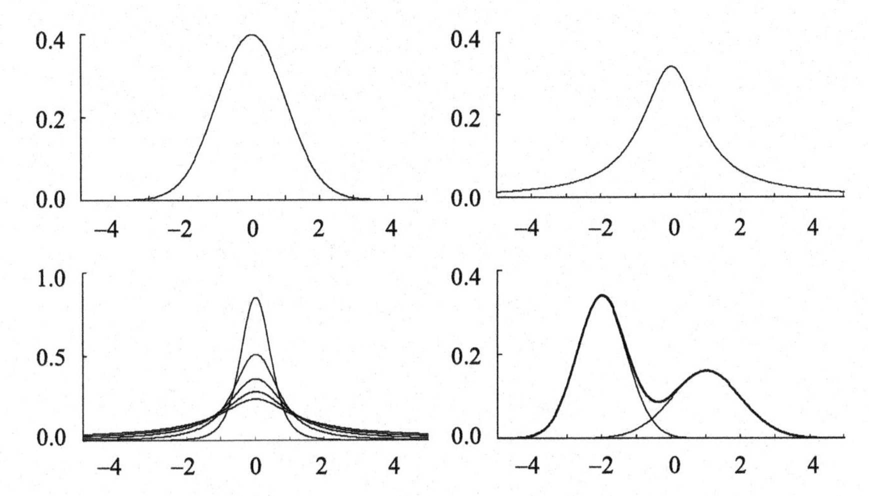

Fig. 2.1. Various examples of probability distributions: standard normal distribution (upper left); Cauchy distribution with $m = 0$ and $\tau^2 = 1$ (upper right); Pearson distributions with $b = 0.6, 0.75, 1, 1.5$, and 3 (lower left); and a mixture of normal distributions (lower right).

This probability distribution is referred to as a *Bernoulli distribution*, and its probability function is given by

$$f(x|p) = p^x(1-p)^{1-x}, \qquad x = 0, 1\,. \tag{2.11}$$

We further assume that the sequence of random variables $X_1, X_2, \ldots, X_n$ is independently distributed having the same Bernoulli distribution. Then the random variable $X = X_1 + X_2 + \cdots + X_n$ denotes the number of occurrences of an event in n trials, and its probability function is given by

$$f(x|p) = {}_nC_x p^x(1-p)^{n-x}, \quad x = 0, 1, 2, \ldots, n\,. \tag{2.12}$$

Such a probability distribution is called a *binomial distribution* with parameters n and p. The mean and variance are $E[X] = np$ and $V(X) = np(1-p)$, respectively.

Example 7 (Poisson distribution model) When very rare events are observed in short intervals, the distribution of the number of events is given by

$$f(x|\lambda) = \frac{\lambda^x}{x!}e^{-\lambda}, \quad x = 0, 1, 2, \ldots \quad (0 < \lambda < \infty)\,. \tag{2.13}$$

This distribution is called a *Poisson distribution.* The mean and variance are $E[X] = \lambda$ and $V(X) = \lambda$. The Poisson distribution is derived as an approximation to the binomial distribution by writing $np = \lambda$ for the probability

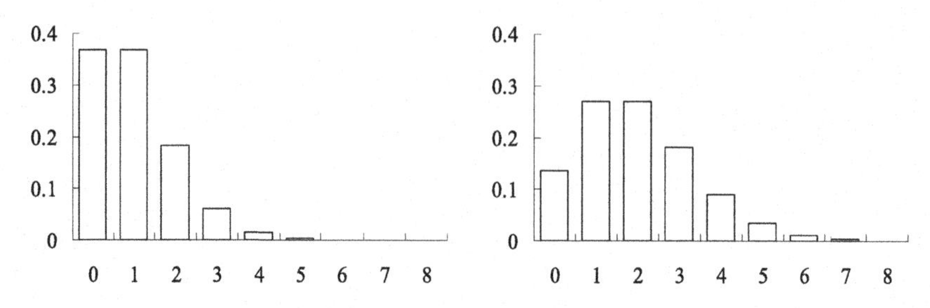

Fig. 2.2. Poisson distributions: left: $\lambda = 1$; right: $\lambda = 2$.

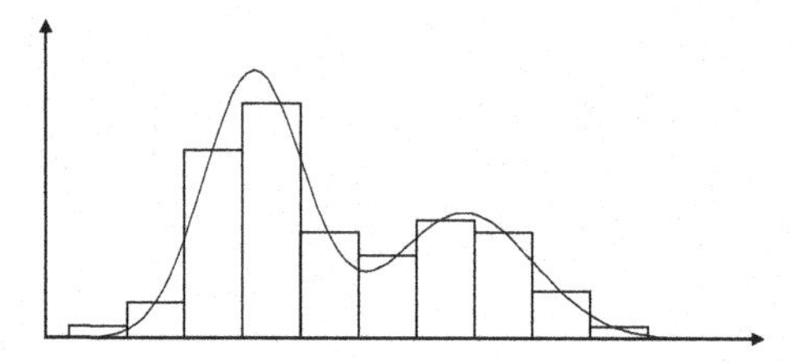

Fig. 2.3. A continuous distribution model and its approximation by a histogram.

function of the binomial distribution, while keeping λ constant. In fact, if n tends to infinity and p approaches 0, then for a fixed integer x,

$$
{}_nC_x p^x (1-p)^{n-x} = \frac{n!}{(n-x)!}\frac{\lambda^x}{x!}\left(1-\frac{\lambda}{n}\right)^n\left(1-\frac{\lambda}{n}\right)^{-x} \quad \longrightarrow \quad \frac{\lambda^x}{x!}e^{-\lambda}. \tag{2.14}
$$

Figure 2.2 shows Poisson distributions for the cases when the parameter λ is 1 and 2. Discrete distributions of various shapes can be represented depending on the value of λ.

Example 8 (Histogram model) A histogram can be obtained by dividing the domain $x_{\min} \le X \le x_{\max}$ of the random variable into appropriate intervals $B_1, \ldots, B_k$, determining the frequencies $n_1, \ldots, n_k$ of the observations that fall in the intervals $B_j = \{x; x_{j-1} \le x < x_j\}$, and graphing the results. If we set $n = n_1 + \cdots + n_k$, and define the relative frequency as $f_j = n_j/n$, a histogram can be thought of as defining the discrete distribution model $f = \{f_1, \ldots, f_k\}$ that is obtained by converting a continuous variable into a discrete variable. On the other hand, if the histogram is thought of as approximating a density function with a stepwise function, the histogram itself can be regarded as a type of continuous distribution model (Figure 2.3).

Example 9 (Probability model) A wide variety of phenomena can be expressed in terms of probability distributions according to the underlying

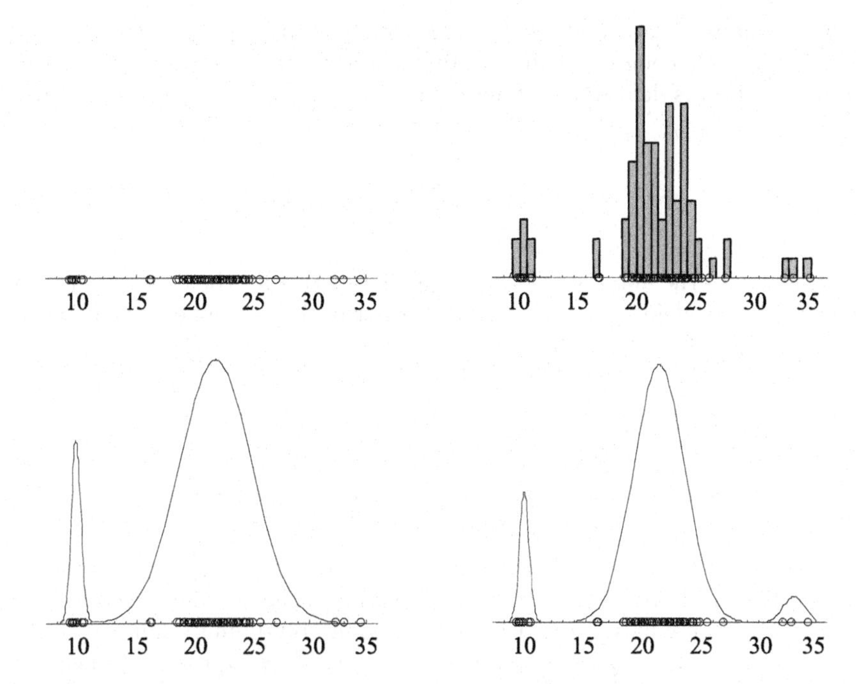

Fig. 2.4. The distribution of the velocities of 82 galaxies [Roeder (1990)]. Data (top left), the histogram (top right), and a mixture of normal distributions model (bottom left: $m = 2$; bottom right: $m = 3$).

problem. The problem is how to construct a probability model based on observed data.

Figure 2.4 shows the observed velocities, x, of 82 galaxies [Roeder (1990)]. Let us approximate the distribution of galaxy velocities using the mixture of normal distributions model in (2.9). If we estimate the parameters for the mixture of normal distributions based on observed data and replace the unknown parameters with estimated values, then the resulting density function $f(x|m, \hat{\boldsymbol{\theta}})$ is a statistical model. A critical issue in fitting the mixture of normal distributions model is the selection of the number of components, m. A two-component model has five parameters, while a three-component model has eight parameters. We must determine which model among the various candidate models best describes the probabilistic structure of the random variable X. Essential to answering this question is the criteria for evaluating the goodness of a statistical model.

Thus far, we have considered univariate random variables. There are many real-world situations, however, in which several variables must be considered simultaneously, for example, temperature and pressure in meteorological data, or interest rate and GDP in economic data. In such cases, $\boldsymbol{X} = (X_1, \ldots, X_p)^T$ becomes a multivariate random vector, for which the distribution function is defined as a function of p variables that are given in terms of $\boldsymbol{x} = (x_1, \ldots, x_p)^T \in \mathrm{R}^p$,

$$\begin{aligned} G(x_1, \ldots, x_p) &= \Pr(\{\omega \in \Omega : X_1(\omega) \leq x_1, \ldots, X_p(\omega) \leq x_p\}) \\ &= \Pr(X_1 \leq x_1, \ldots, X_p \leq x_p) . \end{aligned} \tag{2.15}$$

In parallel with the univariate case, a density function for the multivariate distribution can be defined. For a continuous distribution, a nonnegative function $f(x_1, \ldots, x_p) \geq 0$ that satisfies

$$\int_{-\infty}^{\infty} \cdots \int_{-\infty}^{\infty} f(x_1, \ldots, x_p) dx_1 \cdots dx_p = 1,$$

$$G(x_1, \cdots, x_p) = \int_{-\infty}^{x_1} \cdots \int_{-\infty}^{x_p} f(t_1, \ldots, t_p) dt_1 \cdots dt_p \tag{2.16}$$

is called the *probability density function* of the multivariate random vector $\boldsymbol{X}$.

Consider a discrete case, in which a p-dimensional random vector $\boldsymbol{X} = (X_1, \cdots, X_p)^T$ assumes either a finite or a countably infinite number of discrete values $\boldsymbol{x}_1, \boldsymbol{x}_2, \ldots$, where $\boldsymbol{x}_i = (x_{i1}, \ldots, x_{ip})^T$, $i = 1, 2, \ldots$. Then the probability function of the random vector $\boldsymbol{X}$ is defined by

$$g(\boldsymbol{x}_i) = \mathrm{P_r}(X_1 = x_{i1}, \ldots, X_p = x_{ip}), \quad i = 1, 2, \ldots . \tag{2.17}$$

The probability function satisfies

$$g(\boldsymbol{x}_i) \geq 0, \quad i = 1, 2, \ldots, \quad \text{and} \quad \sum_{i=1}^{\infty} g(\boldsymbol{x}_i) = 1, \tag{2.18}$$

and the distribution function can be expressed as

$$G(x_1, \cdots, x_p) = \sum_{\{i; x_{i1} \leq x_1\}} \cdots \sum_{\{i; x_{ip} \leq x_p\}} g(x_{i1}, \ldots, x_{ip}). \tag{2.19}$$

Example 10 (Multivariate normal distribution) A p-dimensional random vector $\boldsymbol{X} = (X_1, \ldots, X_p)^T$ is said to have a p-variate normal distribution with mean vector $\boldsymbol{\mu}$ and variance covariance matrix Σ if its probability density function is given by

$$f(\boldsymbol{x}|\boldsymbol{\mu}, \Sigma) = \frac{1}{(2\pi)^{p/2}|\Sigma|^{1/2}} \exp\left\{-\frac{1}{2}(\boldsymbol{x} - \boldsymbol{\mu})^T \Sigma^{-1}(\boldsymbol{x} - \boldsymbol{\mu})\right\}, \tag{2.20}$$

where $\boldsymbol{\mu} = (\mu_1, \ldots, \mu_p)^T$ and Σ is a $p \times p$ symmetric positive definite matrix whose $(i, j)^{th}$ component is given by σ_{ij}. We write $\boldsymbol{X} \sim N_p(\boldsymbol{\mu}, \Sigma)$.

Example 11 (Multinomial distribution) Suppose that there exist $k+1$ possible outcomes $E_1, \ldots, E_{k+1}$ in a trial. Let $\mathrm{P}(E_i) = p_i$, where $\sum_{i=1}^{k+1} p_i = 1$, and let X_i $(i = 1, \ldots, k+1)$ denote the number of times outcome E_i occurs in n trials, where $\sum_{i=1}^{k+1} X_i = n$. If the trials are repeated independently, then a multinomial distribution with parameters $n, p_1, \ldots, p_k$ is defined as a discrete distribution having the probability function

$$\Pr(X_1 = x_1, \ldots, X_k = x_k) = \frac{n!}{\prod_{i=1}^{k+1} x_i!} \prod_{i=1}^{k+1} p_i^{x_i}, \tag{2.21}$$

where $x_i = 0, 1, \ldots, n$ (note that $x_{k+1} = n - \sum_{i=1}^{k} x_i$). The mean, variance, and covariance are respectively given by $E[X_i] = np_i,\ i = 1, \ldots, k,\ V(X_i) = np_i(1 - p_i)$, and $\mathrm{Cov}(X_i, X_j) = -np_ip_j \quad (i \neq j)$.

2.3 Conditional Distribution Models

From the viewpoint of statistical modeling, the probability distribution is the most fundamental model in the situation in which the distribution of the random variable X is independent of various other factors. In practice, however, information associated with these variables can be used in various ways. The essence of statistical modeling lies in finding such information and incorporating it into a model in an appropriate form. In the following, we consider cases in which a random variable depends on other variables, on past history, on a spatial pattern, or on prior information. The important thing is that such modeling approaches can be considered as essentially estimating conditional distributions. Thus, the essence of statistical modeling can be thought of as obtaining an appropriate conditional distribution.

In general, if the distribution of the random variable Y is determined in a manner that depends on a p-dimensional variable $\boldsymbol{x} = (x_1, x_2, \ldots, x_p)^T$, then the distribution of Y is expressed as $F(y|\boldsymbol{x})$ or $f(y|\boldsymbol{x})$, and this is called a *conditional distribution model.* There are several ways in which the random variable depends on the other variables $\boldsymbol{x}$. In the following, we consider typical conditional distribution models.

2.3.1 Regression Models

The regression model is used to model the relationship between a response variable y and several explanatory variables $\boldsymbol{x} = (x_1, x_2, \ldots, x_p)^T$. This is equivalent to assuming that the probability distribution of the response variable y varies depending on the explanatory variables $\boldsymbol{x}$ and that a conditional distribution is given in the form of $f(y|\boldsymbol{x})$.

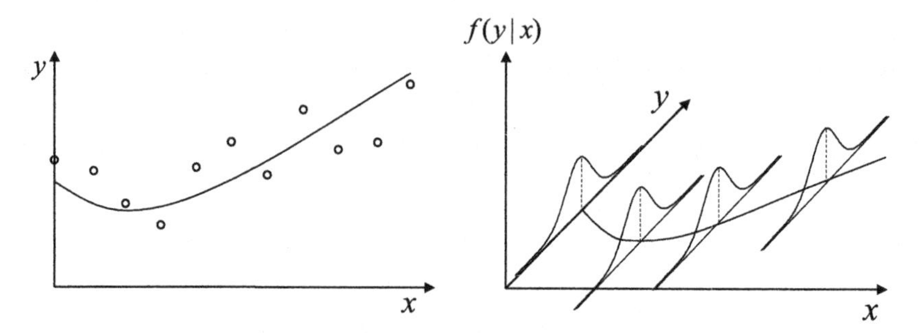

Fig. 2.5. Regression model (left) and conditional distribution model (right) in which the mean of the response variable varies as a function of the explanatory variable x.

Let $\{(y_\alpha, \boldsymbol{x}_\alpha);\ \alpha = 1, 2, \ldots, n\}$ be n sets of data obtained in terms of the response variable y and p explanatory variables $\boldsymbol{x}$. Then the model

$$y_\alpha = u(\boldsymbol{x}_\alpha) + \varepsilon_\alpha, \qquad \alpha = 1, 2, \ldots, n\,, \tag{2.22}$$

of the observed data is called a *regression model*, where $u(\boldsymbol{x})$ is a function of the explanatory variables $\boldsymbol{x}$, and the error terms or noise ε_α are assumed to be independently distributed with mean $E[\varepsilon_\alpha] = 0$ and variance $V(\varepsilon_\alpha) = \sigma^2$. We often assume that the noise ε_α follows the normal distribution $N(0, \sigma^2)$. In such a case, y_α has the normal distribution $N(u(\boldsymbol{x}_\alpha), \sigma^2)$ with mean $u(\boldsymbol{x}_\alpha)$ and variance σ^2, and its density function is given by

$$f(y_\alpha|\boldsymbol{x}_\alpha) = \frac{1}{\sqrt{2\pi\sigma^2}} \exp\left\{ -\frac{(y_\alpha - u(\boldsymbol{x}_\alpha))^2}{2\sigma^2} \right\}, \qquad \alpha = 1, 2, \ldots, n\,. \tag{2.23}$$

This distribution is a type of conditional distribution model in which the mean varies according to $E[Y|\boldsymbol{x}] = u(\boldsymbol{x})$ in a manner that depends on the values of the explanatory variables $\boldsymbol{x}$.

The left panel in Figure 2.5 shows 11 observations and the mean function $u(x)$ of the one-dimensional explanatory variable x and the response variable y. The data y_α at a given point x_α are observed as

$$y_\alpha = \mu_\alpha + \varepsilon_\alpha, \qquad \alpha = 1, 2, \ldots, 11\,, \tag{2.24}$$

with true mean value $E[Y_\alpha|x_\alpha] = \mu_\alpha$ and noise ε_α. The quantity $u(x)$ represents the mean structure of the event, and ε_α is the noise that induces probabilistic fluctuations in the data y_α. The right panel in Figure 2.5 shows a conditional distribution determined using a regression model. Fixing the value of the explanatory variable x gives the probability distribution $f(y|x)$, for which the mean is $u(x)$. Therefore, the regression model in (2.23) determines a class of distributions that move in parallel with the value of x.

Example 12 (Linear regression model) If the regression function or the mean function $u(\boldsymbol{x})$ can be approximated by a linear function of $\boldsymbol{x}$, then the model in (2.22) can be expressed as

$$\begin{aligned} y_\alpha &= \beta_0 + \beta_1 x_{\alpha 1} + \cdots + \beta_p x_{\alpha p} + \varepsilon_\alpha \\ &= \boldsymbol{x}_\alpha^T \boldsymbol{\beta} + \varepsilon_\alpha, \qquad \alpha = 1, 2, \ldots, n, \end{aligned} \tag{2.25}$$

with $\boldsymbol{\beta} = (\beta_0, \beta_1, \ldots, \beta_p)^T$, $\boldsymbol{x}_\alpha = (1, x_{\alpha 1}, x_{\alpha 2}, \ldots, x_{\alpha p})^T$ and is referred to as a *linear regression model.* A linear regression model with Gaussian noise can be expressed by the density function

$$f(y_\alpha | \boldsymbol{x}_\alpha; \boldsymbol{\theta}) = \frac{1}{\sqrt{2\pi\sigma^2}} \exp\left\{ -\frac{(y_\alpha - \boldsymbol{x}_\alpha' \boldsymbol{\beta})^2}{2\sigma^2} \right\}, \quad \alpha = 1, 2, \ldots, n, \tag{2.26}$$

where the unknown parameters in the model are $\boldsymbol{\theta} = (\boldsymbol{\beta}^T, \sigma^2)^T$. In the linear regression model, the critical issue is to determine a set of explanatory variables that appropriately describes changes in the distribution of the response variable y; this problem is referred to as the *variable selection* problem.

Example 13 (Polynomial regression model) A polynomial regression model with Gaussian noise,

$$y_\alpha = \beta_0 + \beta_1 x_\alpha + \cdots + \beta_m x_\alpha^m + \varepsilon_\alpha, \quad \varepsilon_\alpha \sim N(0, \sigma^2), \tag{2.27}$$

assumes that the regression function $u(x)$ can be approximated by $\beta_0 + \beta_1 x + \beta_2 x^2 + \cdots + \beta_m x^m$ with respect to the one-dimensional explanatory variable x. For each order m, the parameters of the polynomial regression model are $\boldsymbol{\beta} = (\beta_0, \beta_1, \ldots, \beta_m)^T$ and the error variance is σ^2. In a polynomial regression model, the crucial task is determining the order m, which is referred to as the *order selection* problem. As shown in Example 16, a model having an order that is too low cannot adequately represent the data structure. On the other hand, a model with an order that is too high causes the model to react excessively to random variations in the data, masking the essential relationship.

Various functions in addition to polynomials are used to represent a regression function. Trigonometric function models are expressed as

$$y_\alpha = a_0 + \sum_{j=1}^{m} \{a_j \cos(j\omega x_\alpha) + b_j \sin(j\omega x_\alpha)\} + \varepsilon_\alpha . \tag{2.28}$$

In addition, various forms of other orthogonal functions can be used to approximate the regression function.

Example 14 (Nonlinear regression models) Thus far, given a regression function $E[Y|\boldsymbol{x}] = u(\boldsymbol{x})$, we have constructed models by assuming functional

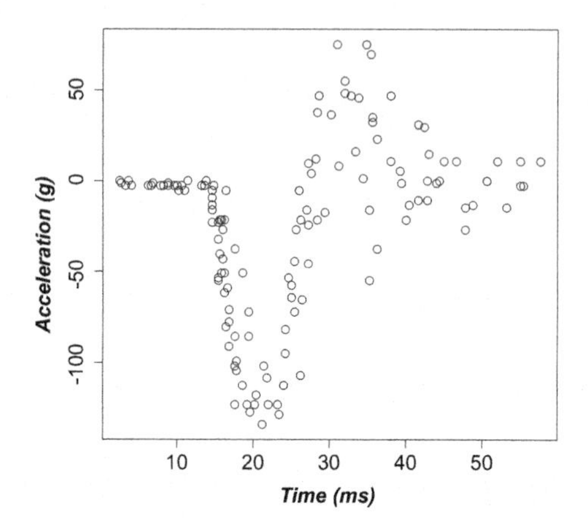

Fig. 2.6. Motorcycle impact data.

forms such as polynomials. The analysis of complex and diverse phenomena, however, requires developing more flexible models. Figure 2.6, for example, plots the measured acceleration $Y(g$; gravity) of the crash dummy's head at a time X (ms, millisecond) from the moment of collision in repeated motorcycle collision experiments [Härdle (1990)]. Neither polynomial models nor models using specific nonlinear functions are adequate for describing the structure of phenomena characterized by data that exhibit this type of complex nonlinear structure.

It is assumed that at each point x_α, y_α is observed as $y_\alpha = \mu_\alpha + \varepsilon_\alpha$, $\alpha = 1, 2, \ldots, n$, with noise ε_α. In order to approximate μ_α, $\alpha = 1, 2, \ldots, n$, in a way that reflects the structure of the phenomenon, we use a regression model

$$y_\alpha = u(x_\alpha; \boldsymbol{\theta}) + \varepsilon_\alpha, \qquad \alpha = 1, 2, \ldots, n. \tag{2.29}$$

For $u(x; \boldsymbol{\theta})$, various models are used depending on the analysis objective, including (1) splines [Green and Silverman (1994)], (2) B-splines [de Boor (1978), Imoto (2001)], (3) kernel functions [Simonoff (1996)], and (4) multi-layer neural network models [Bishop (1995), Ripley (1996)]. Our purpose here is to identify the mean structure of a phenomenon from data based on these flexible models.

Example 15 (Changing variance model) Whereas in the regression models described above, only the mean structure changes as a function of the explanatory variables $\boldsymbol{x}$, in changing variance models the variance of the response variable y also changes as a function of $\boldsymbol{x}$, and such a change is expressed in the form $\sigma^2(\boldsymbol{x})$. In this case, the conditional distribution of y is given by $N(u(\boldsymbol{x}), \sigma^2(\boldsymbol{x}))$. Figure 2.7 shows an example of a conditional distribution determined by a changing variance model in which it has a constant mean. It

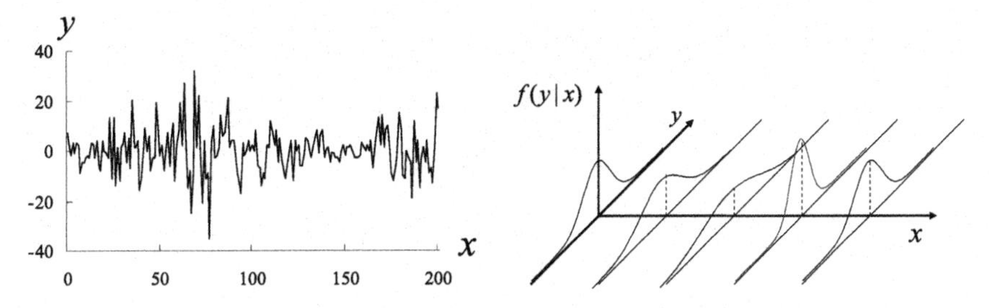

Fig. 2.7. Conditional distributions of changing variance models.

shows that the variance of the distribution changes depending on the value of x. These types of changing variance models are important for analyzing earthquake data and financial data.

Generally, a regression model is composed of a model that approximates the mean function $E[Y|\boldsymbol{x}]$ representing the structure of phenomenon and a probability distribution model that describes the probabilistic fluctuation of the data. Since models that approximate the mean function depend on several parameters, we write $u(\boldsymbol{x};\boldsymbol{\beta})$. Observed data with Gaussian noise are then given as

$$y_\alpha = u(\boldsymbol{x}_\alpha;\boldsymbol{\beta}) + \varepsilon_\alpha, \qquad \alpha = 1, 2, \ldots, n\,, \tag{2.30}$$

and are represented by the density function

$$f(y_\alpha|\boldsymbol{x}_\alpha;\boldsymbol{\theta}) = \frac{1}{\sqrt{2\pi\sigma^2}} \exp\left\{ -\frac{(y_\alpha - u(\boldsymbol{x}_\alpha;\boldsymbol{\beta}))^2}{2\sigma^2} \right\}, \quad \alpha = 1, 2, \ldots, n\,, \tag{2.31}$$

where $\boldsymbol{\theta} = (\boldsymbol{\beta}^T, \sigma^2)^T$.

In the case of a regression model expressed by a density function, we estimate the parameter vector $\boldsymbol{\theta}$ of the model by using the maximum likelihood method, and we denote it as $\hat{\boldsymbol{\theta}} = (\hat{\boldsymbol{\beta}}^T, \hat{\sigma}^2)^T$. Then the density function in which the unknown parameters in (2.31) are replaced with their corresponding estimators,

$$f(y_\alpha|\boldsymbol{x}_\alpha;\hat{\boldsymbol{\theta}}) = \frac{1}{\sqrt{2\pi\hat{\sigma}^2}} \exp\left\{ -\frac{(y_\alpha - u(\boldsymbol{x}_\alpha;\hat{\boldsymbol{\beta}}))^2}{2\hat{\sigma}^2} \right\}, \quad \alpha = 1, 2, \ldots, n\,, \tag{2.32}$$

is called a *statistical model.*

Although the main focus in regression models tends to be modeling for expected values, the distributions of error terms are also important. For a given regression function, different models can be obtained by changing the value of the variance. In addition, models that assume distributions other than

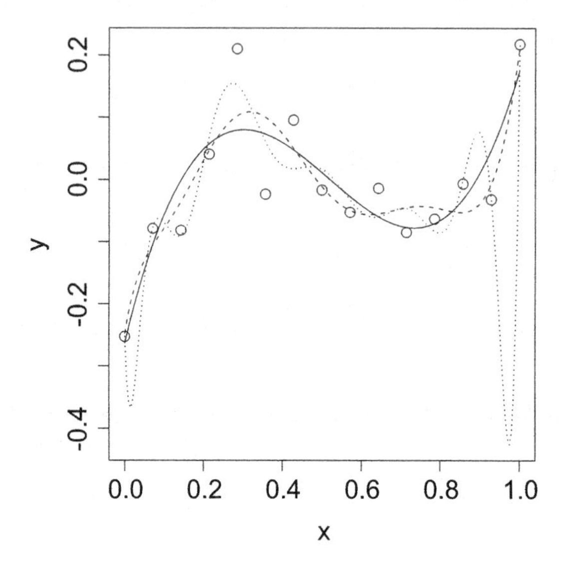

Fig. 2.8. Fitting polynomial regression models of order 3 (solid), 8 (broken), and 12 (dotted).

the normal distribution for the error terms (e.g., Cauchy distribution) are also conceivable.

Example 16 (Fitting a polynomial regression model) Figure 2.8 shows a plot of 15 observations obtained with respect to the explanatory variable x and the response variable y. By ordering the data as $\{(x_\alpha, y_\alpha);\ \alpha = 1, 2, \ldots, 15\}$, we fit the polynomial regression model in (2.27).

For each order m, we estimate the parameters $\boldsymbol{\beta} = (\beta_0, \beta_1, \ldots, \beta_m)^T$ of the polynomial regression model by using either the least square method or the maximum likelihood method that maximizes the log-likelihood function

$$\begin{aligned} &\sum_{\alpha=1}^{n} \log f(y_\alpha | x_\alpha; \boldsymbol{\beta}, \sigma^2) \qquad (2.33) \\ &\quad = -\frac{n}{2}\log(2\pi\sigma^2) - \frac{1}{2\sigma^2}\sum_{\alpha=1}^{n}\{y_\alpha - (\beta_0 + \beta_1 x_\alpha + \cdots + \beta_m x_\alpha^m)\}^2 \end{aligned}$$

and denote the results as $\hat{\boldsymbol{\beta}} = (\hat{\beta}_0, \hat{\beta}_1, \ldots, \hat{\beta}_m)^T$. The figure shows the estimated polynomial regression curves for orders 3, 8, and 12; it shows that estimated polynomials can vary greatly depending on the assumed order. Thus, the problem is deciding the order of the polynomial that should be adopted in the model.

If we consider the problem of order selection from the viewpoint of the goodness of fit of data in an estimated model, that is, from the standpoint of minimizing the squared sum of residuals

$$\sum_{\alpha=1}^{n} (y_\alpha - \hat{y}_\alpha)^2 = \sum_{\alpha=1}^{n} \left\{ y_\alpha - \left(\hat{\beta}_0 + \hat{\beta}_1 x_\alpha + \cdots + \hat{\beta}_m x_\alpha^m \right) \right\}^2 , \quad (2.34)$$

then the higher the order of the model, the smaller the value will be. As a result, we select the highest order [i.e., the $(n-1)^{th}$ order] polynomial that passes through all data points. If the data are free of errors, the error term ε_α in (2.27) will be superfluous, in which case it is sufficient to select the most complex model out of the class of models expressed by a large number of parameters. However, for data that contain intrinsic or observational errors, models that overfit the observed data tend to model the errors excessively and do not adequately approximate the true structure of the phenomenon. Consequently, such models do not predict future events well.

In general, a model that is too complex overadjusts for the random fluctuation in the data, while, on the other hand, overly simplistic models fail to adequately describe the structure of the phenomenon being modeled. Therefore, the key to evaluating a model is to strike a balance between, badness of fit of the data and the model complexity.

Example 17 (Spline functions) Assume that in the data $\{(y_\alpha, x_\alpha); \alpha = 1, 2, \ldots, n\}$ observed with respect to a response variable y and an explanatory variable x, n observations, $x_1, x_2, \ldots, x_n$, are ordered in ascending order in the interval $[a, b]$ as follows:

$$a < x_1 < x_2 < \cdots < x_n < b. \quad (2.35)$$

The essential idea in spline function fitting is to divide the interval containing the data $\{x_1, \ldots, x_n\}$ into several subintervals and to fit a polynomial model in a segment-by-segment manner, rather than fitting a single polynomial model to n sets of observed data.

Let $\xi_1 < \xi_2 < \cdots < \xi_m$ denote the m points that divide (x_1, x_n). These points are referred to as *knots*. A commonly used spline function in practical applications is the cubic spline, in which a third-order polynomial is fitted segment by segment over the subintervals $[a, \xi_1]$, $[\xi_1, \xi_2], \ldots, [\xi_m, b]$, and the polynomials are smoothly connected at the knots. In other words, the model is fitted under the restriction that at each knot, the first and second derivatives of the third-order polynomial are continuous. As a result, the cubic spline function having the knots $\xi_1 < \xi_2 < \cdots < \xi_m$ is given by

$$u(x; \boldsymbol{\theta}) = \beta_0 + \beta_1 x + \beta_2 x^2 + \beta_3 x^3 + \sum_{i=1}^{m} \theta_i (x - \xi_i)_+^3 , \quad (2.36)$$

where $\boldsymbol{\theta} = (\theta_1, \theta_2, \ldots, \theta_m, \beta_0, \beta_1, \beta_2, \beta_3)^T$ and $(x - \xi_i)_+ = \max\{0, x - \xi_i\}$.

It is commonly known, however, that it is not appropriate to fit a cubic polynomial near a boundary since the estimated curve will vary excessively. In order to address this difficulty, the natural cubic spline specifies that the cubic spline be a linear function at the two ends of the interval $(-\infty, \xi_1]$, $[\xi_m, +\infty)$, so that the natural cubic spline is given by

$$u(x; \boldsymbol{\theta}) = \beta_0 + \beta_1 x + \sum_{i=1}^{m-2} \theta_i \left\{ d_i(x) - d_{m-1}(x) \right\}, \tag{2.37}$$

where $\boldsymbol{\theta} = (\theta_1, \theta_2, \ldots, \theta_{m-2}, \beta_0, \beta_1)^T$ and

$$d_i(x) = \frac{(x - \xi_i)_+^3 - (x - \xi_m)_+^3}{\xi_m - \xi_i}.$$

When applying a spline in practical situations, we still need to determine the number of knots and their positions. From a computational standpoint, it is difficult to estimate the positions of knots as parameters. For this reason, we estimate the parameters $\boldsymbol{\theta}$ of the model by using the maximum penalized likelihood method described in Subsection 5.2.4 or the penalized least squares method discussed in Section 6.5. These topics are covered in Chapters 5 and 6. In the B-spline, a basis function is constructed by connecting the segment-wise polynomials, and it can substantially reduce the number of parameters in a model. This topic will be discussed in Section 6.2.

2.3.2 Time Series Model

Observed data, $x_1, \ldots, x_N$, for events that vary with time are referred to as a *time series*. The vast majority of real-world data, including meteorological data, environmental data, financial or economic data, and time-dependent experimental data, constitutes time series. The main aim of time series analysis is to identify the structure of the phenomenon represented by a sequence of measurements and to predict future observations. To analyze such time series data, we consider the conditional distribution

$$f(x_n | x_{n-1}, x_{n-2}, \ldots), \tag{2.38}$$

given observations up to the time $n - 1$.

Example 18 (AR model and ARMA model) In particular, by assuming a linear structure in finite dimensions, we obtain an *autoregressive* (AR) *model* [Akaike (1969, 1970), Brockwell and Davis (1991)];

$$x_n = \sum_{j=1}^{p} a_j x_{n-j} + \varepsilon_n, \qquad \varepsilon_n \sim N(0, \sigma^2), \tag{2.39}$$

where p denotes the order and indicates which information, obtained up to what time in the past, must be used in order to determine a future predictive

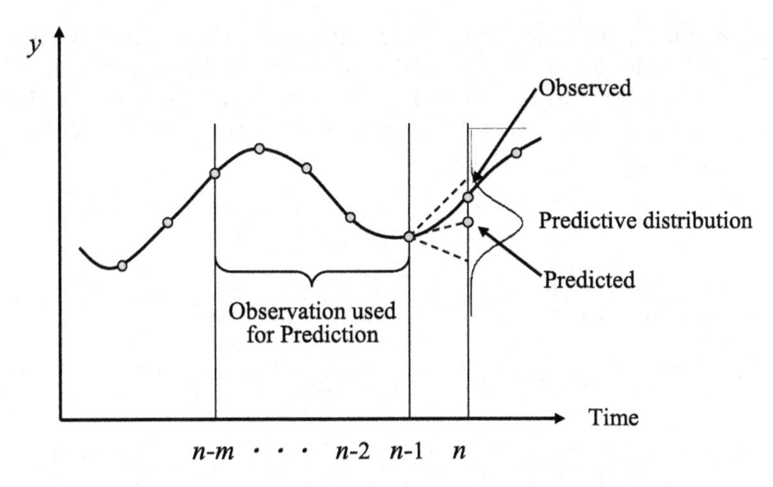

Fig. 2.9. Predictive distribution of time series.

Table 2.1. Residual variances and prediction error variances of AR models with a variety of orders.

p	$\hat{\sigma}_p^2$	PEV_p	p	$\hat{\sigma}_p^2$	PEV_p	p	$\hat{\sigma}_p^2$	PEV_p
0	6.3626	8.0359	7	0.3477	0.3956	14	0.3206	0.3802
1	1.1386	1.3867	8	0.3397	0.3835	15	0.3204	0.3808
2	0.3673	0.4311	9	0.3313	0.3817	16	0.3202	0.3808
3	0.3633	0.4171	10	0.3312	0.3812	17	0.3188	0.3823
4	0.3629	0.4167	11	0.3250	0.3808	18	0.3187	0.3822
5	0.3547	0.4030	12	0.3218	0.3797	19	0.3187	0.3822
6	0.3546	0.4027	13	0.3218	0.3801	20	0.3186	0.3831

distribution. A particular case is that of $p = 0$, which is called *white noise* if x_n is uncorrelated with its own past history. An AR model means that a conditional distribution (also referred to as a predictive distribution) of x_n can be given by the normal distribution having mean $\sum_{j=1}^{p} a_j x_{n-j}$ and variance σ^2.

Similar to the polynomial models, the selection of an appropriate order is an important problem in AR models. When time series data $x_1, \ldots, x_n$ are given, the coefficients a_j and the prediction error variance σ^2 are estimated using the least squares method or the maximum likelihood method. However, the estimated prediction error variance, $\hat{\sigma}_p^2$, of the AR model of order p is a monotonically decreasing function of p. Therefore, if the AR order is determined by this criterion, the maximum order will always be selected, which corresponds to the order selection for the polynomial model in Example 16.

The second column in Table 2.1 indicates the change in $\hat{\sigma}_p^2$ when AR models up to order 20 are fitted to the observations of the rolling angle of a ship [$n = 500$, Kitagawa and Gersch (1996)]. Here, $\hat{\sigma}_p^2$ decreases rapidly up to $p = 2$ and diminishes gradually thereafter. The third column in the table gives the prediction error variance

$$\mathrm{PEV}_p = \frac{1}{500} \sum_{i=501}^{1000} (x_i - \hat{x}_i^p)^2 , \tag{2.40}$$

when the subsequent data $x_{501}, \ldots, x_{1000}$ are predicted by

$$\hat{x}_i^p = \sum_{j=1}^{p} \hat{a}_j^p x_{i-j} \quad (i = 501, \ldots, 1000), \tag{2.41}$$

based on the estimated model of order p, where $\hat{a}_j^p$ is an estimate of the j-th coefficient a_j for the AR model of order p. The value of PEV_p is smallest at $p = 12$, and for higher orders, rather than decreasing, the prediction error variance increases.

Even when the time series has a complex structure and the AR model requires a high order p, in some cases an appropriate model can be obtained with fewer parameters by using past values of ε_n together with past values of the time series. The following model is referred to as an *autoregressive moving average (ARMA) model*:

$$x_n = \sum_{j=1}^{p} a_j x_{n-j} + \varepsilon_n - \sum_{j=1}^{q} b_j \varepsilon_{n-j}. \tag{2.42}$$

In general, if the conditional distribution of a time series x_n is represented by nonlinear functions of the series $x_{n-1}, x_{n-2}, \ldots$ and noise (also called "innovation"), $\varepsilon_n, \varepsilon_{n-1}, \ldots$, then the corresponding model is called a *nonlinear time series model*. If the time series $\boldsymbol{x}_n$ is a vector and the components are interrelated, a multivariate time series model is used for forecasting.

Example 19 (State-space models) A wide variety of time series models such as the ARMA model, trend model, seasonal adjustment model, and time-varying model can be represented using a state-space model. In a state-space model, the time series is expressed by using an unknown m-dimensional state vector $\boldsymbol{\alpha}_n$ as follows:

$$\begin{aligned} \boldsymbol{\alpha}_n &= F_n \boldsymbol{\alpha}_{n-1} + G_n \boldsymbol{v}_n, \\ x_n &= H_n \boldsymbol{\alpha}_n + w_n , \end{aligned} \tag{2.43}$$

where $\boldsymbol{v}_n$ and w_n are white noises that have the normal distributions $N_n(0, Q_n)$ and $N(0, \sigma_n^2)$, respectively. Concerning the state-space model, the Kalman filter algorithm is known to efficiently calculate the conditional distributions

$f(\boldsymbol{\alpha}_n|x_{n-1},x_{n-2},\ldots)$ and $f(\boldsymbol{\alpha}_n|x_n,x_{n-1},\ldots)$ of the unknown state $\boldsymbol{\alpha}_n$ from observed time series; these conditional distributions are referred to as a *state prediction distribution* and a *filter distribution*, respectively. Many important problems in time series analysis, such as prediction and control, computation of likelihood, and decomposition into several components, can be solved by using the estimated state vector.

The generalized state-space model is a generalization of the state-space model [Kitagawa (1987)]. It represents the time series as follows:

$$\begin{aligned} \boldsymbol{\alpha}_n &\sim F(\boldsymbol{\alpha}_n|\boldsymbol{\alpha}_{n-1}), \\ x_n &\sim H(x_n|\boldsymbol{\alpha}_n), \end{aligned} \tag{2.44}$$

where F and H denote appropriately specified conditional probability distributions. In other words, generalized state-space models directly model the two conditional distributions that are essential in time series modeling. This conditional distribution model can also be applied when observed data or states are discrete variables. It can be shown that the hidden Markov model is actually a special case of the generalized state-space model. Recently, a sequential Monte Carlo method for recursive estimation of unknown parameters of the generalized state-space models has been developed [see for example, Durbin and Koopman (2001), Harvey (1989), and Kitagawa and Gersch (1996)].

This method can thus be used to estimate the unknown state vector if the (general) state-space model is specified. Since the log-likelihood of the state-space model can be computed by using the predictive distribution of the state, unknown parameters of the model can be estimated using the maximum likelihood method. However, the state-space model is a very flexible model that is capable of expressing a very wide range of time series models. Therefore, in actual time series modeling, we have to compare a large variety of time series models and select an appropriate one.

2.3.3 Spatial Models

The spatial model represents the distribution of data by associating a spatial arrangement with it. For the case when data are arranged in a regular lattice, as depicted in the left plot of Figure 2.10, a model such as

$$p(x_{ij}|x_{i,j-1},x_{i,j+1},x_{i-1,j},x_{i+1,j}), \tag{2.45}$$

that represents the data x_{ij} at point (i,j), for example, can be constructed as a conditional distribution of the surrounding four points. As a simple example, a model

$$x_{ij} = \frac{1}{4}(x_{i,j-1}+x_{i,j+1}+x_{i-1,j}+x_{i+1,j})+\varepsilon_{ij} \tag{2.46}$$

is conceivable in which ε_{ij} is a normal distribution with mean 0 and variance σ^2.

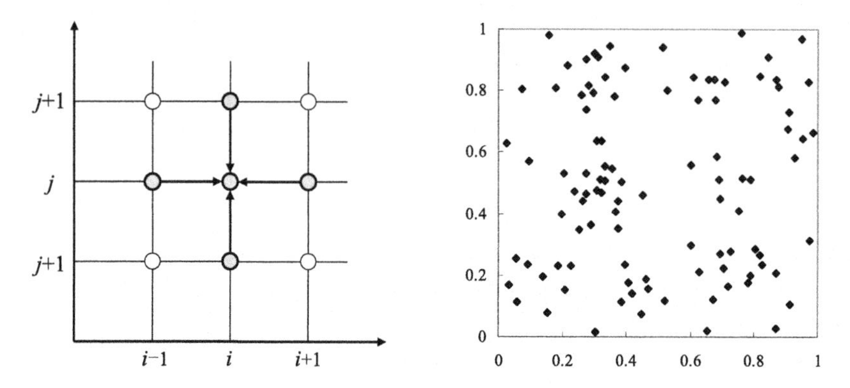

Fig. 2.10. An example of a prediction model for lattice data and spatial data.

On the other hand, in the general case in which the pointwise arrangement of data is not necessarily a lattice pattern, as illustrated in the right plot of Figure 2.10, a model that describes an equilibrium state can be obtained by modeling the local interaction of the points called particles.

Let us assume that the pointwise arrangement $\boldsymbol{x} = \{x_1, x_2, \ldots, x_n\}$ of n particles is given. If we define a potential function $\phi(x, y)$ that models the force acting between two points, the sum of the potential energy at the point arrangement $\boldsymbol{x}$ can be given by

$$H(\boldsymbol{x}) = \sum_{1 \leq i \leq j \leq n} \phi(x_i, x_j) \,. \tag{2.47}$$

Then the Gibbs distribution is defined by

$$f(\boldsymbol{x}) = C \exp\{-H(\boldsymbol{x})\} \,, \tag{2.48}$$

where C is a normalization constant defined such that the integration over the entire space is 1. In this method, models on spatial data can be obtained by establishing concrete forms of the potential function $\phi(x, y)$. For the analysis of spatial data, see Cressie (1991).

3

Information Criterion

In this chapter, we discuss using Kullback–Leibler information as a criterion for evaluating statistical models that approximate the true probability distribution of the data and its properties. We also explain how this criterion for evaluating statistical models leads to the concept of the information criterion, AIC. To this end, we explain the basic framework of model evaluation and the derivation of AIC by adopting a unified approach.

3.1 Kullback–Leibler Information

3.1.1 Definition and Properties

Let $\boldsymbol{x}_n = \{x_1, x_2, \ldots, x_n\}$ be a set of n observations drawn randomly (independently) from an unknown probability distribution function $G(x)$. In the following, we refer to the probability distribution function $G(x)$ that generates data as the true model or the true distribution. In contrast, let $F(x)$ be an arbitrarily specified model. If the probability distribution functions $G(x)$ and $F(x)$ have density functions $g(x)$ and $f(x)$, respectively, then they are called *continuous models* (or *continuous distribution models*). If, given either a finite set or a countably infinite set of discrete points $\{x_1, x_2, \ldots, x_k, \ldots\}$, they are expressed as probabilities of events

$$\begin{aligned} g_i &= g(x_i) \equiv \Pr(\{\omega;\ X(\omega) = x_i\}), \\ f_i &= f(x_i) \equiv \Pr(\{\omega;\ X(\omega) = x_i\}), \quad i = 1, 2, \ldots, \end{aligned} \tag{3.1}$$

then these models are called *discrete models* (*discrete distribution models*).

We assume that the goodness of the model $f(x)$ is assessed in terms of the closeness as a probability distribution to the true distribution $g(x)$. As a measure of this closeness, Akaike (1973) proposed the use of the following *Kullback–Leibler information* [or Kullback–Leibler divergence, Kullback–Leibler (1951), hereinafter abbreviated as "K-L information"]:

$$I(G;F) = E_G \left[\log \left\{ \frac{G(X)}{F(X)} \right\} \right], \tag{3.2}$$

where E_G represents the expectation with respect to the probability distribution G.

If the probability distribution functions are continuous models that have the density functions $g(x)$ and $f(x)$, then the K-L information can be expressed as

$$I(g;f) = \int_{-\infty}^{\infty} \log \left\{ \frac{g(x)}{f(x)} \right\} g(x) dx. \tag{3.3}$$

If the probability distribution functions are discrete models for which the probabilities are given by $\{g(x_i); i = 1, 2, \ldots\}$ and $\{f(x_i); i = 1, 2, \ldots\}$, then the K-L information can be expressed as

$$I(g;f) = \sum_{i=1}^{\infty} g(x_i) \log \left\{ \frac{g(x_i)}{f(x_i)} \right\}. \tag{3.4}$$

By unifying the continuous and discrete models, we can express the K-L information as follows:

$$\begin{aligned} I(g\,;f) &= \int \log \left\{ \frac{g(x)}{f(x)} \right\} dG(x) \\ &= \begin{cases} \displaystyle\int_{-\infty}^{\infty} \log \left\{ \frac{g(x)}{f(x)} \right\} g(x) dx, & \text{for continuous model,} \\ \displaystyle\sum_{i=1}^{\infty} g(x_i) \log \left\{ \frac{g(x_i)}{f(x_i)} \right\}, & \text{for discrete model.} \end{cases} \end{aligned} \tag{3.5}$$

Properties of K-L information. The K-L information has the following properties:

(i) $I(g;f) \geq 0$,
(ii) $I(g;f) = 0 \iff g(x) = f(x)$.

In view of these properties, we consider that the smaller the quantity of K-L information, the closer the model $f(x)$ is to $g(x)$.

Proof. First, let us consider the function $K(t) = \log t - t + 1$, which is defined for $t > 0$. In this case, the derivative of $K(t)$, $K'(t) = t^{-1} - 1$, satisfies the condition $K'(1) = 0$, and $K(t)$ takes its maximum, $K(1) = 0$, at $t = 1$. Therefore, the inequality $K(t) \leq 0$ holds for all t such that $t > 0$. The equality holds only for $t = 1$, which means that the relationship

$$\log t \leq t - 1 \quad \text{(the equality holds only when } t = 1\text{)}$$

holds.

For the continuous model, by substituting $t = f(x)/g(x)$ into this expression, we obtain

$$\log \frac{f(x)}{g(x)} \leq \frac{f(x)}{g(x)} - 1.$$

By multiplying both sides of the equation by $g(x)$ and integrating them, we obtain

$$\begin{aligned} \int \log \left\{ \frac{f(x)}{g(x)} \right\} g(x) dx &\leq \int \left\{ \frac{f(x)}{g(x)} - 1 \right\} g(x) dx \\ &= \int f(x) dx - \int g(x) dx = 0. \end{aligned}$$

This gives

$$\int \log \left\{ \frac{g(x)}{f(x)} \right\} g(x) dx = - \int \log \left\{ \frac{f(x)}{g(x)} \right\} g(x) dx \geq 0,$$

thus demonstrating (i). Clearly, the equality holds only when $g(x) = f(x)$.

For the discrete model, it suffices to replace the density functions $g(x)$ and $f(x)$ by the probability functions $g(x_i)$ and $f(x_i)$, respectively, and sum the terms over $i = 1, 2, \ldots$ instead of integrating.

Measures of the similarity between distributions. As a measure of the closeness between distributions, the following quantities have been proposed in addition to the K-L information [Kawada (1987)]:

$$\begin{aligned} \chi^2(g; f) &= \sum_{i=1}^{k} \frac{g_i^2}{f_i} - 1 = \sum_{i=1}^{k} \frac{(f_i - g_i)^2}{f_i} && \chi^2\text{-statistics}, \\ I_K(g; f) &= \int \left\{ \sqrt{f(x)} - \sqrt{g(x)} \right\}^2 dx && \text{Hellinger distance}, \\ I_\lambda(g; f) &= \frac{1}{\lambda} \int \left\{ \left(\frac{g(x)}{f(x)} \right)^\lambda - 1 \right\} g(x) dx && \text{Generalized information}, \\ D(g; f) &= \int u \left(\frac{g(x)}{f(x)} \right) g(x) dx && \text{Divergence}, \\ L_1(g; f) &= \int |g(x) - f(x)| dx && L^1\text{-norm}, \\ L_2(g; f) &= \int \{g(x) - f(x)\}^2 dx && L^2\text{-norm}. \end{aligned}$$

In the above divergence, $D(g; f)$, letting $u(x) = \log x$ produces K-L information $I(g; f)$; similarly, letting $u(x) = \lambda^{-1}(x^\lambda - 1)$ produces generalized information $I_\lambda(g; f)$. In $I_\lambda(g; f)$, when $\lambda \to 0$, we obtain K-L information $I(g; f)$. In this book, following Akaike (1973), the model evaluation criterion based on the K-L information will be referred to generically as an *information criterion*.

3.1.2 Examples of K-L Information

We illustrate K-L information by using several specific examples.

Example 1 (K-L information for normal models) Suppose that the true model $g(x)$ and the specified model $f(x)$ have normal distributions $N(\xi, \tau^2)$ and $N(\mu, \sigma^2)$, respectively. If E_G is an expectation with respect to the true model, the random variable X is distributed according to $N(\xi, \tau^2)$, and therefore, the following equation holds:

$$\begin{aligned} E_G\left[(X-\mu)^2\right] &= E_G\left[(X-\xi)^2 + 2(X-\xi)(\xi-\mu) + (\xi-\mu)^2\right] \\ &= \tau^2 + (\xi-\mu)^2. \end{aligned} \tag{3.6}$$

Thus, for the normal distribution $f(x) = (2\pi\sigma^2)^{-\frac{1}{2}} \exp\left\{-(x-\mu)^2/(2\sigma^2)\right\}$, we obtain

$$\begin{aligned} E_G\left[\log f(X)\right] &= E_G\left[-\frac{1}{2}\log(2\pi\sigma^2) - \frac{(X-\mu)^2}{2\sigma^2}\right] \\ &= -\frac{1}{2}\log(2\pi\sigma^2) - \frac{\tau^2 + (\xi-\mu)^2}{2\sigma^2}. \end{aligned} \tag{3.7}$$

In particular, if we let $\mu = \xi$ and $\sigma^2 = \tau^2$ in this expression, it follows that

$$E_G\left[\log g(X)\right] = -\frac{1}{2}\log(2\pi\tau^2) - \frac{1}{2}. \tag{3.8}$$

Therefore, the K-L information of the model $f(x)$ with respect to $g(x)$ is given by

$$\begin{aligned} I(g\,;f) &= E_G\left[\log g(X)\right] - E_G\left[\log f(X)\right] \\ &= \frac{1}{2}\left\{\log\frac{\sigma^2}{\tau^2} + \frac{\tau^2 + (\xi-\mu)^2}{\sigma^2} - 1\right\}. \end{aligned} \tag{3.9}$$

Example 2 (K-L information for normal and Laplace models) Assume that the true model is a two-sided exponential (Laplace) distribution $g(x) = \frac{1}{2}\exp(-|x|)$ and that the specified model $f(x)$ is $N(\mu, \sigma^2)$. In this case, we obtain

$$\begin{aligned} E_G\left[\log g(X)\right] &= -\log 2 - \frac{1}{2}\int_{-\infty}^{\infty} |x| e^{-|x|} dx \\ &= -\log 2 - \int_0^{\infty} x e^{-x} dx \\ &= -\log 2 - 1, \end{aligned} \tag{3.10}$$

$$\begin{aligned} E_G\left[\log f(X)\right] &= -\frac{1}{2}\log(2\pi\sigma^2) - \frac{1}{4\sigma^2}\int_{-\infty}^{\infty}(x-\mu)^2 e^{-|x|} dx \\ &= -\frac{1}{2}\log(2\pi\sigma^2) - \frac{1}{4\sigma^2}(4 + 2\mu^2). \end{aligned} \tag{3.11}$$

Then the K-L information of the model $f(x)$ with respect to $g(x)$ is given by

$$I(g\,;f) = \frac{1}{2}\log(2\pi\sigma^2) + \frac{2+\mu^2}{2\sigma^2} - \log 2 - 1. \tag{3.12}$$

Example 3 (K-L information for two discrete models) Assume that two dice have the following probabilities for rolling the numbers one to six:

$$\begin{aligned} f_a &= \{0.2,\ 0.12,\ 0.18,\ 0.12,\ 0.20,\ 0.18\}, \\ f_b &= \{0.18,\ 0.12,\ 0.14,\ 0.19,\ 0.22,\ 0.15\}. \end{aligned}$$

In this case, which is the fairer die? Since an ideal die has the probabilities $g = \{1/6,\ 1/6,\ 1/6,\ 1/6,\ 1/6,\ 1/6\}$, we take this to be the true model. When we calculate the K-L information, $I(g; f)$, the die that gives the smaller value must be closer to the ideal fair die. Calculating the value of

$$I(g; f) = \sum_{i=1}^{6} g_i \log \frac{g_i}{f_i}, \tag{3.13}$$

we obtain $I(g\,;f_a) = 0.023$ and $I(g\,;f_b) = 0.020$. Thus, in terms of K-L information, it must be concluded that die f_b is the fairer of the two.

3.1.3 Topics on K-L Information

Boltzmann's entropy. The negative of the K-L information, $B(g\,;f) = -I(g\,;f)$, is referred to as *Boltzmann's entropy.* In the case of the discrete distribution model $f = \{f_1, \ldots, f_k\}$, the entropy can be interpreted as a quantity that varies proportionally with the logarithm of the probability W in which the relative frequency of the sample obtained from the specified model agrees with the true distribution.

Proof. Suppose that we have n independent samples from a distribution that follows the model f, and assume that either a frequency distribution $\{n_1, \ldots, n_k\}$ $(n_1 + n_2 + \cdots + n_k = n)$ or a relative frequency $\{g_1, g_2, \ldots, g_k\}$ $(g_i = n_i/n)$ is obtained. Since the probability with which such a frequency distribution $\{n_1, \ldots, n_k\}$ is obtained is

$$W = \frac{n!}{n_1! \cdots n_k!} f_1^{n_1} \cdots f_k^{n_k}, \tag{3.14}$$

we take the logarithm of this quantity, and, using Stirling's approximation $(\log n! \sim n \log n - n)$, we obtain

$$\log W = \log n! - \sum_{i=1}^{k} \log n_i! + \sum_{i=1}^{k} n_i \log f_i$$

$$
\begin{aligned}
&\sim n\log n - n - \sum_{i=1}^{k} n_i \log n_i + \sum_{i=1}^{k} n_i + \sum_{i=1}^{k} n_i \log f_i \\
&= -\sum_{i=1}^{k} n_i \log\left\{\frac{n_i}{n}\right\} + \sum_{i=1}^{k} n_i \log f_i \\
&= \sum_{i=1}^{k} n_i \log\left\{\frac{f_i}{g_i}\right\} = n\sum_{i=1}^{k} g_i \log\left\{\frac{f_i}{g_i}\right\} \\
&= n \cdot B(g\,;f).
\end{aligned}
$$

Hence, it follows that $B(g\,;f) \sim n^{-1}\log W$; that is, $B(g;f)$ is approximately proportional to the logarithm of the probability of which the relative frequency of the sample obtained from the specified model agrees with the true distribution.

We notice that, in the above statement, the K-L information is not the probability of obtaining the distribution defined by a model from the true distribution. Rather, it is thought of as the probability of obtaining the observed data from the model.

On the functional form of K-L information. If the differentiable function F defined on $(0, \infty)$ satisfies the relationship

$$
\sum_{i=1}^{k} g_i F(f_i) \le \sum_{i=1}^{k} g_i F(g_i) \tag{3.15}
$$

for any two probability functions $\{g_1, \ldots, g_k\}$ and $\{f_1, \ldots, f_k\}$, then $F(g) = \alpha + \beta \log g$ for some α, β with $\beta > 0$.

Proof. In order to demonstrate that $F(g) = \alpha + \beta\log g$, it suffices to show that $gF'(g) = \beta > 0$ and hence that $\partial F/\partial g = \beta/g$. Let $h = (h_1, \ldots, h_k)^T$ be an arbitrary vector that satisfies $\sum_{i=1}^{k} h_i = 0$ and $|h_i| \le \max\{g_i, 1-g_i\}$. Since $g + \lambda h$ is a probability distribution, it follows from the assumption that

$$
\varphi(\lambda) \equiv \sum_{i=1}^{k} g_i F(g_i + \lambda h_i) \le \sum_{i=1}^{k} g_i F(g_i) = \varphi(0).
$$

Therefore, since

$$
\varphi'(\lambda) = \sum_{i=1}^{k} g_i F'(g_i + \lambda h_i) h_i, \quad \varphi'(0) = \sum_{i=1}^{k} g_i F'(g_i) h_i = 0
$$

are always true, by writing $h_1 = C,\ h_2 = -C,\ h_i = 0\ (i = 3, \ldots, k)$, we have

$$
g_1 F'(g_1) = g_2 F'(g_2) = \text{const} = \beta.
$$

The equality for other values of i can be shown in a similar manner.

This result does not imply that the measure that satisfies $I(g : f) \geq 0$ is intrinsically limited to the K-L information. Rather, as indicated by (3.16) in the next section, the result shows that any measure that can be decomposed into two additive terms is limited to the K-L information.

3.2 Expected Log-Likelihood and Corresponding Estimator

The preceding section showed that we can evaluate the appropriateness of a given model by calculating the K-L information. However, K-L information can be used in actual modeling only in limited cases, since K-L information contains the unknown distribution g, so that its value cannot be calculated directly.

K-L information can be decomposed into

$$I(g\,;f) = E_G\left[\log\left\{\frac{g(X)}{f(X)}\right\}\right] = E_G\left[\log g(X)\right] - E_G\left[\log f(X)\right]. \tag{3.16}$$

Moreover, because the first term on the right-hand side is a constant that depends solely on the true model g, it is clear that in order to compare different models, it is sufficient to consider only the second term on the right-hand side. This term is called the *expected log-likelihood.* The larger this value is for a model, the smaller its K-L information is and the better the model is.

Since the expected log-likelihood can be expressed as

$$\begin{aligned} E_G\left[\log f(X)\right] &= \int \log f(x) dG(x) \\ &= \begin{cases} \displaystyle\int_{-\infty}^{\infty} g(x)\log f(x)dx, & \text{for continuous models,} \\ \displaystyle\sum_{i=1}^{\infty} g(x_i)\log f(x_i), & \text{for discrete models,} \end{cases} \end{aligned} \tag{3.17}$$

it still depends on the true distribution g and is an unknown quantity that eludes explicit computation. However, if a good estimate of the expected log-likelihood can be obtained from the data, this estimate can be used as a criterion for comparing models. Let us now consider the following problem.

Let $\boldsymbol{x}_n = \{x_1, x_2, \ldots, x_n\}$ be data observed from the true distribution $G(x)$ or $g(x)$. An estimate of the expected log-likelihood can be obtained by replacing the unknown probability distribution G contained in (3.17) with an empirical distribution function $\hat{G}$ based on data $\boldsymbol{x}_n$. The empirical distribution function is the distribution function for the probability function $\hat{g}(x_\alpha) = 1/n$ $(\alpha = 1, 2, \ldots, n)$ that has the equal probability $1/n$ for each of n observations

$\{x_1, x_2, \ldots, x_n\}$ (see Section 5.1). In fact, by replacing the unknown probability distribution G contained in (3.17) with the empirical distribution function $\hat{G}(x)$, we obtain

$$\begin{aligned} E_{\hat{G}}\left[\log f(X)\right] &= \int \log f(x) d\hat{G}(x) \\ &= \sum_{\alpha=1}^{n} \hat{g}(x_\alpha) \log f(x_\alpha) \qquad (3.18) \\ &= \frac{1}{n} \sum_{\alpha=1}^{n} \log f(x_\alpha). \end{aligned}$$

According to the law of large numbers, when the number of observations, n, tends to infinity, the mean of the random variables $Y_\alpha = \log f(X_\alpha)$ $(\alpha = 1, 2, \ldots, n)$ converges in probability to its expectation, that is, the convergence

$$\frac{1}{n} \sum_{\alpha=1}^{n} \log f(X_\alpha) \longrightarrow E_G\left[\log f(X)\right], \qquad n \to +\infty, \qquad (3.19)$$

holds. Therefore, it is clear that the estimate based on the empirical distribution function in (3.18) is a natural estimate of the expected log-likelihood. The estimate of the expected log-likelihood multiplied by n, i.e.,

$$n \int \log f(x) d\hat{G}(x) = \sum_{\alpha=1}^{n} \log f(x_\alpha), \qquad (3.20)$$

is the log-likelihood of the model $f(x)$. This means that the *log-likelihood*, frequently used in statistical analyses, is clearly understood as being an approximation to the K-L information.

Example 4 (Expected log-likelihood for normal models) Let both of the continuous models $g(x)$ and $f(x)$ be the standard normal distribution $N(0, 1)$ with mean 0 and variance 1. Let us generate n observations, $\{x_1, x_2, \ldots, x_n\}$, from the true model $g(x)$ to construct the empirical distribution function $\hat{G}$. In the next step, we calculate the value of (3.18),

$$E_{\hat{G}}\left[\log f(X)\right] = -\frac{1}{2} \log(2\pi) - \frac{1}{2n} \sum_{\alpha=1}^{n} x_\alpha^2.$$

Table 3.1 shows the results of obtaining the mean and the variance of $E_{\hat{G}}[\log f(X)]$ by repeating this process 1,000 times.

Since the average of the 1,000 trials is very close to the true value, that is, the expected log-likelihood

$$E_G[\log f(X)] = \int g(x) \log f(x) dx = -\frac{1}{2} \log(2\pi) - \frac{1}{2} = -1.4189,$$

Table 3.1. Distribution of the log-likelihood of a normal distribution model. The mean, variance, and standard deviation are obtained by running 1,000 Monte Carlo trials. The expression $E_G[\log f(X)]$ represents the expected log-likelihood.

n	10	100	1,000	10,000	$E_G[\log f(X)]$
Mean	−1.4188	−1.4185	−1.4191	−1.4189	−1.4189
Variance	0.05079	0.00497	0.00050	0.00005	—
Standard deviation	0.22537	0.07056	0.02232	0.00696	—

the results suggest that even for a small number of observations, the log-likelihood has little bias. By contrast, the variance decreases in inverse proportion to n.

3.3 Maximum Likelihood Method and Maximum Likelihood Estimators

3.3.1 Log-Likelihood Function and Maximum Likelihood Estimators

Let us consider the case in which a model is given in the form of a probability distribution $f(x|\boldsymbol{\theta})(\boldsymbol{\theta} \in \Theta \subset R^p)$, having unknown p-dimensional parameters $\boldsymbol{\theta} = (\theta_1, \theta_2, \ldots, \theta_p)^T$. In this case, given data $\boldsymbol{x}_n = \{x_1, x_2, \ldots, x_n\}$, the log-likelihood can be determined for each $\boldsymbol{\theta} \in \Theta$. Therefore, by regarding the log-likelihood as a function of $\boldsymbol{\theta} \in \Theta$, and representing it as

$$\ell(\boldsymbol{\theta}) = \sum_{\alpha=1}^{n} \log f(x_\alpha|\boldsymbol{\theta}), \tag{3.21}$$

the log-likelihood is referred to as the *log-likelihood function*. A natural estimator of $\boldsymbol{\theta}$ is defined by finding the maximizer $\boldsymbol{\theta} \in \Theta$ of the $\ell(\boldsymbol{\theta})$, that is, by determining $\boldsymbol{\theta}$ that satisfies the equation

$$\ell(\hat{\boldsymbol{\theta}}) = \max_{\boldsymbol{\theta} \in \Theta} \ell(\boldsymbol{\theta}). \tag{3.22}$$

This method is called the *maximum likelihood method*, and $\hat{\boldsymbol{\theta}}$ is called the *maximum likelihood estimator*. If the data used in the estimation must be specified explicitly, then the maximum likelihood estimator is denoted by $\hat{\boldsymbol{\theta}}(\boldsymbol{x}_n)$. The model $f(x|\hat{\boldsymbol{\theta}})$ determined by $\hat{\boldsymbol{\theta}}$ is called the *maximum likelihood model*, and the term $\ell(\hat{\boldsymbol{\theta}}) = \sum_{\alpha=1}^{n} \log f(x_\alpha|\hat{\boldsymbol{\theta}})$ is called the *maximum log-likelihood*.

3.3.2 Implementation of the Maximum Likelihood Method by Means of Likelihood Equations

If the log-likelihood function $\ell(\boldsymbol{\theta})$ is continuously differentiable, the maximum likelihood estimator $\hat{\boldsymbol{\theta}}$ is given as a solution of the likelihood equation

$$\frac{\partial \ell(\boldsymbol{\theta})}{\partial \theta_i} = 0, \quad i = 1, 2, \ldots, p \quad \text{or} \quad \frac{\partial \ell(\boldsymbol{\theta})}{\partial \boldsymbol{\theta}} = \mathbf{0}, \tag{3.23}$$

where $\partial \ell(\boldsymbol{\theta})/\partial \boldsymbol{\theta}$ is a p-dimensional vector, the ith component of which is given by $\partial \ell(\boldsymbol{\theta})/\partial \theta_i$, and $\mathbf{0}$ is the p-dimensional zero vector, all the components of which are 0. In particular, if the likelihood equation is a linear equation having p-dimensional parameters, the maximum likelihood estimator can be expressed explicitly.

Example 5 (Normal model) Let us consider the normal distribution model $N(\mu, \sigma^2)$ with respect to the data $\{x_1, x_2, \ldots, x_n\}$. Since the log-likelihood function is given by

$$\ell(\mu, \sigma^2) = -\frac{n}{2} \log(2\pi\sigma^2) - \frac{1}{2\sigma^2} \sum_{\alpha=1}^{n} (x_\alpha - \mu)^2, \tag{3.24}$$

the likelihood equation takes the form

$$\begin{aligned} \frac{\partial \ell(\mu, \sigma^2)}{\partial \mu} &= \frac{1}{\sigma^2} \sum_{\alpha=1}^{n} (x_\alpha - \mu) = \frac{1}{\sigma^2} \left(\sum_{\alpha=1}^{n} x_\alpha - n\mu \right) = 0, \\ \frac{\partial \ell(\mu, \sigma^2)}{\partial \sigma^2} &= -\frac{n}{2\sigma^2} + \frac{1}{2(\sigma^2)^2} \sum_{\alpha=1}^{n} (x_\alpha - \mu)^2 = 0. \end{aligned}$$

It follows, then, that the maximum likelihood estimators for μ and σ^2 are

$$\hat{\mu} = \frac{1}{n} \sum_{\alpha=1}^{n} x_\alpha, \quad \hat{\sigma}^2 = \frac{1}{n} \sum_{\alpha=1}^{n} (x_\alpha - \hat{\mu})^2. \tag{3.25}$$

For the following 20 observations

−7.99	−4.01	−1.56	−0.99	−0.93	−0.80	−0.77	−0.71	−0.42	−0.02
0.65	0.78	0.80	1.14	1.15	1.24	1.29	2.81	4.84	6.82

the maximum likelihood estimates of μ and σ^2 are calculated as

$$\hat{\mu} = \frac{1}{n} \sum_{\alpha=1}^{n} x_\alpha = 0.166, \qquad \hat{\sigma}^2 = \frac{1}{n} \sum_{\alpha=1}^{n} (x_\alpha - \hat{\mu})^2 = 8.545, \tag{3.26}$$

and the maximum log-likelihood is

$$\ell(\hat{\mu}, \hat{\sigma}^2) = -\frac{n}{2} \log(2\pi\hat{\sigma}^2) - \frac{n}{2} = -49.832. \tag{3.27}$$

Example 6 (Bernoulli model) The log-likelihood function based on n observations $\{x_1, x_2, \ldots, x_n\}$ drawn from the Bernoulli distribution $f(x|p) = p^x(1-p)^{1-x}$ $(x = 0, 1)$ is

$$\begin{aligned} \ell(p) &= \log\left\{\prod_{\alpha=1}^{n} p^{x_\alpha}(1-p)^{1-x_\alpha}\right\} \\ &= \sum_{\alpha=1}^{n} x_\alpha \log p + \left(n - \sum_{\alpha=1}^{n} x_\alpha\right)\log(1-p). \end{aligned} \tag{3.28}$$

Consequently, the likelihood equation is

$$\frac{\partial \ell(p)}{\partial p} = \frac{1}{p}\sum_{\alpha=1}^{n} x_\alpha - \frac{1}{1-p}\left(n - \sum_{\alpha=1}^{n} x_\alpha\right) = 0. \tag{3.29}$$

Thus, the maximum likelihood estimator for p is given by

$$\hat{p} = \frac{1}{n}\sum_{\alpha=1}^{n} x_\alpha. \tag{3.30}$$

Example 7 (Linear regression model) Let $\{y_\alpha, x_{\alpha 1}, x_{\alpha 2}, \ldots, x_{\alpha p}\}$ $(\alpha = 1, 2, \ldots, n)$ be n sets of data that are observed with respect to a response variable y and p explanatory variables $\{x_1, x_2, \ldots, x_p\}$. In order to describe the relationship between the variables, we assume the following linear regression model with Gaussian noise:

$$y_\alpha = \boldsymbol{x}_\alpha^T \boldsymbol{\beta} + \varepsilon_\alpha, \quad \varepsilon_\alpha \sim N(0, \sigma^2), \quad \alpha = 1, 2, \ldots, n, \tag{3.31}$$

where $\boldsymbol{x}_\alpha = (1, x_{\alpha 1}, x_{\alpha 2}, \ldots, x_{\alpha p})^T$ and $\boldsymbol{\beta} = (\beta_0, \beta_1, \ldots, \beta_p)^T$. Since the probability density function of y_α is

$$f(y_\alpha|\boldsymbol{x}_\alpha; \boldsymbol{\theta}) = \frac{1}{\sqrt{2\pi\sigma^2}}\exp\left\{-\frac{1}{2\sigma^2}\left(y_\alpha - \boldsymbol{x}_\alpha^T\boldsymbol{\beta}\right)^2\right\}, \tag{3.32}$$

the log-likelihood function is expressed as

$$\begin{aligned} \ell(\boldsymbol{\theta}) &= \sum_{\alpha=1}^{n} \log f(y_\alpha|\boldsymbol{x}_\alpha; \boldsymbol{\theta}) \\ &= -\frac{n}{2}\log(2\pi\sigma^2) - \frac{1}{2\sigma^2}\sum_{\alpha=1}^{n}\left(y_\alpha - \boldsymbol{x}_\alpha^T\boldsymbol{\beta}\right)^2 \\ &= -\frac{n}{2}\log(2\pi\sigma^2) - \frac{1}{2\sigma^2}(\boldsymbol{y} - X\boldsymbol{\beta})^T(\boldsymbol{y} - X\boldsymbol{\beta}), \end{aligned} \tag{3.33}$$

where $\boldsymbol{y} = (y_1, y_2, \ldots, y_n)^T$ and $X = (\boldsymbol{x}_1, \boldsymbol{x}_2, \ldots, \boldsymbol{x}_n)^T$. By taking partial derivatives of the above equation with respect to the parameter vector $\boldsymbol{\theta} = (\boldsymbol{\beta}^T, \sigma^2)^T$, the likelihood equation is given by

$$\frac{\partial \ell(\boldsymbol{\theta})}{\partial \boldsymbol{\beta}} = -\frac{1}{2\sigma^2}\left(-2X^T\boldsymbol{y} + 2X^TX\boldsymbol{\beta}\right) = \mathbf{0},$$

$$\frac{\partial \ell(\boldsymbol{\theta})}{\partial \sigma^2} = -\frac{n}{2\sigma^2} + \frac{1}{2\sigma^4}(\boldsymbol{y} - X\boldsymbol{\beta})^T(\boldsymbol{y} - X\boldsymbol{\beta}) = 0. \tag{3.34}$$

Consequently, the maximum likelihood estimators for $\boldsymbol{\beta}$ and σ^2 are given by

$$\hat{\boldsymbol{\beta}} = (X^TX)^{-1}X^T\boldsymbol{y}, \qquad \hat{\sigma}^2 = \frac{1}{n}(\boldsymbol{y} - X\hat{\boldsymbol{\beta}})^T(\boldsymbol{y} - X\hat{\boldsymbol{\beta}}). \tag{3.35}$$

3.3.3 Implementation of the Maximum Likelihood Method by Numerical Optimization

Although in the preceding section we showed cases in which it was possible to obtain an explicit solution to the likelihood equations, in general likelihood equations are complex nonlinear functions of the parameter vector $\boldsymbol{\theta}$. In this subsection, we describe how to obtain the maximum likelihood estimator in such situations.

When a given likelihood equation cannot be solved explicitly, a numerical optimization method is frequently employed, which involves starting from an appropriately chosen initial value $\boldsymbol{\theta}_0$ and successively generating quantities $\boldsymbol{\theta}_1, \boldsymbol{\theta}_2, \ldots$, in order to cause convergence to the solution $\hat{\boldsymbol{\theta}}$. Assuming that the estimated value $\boldsymbol{\theta}_k$ can be determined at some stage, we determine the next point, $\boldsymbol{\theta}_{k+1}$, which yields a larger likelihood, using the method described below.

In the maximum likelihood method, in order to determine the $\hat{\boldsymbol{\theta}}$ that maximizes $\ell(\boldsymbol{\theta})$, we find $\boldsymbol{\theta}$ that satisfies the necessary condition, namely the likelihood equation $\partial \ell(\boldsymbol{\theta})/\partial \boldsymbol{\theta} = \mathbf{0}$. However, since $\boldsymbol{\theta}_k$ does not exactly satisfy $\partial \ell(\boldsymbol{\theta})/\partial \boldsymbol{\theta} = \mathbf{0}$, we generate the next point, $\boldsymbol{\theta}_{k+1}$, in order to approximate 0 closer. For this purpose, we first perform a Taylor series expansion of $\partial \ell(\boldsymbol{\theta})/\partial \boldsymbol{\theta}$ in the neighborhood of $\boldsymbol{\theta}_k$,

$$\frac{\partial \ell(\boldsymbol{\theta})}{\partial \boldsymbol{\theta}} \approx \frac{\partial \ell(\boldsymbol{\theta}_k)}{\partial \boldsymbol{\theta}} + \frac{\partial^2 \ell(\boldsymbol{\theta}_k)}{\partial \boldsymbol{\theta} \partial \boldsymbol{\theta}^T}(\boldsymbol{\theta} - \boldsymbol{\theta}_k). \tag{3.36}$$

Then by writing

$$\boldsymbol{g}(\boldsymbol{\theta}) = \left(\frac{\partial \ell(\boldsymbol{\theta})}{\partial \theta_1}, \frac{\partial \ell(\boldsymbol{\theta})}{\partial \theta_2}, \ldots, \frac{\partial \ell(\boldsymbol{\theta})}{\partial \theta_p}\right)^T,$$

$$H(\boldsymbol{\theta}) = \frac{\partial^2 \ell(\boldsymbol{\theta})}{\partial \boldsymbol{\theta} \partial \boldsymbol{\theta}^T} = \left(\frac{\partial^2 \ell(\boldsymbol{\theta})}{\partial \theta_i \partial \theta_j}\right), \quad i, j = 1, 2, \ldots, p, \tag{3.37}$$

in terms of $\boldsymbol{\theta}$ that satisfies $\partial \ell(\boldsymbol{\theta})/\partial \boldsymbol{\theta} = \mathbf{0}$, we obtain

$$\mathbf{0} = \boldsymbol{g}(\boldsymbol{\theta}) \approx \boldsymbol{g}(\boldsymbol{\theta}_k) + H(\boldsymbol{\theta}_k)(\boldsymbol{\theta} - \boldsymbol{\theta}_k), \tag{3.38}$$

where the quantity $\boldsymbol{g}(\boldsymbol{\theta}_k)$ is a gradient vector and $H(\boldsymbol{\theta}_k)$ is a Hessian matrix. By virtue of (3.38), it follows that $\boldsymbol{\theta} \approx \boldsymbol{\theta}_k - H(\boldsymbol{\theta}_k)^{-1}\boldsymbol{g}(\boldsymbol{\theta}_k)$. Therefore, using

$$\boldsymbol{\theta}_{k+1} \equiv \boldsymbol{\theta}_k - H(\boldsymbol{\theta}_k)^{-1}\boldsymbol{g}(\boldsymbol{\theta}_k),$$

we determine the next point, $\boldsymbol{\theta}_{k+1}$. This technique, called *the Newton–Raphson method*, is known to converge rapidly near the root, or in other words, provided an appropriate initial value is chosen.

Thus, while the Newton–Raphson method is considered to be an efficient technique, several difficulties may be encountered when it is applied to maximum likelihood estimation: (1) in many cases, it may prove difficult to calculate the Hessian matrix, which is the 2nd-order partial derivative of the log-likelihood; (2) for each matrix, the method requires calculating the inverse matrix of $H(\boldsymbol{\theta}_k)$; and (3) depending on how the initial value is selected, the method may converge very slowly or even diverge.

In order to mitigate these problems, a *quasi-Newton method* is employed. This method does not involve calculating the Hessian matrix and automatically generates the inverse matrix, $H^{-1}(\boldsymbol{\theta}_k)$. In addition, step widths can be introduced either to accelerate convergence or to prevent divergence. Specifically, the following algorithm is employed in order to successively generate $\boldsymbol{\theta}_{k+1}$:

(i) Determine a search (descending) direction vector $\boldsymbol{d}_k = -H_k^{-1}\boldsymbol{g}_k$.
(ii) Determine the optimum step width λ_k that maximizes $\ell(\boldsymbol{\theta}_k + \lambda \boldsymbol{d}_k)$.
(iii) By taking $\boldsymbol{\theta}_{k+1} \equiv \boldsymbol{\theta}_k + \lambda_k \boldsymbol{d}_k$, determine the next point, $\boldsymbol{\theta}_{k+1}$, and set $\boldsymbol{y}_k \equiv \boldsymbol{g}(\boldsymbol{\theta}_{k+1}) - \boldsymbol{g}(\boldsymbol{\theta}_k)$.
(iv) Update an estimate of $H(\boldsymbol{\theta}_k)^{-1}$ by using either the Davidon–Fletcher–Powell (DFP) algorithm or the Broyden–Fletcher–Goldfarb–Shanno (BFGS) algorithm:

$$H_{k+1}^{-1} = H_k^{-1} + \frac{\boldsymbol{s}_k \boldsymbol{s}_k^T}{\boldsymbol{s}_k^T \boldsymbol{y}_k} - \frac{H_k^{-1}\boldsymbol{y}_k \boldsymbol{y}_k^T H_k^{-1}}{\boldsymbol{y}_k^T H_k^{-1} \boldsymbol{y}_k}, \tag{3.39}$$

$$H_{k+1}^{-1} = H_k^{-1} + \frac{\boldsymbol{s}_k \boldsymbol{y}_k^T H_k^{-1}}{\boldsymbol{s}_k^T \boldsymbol{y}_k} - \frac{H_k^{-1}\boldsymbol{y}_k \boldsymbol{s}_k^T}{\boldsymbol{s}_k^T \boldsymbol{y}_k} + \left\{1 + \frac{\boldsymbol{y}_k H_k^{-1} \boldsymbol{y}_k^T}{\boldsymbol{s}_k^T \boldsymbol{y}_k}\right\} \frac{\boldsymbol{s}_k \boldsymbol{s}_k^T}{\boldsymbol{s}_k^T \boldsymbol{y}_k},$$

where $\boldsymbol{s}_k = \boldsymbol{\theta}_{k+1} - \boldsymbol{\theta}_k$.

When applying the quasi-Newton method, one starts with appropriate initial values, $\boldsymbol{\theta}_0$ and H_0^{-1}, and successively determines $\boldsymbol{\theta}_k$ and H_k^{-1}. As an initial value for H_0^{-1}, the identity matrix I, an appropriately scaled matrix of the unit matrix, or an approximate value of $H(\boldsymbol{\theta}_0)^{-1}$ is used. In situations in which it is also difficult to calculate the gradient vector $\boldsymbol{g}(\boldsymbol{\theta})$ of a log-likelihood function, $\boldsymbol{g}(\boldsymbol{\theta})$ can be determined solely from the log-likelihood by numerical differentiation.

Other methods besides the Newton–Raphson method and the quasi-Newton method described above (for example, the simplex method) can be used to obtain the maximum likelihood estimate, since it suffices to determine $\boldsymbol{\theta}$ that maximizes the log-likelihood function.

Example 8 (Cauchy distribution model) Consider the Cauchy distribution model expressed by

$$f(x|\mu,\tau^2) = \sum_{\alpha=1}^{n} \log f(x_\alpha|\mu,\tau^2) = \frac{1}{\pi}\frac{\tau}{(y-\mu)^2+\tau^2} \tag{3.40}$$

for the data shown in Example 5. The log-likelihood of the Cauchy distribution model is given by

$$\ell(\mu,\tau^2) = \frac{n}{2}\log\tau^2 - n\log\pi - \sum_{\alpha=1}^{n}\log\left\{(x_\alpha-\mu)^2+\tau^2\right\}. \tag{3.41}$$

Then the first derivatives of $\ell(\mu,\tau^2)$ with respect to μ and τ^2 are

$$\begin{aligned}\frac{\partial\ell}{\partial\mu} &= 2\sum_{\alpha=1}^{n}\frac{x_\alpha-\mu}{(x_\alpha-\mu)^2+\tau^2},\\ \frac{\partial\ell}{\partial\tau^2} &= \frac{n}{2\tau^2} - \sum_{\alpha=1}^{n}\frac{1}{(x_\alpha-\mu)^2+\tau^2}.\end{aligned} \tag{3.42}$$

The maximum likelihood estimates of the parameters μ and τ^2 are then obtained by maximizing the log-likelihood using the quasi-Newton method. Table 3.2 shows the results of the quasi-Newton method when the initial estimates are set to $\theta_0 = (\mu_0, \tau_0^2)^T = (0,1)^T$. The quasi-Newton method only required five iterations to find the maximum likelihood estimates.

Table 3.2. Estimation of the parameters of the Cauchy distribution model by a quasi-Newton algorithm.

k	μ_k	τ_k^2	$\ell(\theta_k)$	$\partial\ell/\partial\mu$	$\partial\ell/\partial\tau^2$
0	0.00000	1.00000	48.12676	−0.83954	−1.09776
1	0.23089	1.30191	47.87427	0.18795	−0.14373
2	0.17969	1.35705	47.86554	−0.04627	−0.04276
3	0.18940	1.37942	47.86484	0.00244	−0.00106
4	0.18886	1.38004	47.86484	−0.00003	−0.00002
5	0.18887	1.38005	47.86484	0.00000	0.00000

Example 9 (Time series model) In general, the time series are mutually correlated and the log-likelihood of the time series model cannot be expressed as the sum of the logarithms of the density function of each observation.

However, the likelihood can generally be expressed by using the conditional distributions as follows:

$$L(\theta) = f(y_1, \ldots, y_N|\theta) = \prod_{n=1}^{N} f(y_n|y_1, \ldots, y_{n-1}). \tag{3.43}$$

Here, for some simple models, each conditional distribution on the right-hand side of the above expression can be obtained from the specified model. For example, for the autoregressive model,

$$y_n = \sum_{j=1}^{m} a_j y_{n-j} + \varepsilon_n, \quad \varepsilon_n \sim N(0, \sigma^2), \tag{3.44}$$

for $n > m$, the conditional distribution is obtained by

$$f(y_n|y_1, \ldots, y_{n-1}) = \frac{1}{\sqrt{2\pi\sigma^2}} \exp\left\{ -\frac{1}{2\sigma^2}\left(y_n - \sum_{j=1}^{m} a_j y_{n-j} \right)^2 \right\}. \tag{3.45}$$

By ignoring the first m conditional distributions, the log-likelihood of an AR model can be approximated by

$$\ell(\theta) = -\frac{N-m}{2}\log(2\pi\sigma^2) - \frac{1}{2\sigma^2}\sum_{n=m+1}^{N}\left(y_n - \sum_{j=1}^{m} a_j y_{n-j} \right)^2, \tag{3.46}$$

where $\theta = (a_1, \ldots, a_m, \sigma^2)^T$. The least squares estimates of the parameters of the AR model are easily obtained by maximizing the approximate log-likelihood. However, for exact maximum likelihood estimation, we need to use the state-space representation of the model shown below.

In general, we assume that the time series y_n is expressed by a state-space model

$$\begin{aligned} \boldsymbol{x}_n &= F_n \boldsymbol{x}_{n-1} + G_n \boldsymbol{v}_n, \\ y_n &= H_n \boldsymbol{x}_n + w_n, \end{aligned} \tag{3.47}$$

where $\boldsymbol{x}_n$ is a properly defined k-dimensional state vector; F_n, G_n, and H_n are $k \times k$, $k \times \ell$, and $1 \times k$ matrices; and $\boldsymbol{v}_n \sim N_\ell(0, Q_n)$ and $w_n \sim N(0, \sigma^2)$. Then the one-step-ahead predictor $\boldsymbol{x}_{n|n-1}$ and its variance covariance matrix $V_{n|n-1}$ of the state vector $\boldsymbol{x}_n$ given the observations $y_1, \ldots, y_{n-1}$ can be obtained very efficiently by using the Kalman filter recursive algorithm as follows [Anderson and Moore (1979) and Kitagawa and Gersch (1996)]:

One-step-ahead prediction

$$\begin{aligned} \boldsymbol{x}_{n|n-1} &= F_n \boldsymbol{x}_{n-1|n-1}, \\ V_{n|n-1} &= F_n V_{n-1|n-1} F_n^T + G_n Q_n G_n^T. \end{aligned} \tag{3.48}$$

Filter

$$\begin{aligned}K_n &= V_{n|n-1}H_n^T(H_nV_{n|n-1}H_n^T + \sigma^2)^{-1},\\ \boldsymbol{x}_{n|n} &= \boldsymbol{x}_{n|n-1} + K_n(y_n - H_n\boldsymbol{x}_{n|n-1}),\\ V_{n|n} &= (I - K_nH_n)V_{n|n-1}.\end{aligned} \tag{3.49}$$

Then the one-step-ahead predictive distribution of the observation y_n given $\{y_1, \ldots, y_{n-1}\}$ can be expressed as

$$p(y_n|y_1, \ldots, y_{n-1}) = \frac{1}{\sqrt{2\pi r_n}} \exp\left\{-\frac{(y_n - H_n\boldsymbol{x}_{n|n-1})^2}{2r_n}\right\} \tag{3.50}$$

with $r_n = H_nV_{n|n-1}H_n^T + R_n$. Therefore, if the model contains some unknown parameter vector $\boldsymbol{\theta}$, the log-likelihood of the time series model expressed in the state-space model is given by

$$\ell(\boldsymbol{\theta}) = -\frac{1}{2}\left\{N\log 2\pi + \sum_{n=1}^{N}\log r_n + \sum_{n=1}^{N}\frac{(y_n - H_n\boldsymbol{x}_{n|n-1})^2}{r_n}\right\}. \tag{3.51}$$

The maximum likelihood estimate of the parameter $\hat{\boldsymbol{\theta}}$ is obtained by maximizing (3.51) with respect to those parameters used in a numerical optimization method.

3.3.4 Fluctuations of the Maximum Likelihood Estimators

Assume that the true distribution $g(x)$ that generates data is the standard normal distribution $N(0, 1)$ with mean 0 and variance 1 and that the specified model $f(x|\theta)$ is a normal distribution in which either the mean μ or the variance σ^2 is unknown. Figures 3.1 and 3.2 are plots of the log-likelihood function

$$\ell(\mu) = -\frac{n}{2}\log(2\pi) - \frac{1}{2}\sum_{\alpha=1}^{n}(x_\alpha - \mu)^2, \tag{3.52}$$

based on n observations with an unknown mean μ and the variance $\sigma^2 = 1$. The horizontal axis represents the value of μ, and the vertical axis represents the corresponding value of $\ell(\mu)$. Figures 3.1 and 3.2 show log-likelihood functions based on $n = 10$ and $n = 100$ observations, respectively. In these figures, random numbers are used to generate 10 sets of observations $\{x_1, x_2, \ldots, x_n\}$ following the distribution $N(0\ ,1)$, and the log-likelihood functions $\ell(\mu)$ $(-2 \le \mu \le 2)$ calculated from the observation sets are overlaid. The value of μ that maximizes these functions is the maximum likelihood estimate of the mean, which is plotted on the horizontal axis with lines pointing downward from the axis. The estimator has a scattered profile depending on the data involved. In the figures, the bold curves represent the expected log-likelihood function

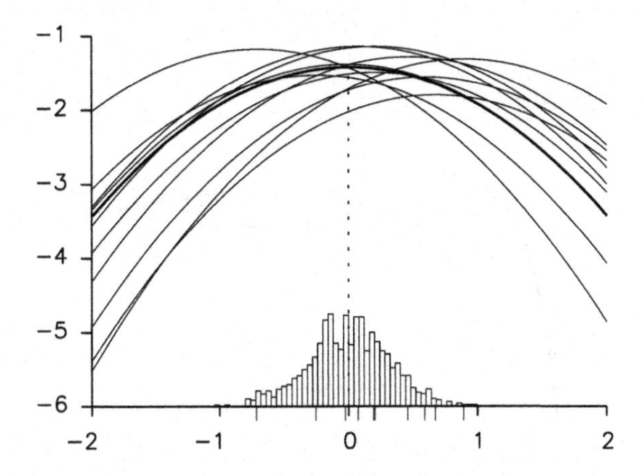

Fig. 3.1. Distributions of expected log-likelihood (bold lines), log-likelihood (thin lines), and maximum likelihood estimators with respect to the mean μ of normal distributions; $n = 10$.

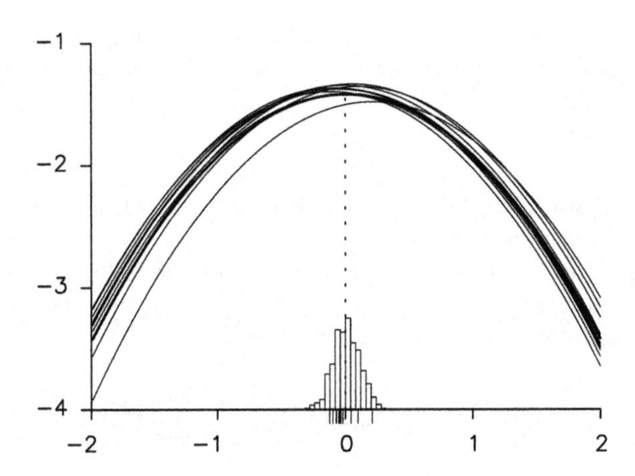

Fig. 3.2. Distributions of the expected log-likelihood (bold), log-likelihood (thin), and maximum likelihood estimator with respect to the mean μ of the normal distribution; $n = 100$.

$$nE_G\left[\log f(X|\mu)\right] = n\int g(x)\log f(x|\mu)dx = -\frac{n}{2}\log(2\pi) - \frac{n(1+\mu^2)}{2},$$

and the values of the true parameter μ_0 corresponding to the function are plotted as dotted lines. The difference between these values and the maximum likelihood estimate is the estimation error of μ. The histogram in the figure, which shows the distribution of the maximum likelihood estimates resulting from similar calculations repeated 1,000 times, indicates that the maximum

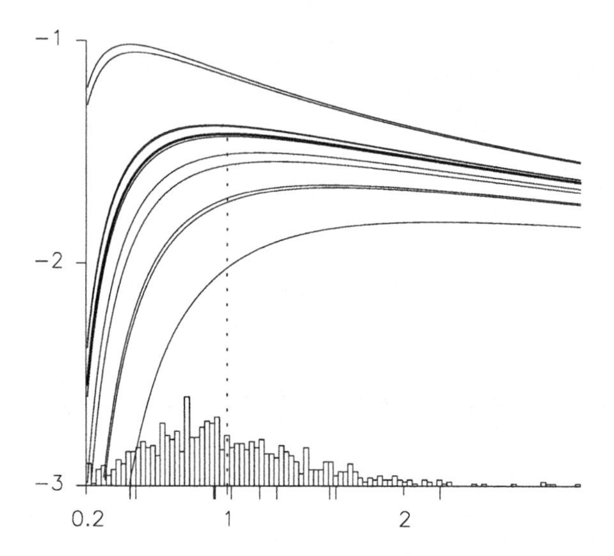

Fig. 3.3. Distributions of the expected log-likelihood (bold), log-likelihood (thin), and maximum likelihood estimator with respect to the variance σ^2 of the normal distribution; $n = 10$.

likelihood estimator has a distribution over a range of ± 1 in the case of $n = 10$, and ± 0.3 in the case of $n = 100$.

Figures 3.3 and 3.4 show 10 overlaid plots of the following log-likelihood function, obtained from $n = 10$ and $n = 100$ observations, respectively, with unknown variance σ^2 and the mean $\mu = 0$:

$$\ell(\sigma^2) = -\frac{n}{2} \log(2\pi\sigma^2) - \frac{1}{2\sigma^2} \sum_{\alpha=1}^{n} x_\alpha^2.$$

In this case, $\ell(\sigma^2)$ is an asymmetric function of σ^2, and the corresponding distribution of the maximum likelihood estimator is also asymmetric. In this case, too, the figures suggest that the distribution of the estimators converges to the true value as n increases. In the figures, the bold curve represents the expected log-likelihood function

$$nE_G \left[\log f(X|\sigma^2) \right] = n \int g(x) \log f(x|\sigma^2) dx = -\frac{n}{2} \log(2\pi\sigma^2) - \frac{n}{2\sigma^2},$$

and the value of the corresponding true parameter is shown by the dotted line. The difference between this value and the maximum likelihood estimator is the estimation error of σ^2. The histograms in the figures show the distribution of the maximum likelihood estimator when the same calculations are repeated 1,000 times.

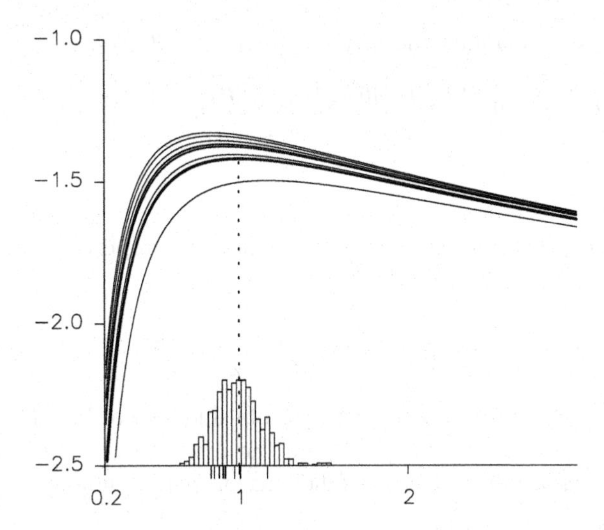

Fig. 3.4. Distributions of the expected log-likelihood (bold), log-likelihood (thin), and maximum likelihood estimator with respect to the variance σ^2 of the normal distribution; $n = 100$.

3.3.5 Asymptotic Properties of the Maximum Likelihood Estimators

This section discusses the asymptotic properties of the maximum likelihood estimator of a continuous parametric model $\{f(x|\boldsymbol{\theta});\ \boldsymbol{\theta} \in \Theta \subset R^p\}$ with p-dimensional parameter vector $\boldsymbol{\theta}$.

Asymptotic normality. Assume that the following regularity condition holds for the density function $f(x|\boldsymbol{\theta})$:

(1) The function $\log f(x|\boldsymbol{\theta})$ is three times continuously differentiable with respect to $\boldsymbol{\theta} = (\theta_1, \theta_2, \ldots, \theta_p)^T$.
(2) There exist integrable functions on R, $F_1(x)$, and $F_2(x)$ and a function $H(x)$ such that

$$\int_{-\infty}^{\infty} H(x) f(x|\boldsymbol{\theta}) dx < M,$$

for an appropriate real value M, and the following inequalities hold for any $\boldsymbol{\theta} \in \Theta$:

$$\left| \frac{\partial \log f(x|\boldsymbol{\theta})}{\partial \theta_i} \right| < F_1(x), \qquad \left| \frac{\partial^2 \log f(x|\boldsymbol{\theta})}{\partial \theta_i \partial \theta_j} \right| < F_2(x),$$

$$\left| \frac{\partial^3 \log f(x|\boldsymbol{\theta})}{\partial \theta_i \partial \theta_j \partial \theta_k} \right| < H(x), \quad i, j, k = 1, 2, \ldots, p.$$

(3) The following inequality holds for an arbitrary $\boldsymbol{\theta} \in \Theta$:

$$0 < \int_{-\infty}^{\infty} f(x|\boldsymbol{\theta}) \frac{\partial \log f(x|\boldsymbol{\theta})}{\partial \theta_i} \frac{\partial \log f(x|\boldsymbol{\theta})}{\partial \theta_j} dx < \infty, \quad i, j = 1, \ldots, p. \tag{3.53}$$

Then, under the above conditions the following properties can be derived:
(a) Assume that $\boldsymbol{\theta}_0$ is a solution of

$$\int f(x|\boldsymbol{\theta}) \frac{\partial \log f(x|\boldsymbol{\theta})}{\partial \boldsymbol{\theta}} dx = \mathbf{0} \tag{3.54}$$

and that data $\boldsymbol{x}_n = \{x_1, x_2, \ldots, x_n\}$ are obtained according to the density function $f(x|\boldsymbol{\theta}_0)$. In addition, let $\hat{\boldsymbol{\theta}}_n$ be the maximum likelihood estimator based on n observations. Then the following properties hold:

(i) The likelihood equation

$$\frac{\partial \ell(\boldsymbol{\theta})}{\partial \boldsymbol{\theta}} = \sum_{\alpha=1}^{n} \frac{\partial \log f(x_\alpha|\boldsymbol{\theta})}{\partial \boldsymbol{\theta}} = \mathbf{0} \tag{3.55}$$

has a solution that converges to $\boldsymbol{\theta}_0$.

(ii) The maximum likelihood estimator $\hat{\boldsymbol{\theta}}_n$ converges in probability to $\boldsymbol{\theta}_0$ when $n \to +\infty$.

(iii) The maximum likelihood estimator $\hat{\boldsymbol{\theta}}_n$ has asymptotic normality, that is, the distribution of $\sqrt{n}(\hat{\boldsymbol{\theta}}_n - \boldsymbol{\theta}_0)$ converges in law to the p-dimensional normal distribution $N_p(\mathbf{0}, I(\boldsymbol{\theta}_0)^{-1})$ with the mean vector $\mathbf{0}$ and the variance covariance matrix $I(\boldsymbol{\theta}_0)^{-1}$, where the matrix $I(\boldsymbol{\theta}_0)$ is the value of the matrix $I(\boldsymbol{\theta})$ at $\boldsymbol{\theta} = \boldsymbol{\theta}_0$, which is given by

$$I(\boldsymbol{\theta}) = \int f(x|\boldsymbol{\theta}) \frac{\partial \log f(x|\boldsymbol{\theta})}{\partial \boldsymbol{\theta}} \frac{\partial \log f(x|\boldsymbol{\theta})}{\partial \boldsymbol{\theta}^T} dx. \tag{3.56}$$

This matrix $I(\boldsymbol{\theta})$, with $(i,j)^{th}$ component given as (3.53) under condition (3), is called the *Fisher information matrix.*

Although the asymptotic normality stated above assumes the existence of $\boldsymbol{\theta}_0 \in \Theta$ that satisfies the assumption $g(x) = f(x|\boldsymbol{\theta}_0)$, similar results, given below, can also be obtained even when the assumption does not hold:

(b) Assume that $\boldsymbol{\theta}_0$ is a solution of

$$\int g(x) \frac{\partial \log f(x|\boldsymbol{\theta})}{\partial \boldsymbol{\theta}} dx = \mathbf{0} \tag{3.57}$$

and that data $\boldsymbol{x}_n = \{x_1, x_2, \cdots, x_n\}$ are observed according to the distribution $g(x)$. In this case, the following statements hold with respect to the maximum likelihood estimator $\hat{\boldsymbol{\theta}}_n$:

(i) The maximum likelihood estimator $\hat{\boldsymbol{\theta}}_n$ converges in probability to $\boldsymbol{\theta}_0$ as $n \to +\infty$.

(ii) The distribution of $\sqrt{n}(\hat{\boldsymbol{\theta}}_n - \boldsymbol{\theta}_0)$ with respect to the maximum likelihood estimator $\hat{\boldsymbol{\theta}}_n$ converges in law to the p-dimensional normal distribution with the mean vector $\mathbf{0}$ and the variance covariance matrix $J^{-1}(\boldsymbol{\theta}_0)I(\boldsymbol{\theta}_0)J^{-1}(\boldsymbol{\theta}_0)$ as $n \to +\infty$. In other words, when $n \to +\infty$, the following holds:

$$\sqrt{n}(\hat{\boldsymbol{\theta}}_n - \boldsymbol{\theta}_0) \to N_p\left(\mathbf{0}, J^{-1}(\boldsymbol{\theta}_0)I(\boldsymbol{\theta}_0)J^{-1}(\boldsymbol{\theta}_0)\right), \tag{3.58}$$

where the matrices $I(\boldsymbol{\theta}_0)$ and $J(\boldsymbol{\theta}_0)$ are the $p \times p$ matrices evaluated at $\boldsymbol{\theta}= \boldsymbol{\theta}_0$ and are given by the following equations:

$$\begin{aligned} I(\boldsymbol{\theta}) &= \int g(x)\frac{\partial \log f(x|\boldsymbol{\theta})}{\partial \boldsymbol{\theta}}\frac{\partial \log f(x|\boldsymbol{\theta})}{\partial \boldsymbol{\theta}^T}dx \\ &= \left(\int g(x)\frac{\partial \log f(x|\boldsymbol{\theta})}{\partial \theta_i}\frac{\partial \log f(x|\boldsymbol{\theta})}{\partial \theta_j}dx\right), \end{aligned} \tag{3.59}$$

$$\begin{aligned} J(\boldsymbol{\theta}) &= -\int g(x)\frac{\partial^2 \log f(x|\boldsymbol{\theta})}{\partial \boldsymbol{\theta}\partial \boldsymbol{\theta}^T}dx \\ &= -\left(\int g(x)\frac{\partial^2 \log f(x|\boldsymbol{\theta})}{\partial \theta_i \partial \theta_j}dx\right), \quad i,j = 1,\ldots,p. \end{aligned} \tag{3.60}$$

Outline of the Proof. By using a Taylor expansion of the first derivative of the maximum log-likelihood $\ell(\hat{\boldsymbol{\theta}}_n) = \sum_{\alpha=1}^n \log f(x_\alpha|\hat{\boldsymbol{\theta}}_n)$ around $\boldsymbol{\theta}_0$, we obtain

$$\mathbf{0} = \frac{\partial \ell(\hat{\boldsymbol{\theta}}_n)}{\partial \boldsymbol{\theta}} = \frac{\partial \ell(\boldsymbol{\theta}_0)}{\partial \boldsymbol{\theta}} + \frac{\partial^2 \ell(\boldsymbol{\theta}_0)}{\partial \boldsymbol{\theta}\partial \boldsymbol{\theta}^T}(\hat{\boldsymbol{\theta}}_n - \boldsymbol{\theta}_0) + \cdots. \tag{3.61}$$

From the Taylor series expansion formula, the following approximation for the maximum likelihood estimator $\hat{\boldsymbol{\theta}}_n$ can be obtained:

$$-\frac{\partial^2 \ell(\boldsymbol{\theta}_0)}{\partial \boldsymbol{\theta}\partial \boldsymbol{\theta}^T}(\hat{\boldsymbol{\theta}}_n - \boldsymbol{\theta}_0) = \frac{\partial \ell(\boldsymbol{\theta}_0)}{\partial \boldsymbol{\theta}}. \tag{3.62}$$

By the law of large numbers, when $n \to +\infty$, it can be shown that

$$-\frac{1}{n}\frac{\partial^2 \ell(\boldsymbol{\theta}_0)}{\partial \boldsymbol{\theta}\partial \boldsymbol{\theta}^T} = -\frac{1}{n}\sum_{\alpha=1}^n \frac{\partial^2}{\partial \boldsymbol{\theta}\partial \boldsymbol{\theta}^T}\log f(x_\alpha|\boldsymbol{\theta})\bigg|_{\boldsymbol{\theta}_0} \to J(\boldsymbol{\theta}_0), \tag{3.63}$$

where $|_{\boldsymbol{\theta}_0}$ is the value of the derivative at $\boldsymbol{\theta} = \boldsymbol{\theta}_0$.

By virtue of the fact that when the p-dimensional random vector is written as $\boldsymbol{X}_\alpha = \partial \log f(X_\alpha|\boldsymbol{\theta})/\partial \boldsymbol{\theta}|_{\boldsymbol{\theta}_0}$ in the multivariate central limit theorem of

Remark 1 below and the right-hand side of (3.62) is $E_G[\boldsymbol{X}_\alpha] = 0$, $E_G[\boldsymbol{X}_\alpha \boldsymbol{X}_\alpha^T] = I(\boldsymbol{\theta}_0)$, it follows that

$$\sqrt{n}\frac{1}{n}\frac{\partial \ell(\boldsymbol{\theta}_0)}{\partial \boldsymbol{\theta}} = \sqrt{n}\frac{1}{n}\sum_{\alpha=1}^{n} \frac{\partial}{\partial \boldsymbol{\theta}} \log f(x_\alpha|\boldsymbol{\theta})\bigg|_{\boldsymbol{\theta}_0} \to N_p(\mathbf{0}, I(\boldsymbol{\theta}_0)). \quad (3.64)$$

Then it follows from (3.62), (3.63), and (3.64) that, when $n \to +\infty$, we obtain

$$\sqrt{n}J(\boldsymbol{\theta}_0)(\hat{\boldsymbol{\theta}} - \boldsymbol{\theta}_0) \longrightarrow N_p(\mathbf{0}, I(\boldsymbol{\theta}_0)). \quad (3.65)$$

Therefore, the convergence in law

$$\sqrt{n}(\hat{\boldsymbol{\theta}} - \boldsymbol{\theta}_0) \longrightarrow N_p\left(\mathbf{0}, J^{-1}(\boldsymbol{\theta}_0)I(\boldsymbol{\theta}_0)J^{-1}(\boldsymbol{\theta}_0)\right) \quad (3.66)$$

holds as n tends to infinity. In fact, it has been shown that this asymptotic normality holds even when the existence of higher-order derivatives is not assumed [Huber (1967)].

If the distribution $g(x)$ that generated the data is included in the class of parametric models $\{f(x|\boldsymbol{\theta}); \boldsymbol{\theta} \in \Theta \subset R^p\}$, from Remark 2 shown below, the equality $I(\boldsymbol{\theta}_0) = J(\boldsymbol{\theta}_0)$ holds, and the asymptotic variance covariance matrix for $\sqrt{n}(\hat{\boldsymbol{\theta}} - \boldsymbol{\theta}_0)$ becomes

$$J^{-1}(\boldsymbol{\theta}_0)I(\boldsymbol{\theta}_0)J^{-1}(\boldsymbol{\theta}_0) = I(\boldsymbol{\theta}_0)^{-1}, \quad (3.67)$$

and the result (a) (iii) falls out.

Remark 1 (Multivariate central limit theorem) Let $\{\boldsymbol{X}_1, \boldsymbol{X}_2, \ldots, \boldsymbol{X}_n, \ldots\}$ be a sequence of mutually independent random vectors drawn from a p-dimensional probability distribution and that have mean vector $E[\boldsymbol{X}_\alpha] = \boldsymbol{\mu}$ and variance covariance matrix $E[(\boldsymbol{X}_\alpha - \boldsymbol{\mu})(\boldsymbol{X}_\alpha - \boldsymbol{\mu})^T] = \Sigma$. Then the distribution of $\sqrt{n}(\overline{\boldsymbol{X}} - \boldsymbol{\mu})$ with respect to the sample mean vector $\overline{\boldsymbol{X}} = \frac{1}{n}\sum_{\alpha=1}^{n} \boldsymbol{X}_\alpha$ converges in law to a p-dimensional normal distribution with mean vector $\mathbf{0}$ and variance covariance matrix Σ when $n \to +\infty$. In other words, when $n \to +\infty$, it holds that

$$\frac{1}{\sqrt{n}}\sum_{\alpha=1}^{n}(\boldsymbol{X}_\alpha - \boldsymbol{\mu}) = \sqrt{n}(\overline{\boldsymbol{X}} - \boldsymbol{\mu}) \;\to\; N_p(\mathbf{0}, \Sigma). \quad (3.68)$$

Remark 2 (Relationship between the matrices $I(\boldsymbol{\theta})$ and $J(\boldsymbol{\theta})$) The following equality holds with respect to the second derivative of the log-likelihood function:

$$\begin{aligned}
&\frac{\partial^2}{\partial \theta_i \partial \theta_j} \log f(x|\boldsymbol{\theta}) \\
&\quad = \frac{\partial}{\partial \theta_i}\left\{\frac{\partial}{\partial \theta_j} \log f(x|\boldsymbol{\theta})\right\}
\end{aligned}$$

$$= \frac{\partial}{\partial \theta_i} \left\{ \frac{1}{f(x|\boldsymbol{\theta})} \frac{\partial}{\partial \theta_j} f(x|\boldsymbol{\theta}) \right\}$$
$$= \frac{1}{f(x|\boldsymbol{\theta})} \frac{\partial^2}{\partial \theta_i \partial \theta_j} f(x|\boldsymbol{\theta}) - \frac{1}{f(x|\boldsymbol{\theta})^2} \frac{\partial}{\partial \theta_i} f(x|\boldsymbol{\theta}) \frac{\partial}{\partial \theta_j} f(x|\boldsymbol{\theta})$$
$$= \frac{1}{f(x|\boldsymbol{\theta})} \frac{\partial^2}{\partial \theta_i \partial \theta_j} f(x|\boldsymbol{\theta}) - \frac{\partial}{\partial \theta_i} \log f(x|\boldsymbol{\theta}) \frac{\partial}{\partial \theta_j} \log f(x|\boldsymbol{\theta}).$$

By taking the expectation of the both sides with respect to the distribution $G(x)$, we obtain

$$E_G \left[\frac{\partial^2}{\partial \theta_i \partial \theta_j} \log f(x|\boldsymbol{\theta}) \right]$$
$$= E_G \left[\frac{1}{f(x|\boldsymbol{\theta})} \frac{\partial^2}{\partial \theta_i \partial \theta_j} f(x|\boldsymbol{\theta}) \right] - E_G \left[\frac{\partial}{\partial \theta_i} \log f(x|\boldsymbol{\theta}) \frac{\partial}{\partial \theta_j} \log f(x|\boldsymbol{\theta}) \right].$$

Hence, in general, we know that $I(\boldsymbol{\theta}) \neq J(\boldsymbol{\theta})$. However, if there exists a parameter vector $\boldsymbol{\theta}_0 \in \Theta$ such that $g(x) = f(x|\boldsymbol{\theta}_0)$, the first term on the right-hand side becomes

$$E_G \left[\frac{1}{f(x|\boldsymbol{\theta}_0)} \frac{\partial^2}{\partial \theta_i \partial \theta_j} f(x|\boldsymbol{\theta}_0) \right] = \int \frac{\partial^2}{\partial \theta_i \partial \theta_j} f(x|\boldsymbol{\theta}_0) dx$$
$$= \frac{\partial^2}{\partial \theta_i \partial \theta_j} \int f(x|\boldsymbol{\theta}_0) dx = 0,$$

and therefore the equality $I_{ij}(\theta_0) = J_{ij}(\theta_0)$ $(i, j = 1, 2, \ldots, p)$ holds; hence, we have $I(\boldsymbol{\theta}_0) = J(\boldsymbol{\theta}_0)$.

3.4 Information Criterion AIC

3.4.1 Log-Likelihood and Expected Log-Likelihood

The argument that has been presented thus far can be summarized as follows. When we build a model using data, we assume that the data $\boldsymbol{x}_n = \{x_1, x_2, \ldots, x_n\}$ are generated according to the true distribution $G(x)$ or $g(x)$. In order to capture the structure of the given phenomena, we assume a parametric model $\{f(x|\boldsymbol{\theta});\ \boldsymbol{\theta} \in \Theta \subset R^p\}$ having p-dimensional parameters, and we estimate it by using the maximum likelihood method. In other words, we construct a statistical model $f(x|\hat{\boldsymbol{\theta}})$ by replacing the unknown parameter $\boldsymbol{\theta}$ contained in the probability distribution by the maximum likelihood estimator $\hat{\boldsymbol{\theta}}$. Our purpose here is to evaluate the goodness or badness of the statistical model $f(x|\hat{\boldsymbol{\theta}})$ thus constructed. We now consider the evaluation of a model from the standpoint of making a prediction.

Our task is to evaluate the expected goodness or badness of the estimated model $f(z|\hat{\boldsymbol{\theta}})$ when it is used to predict the independent future data $Z = z$

generated from the unknown true distribution $g(z)$. The K-L information described below is used to measure the closeness of the two distributions:

$$\begin{aligned} I\{g(z); f(z|\hat{\boldsymbol{\theta}})\} &= E_G\left[\log\left\{\frac{g(Z)}{f(Z|\hat{\boldsymbol{\theta}})}\right\}\right] \\ &= E_G\left[\log g(Z)\right] - E_G\left[\log f(Z|\hat{\boldsymbol{\theta}})\right], \end{aligned} \tag{3.69}$$

where the expectation is taken with respect to the unknown probability distribution $G(z)$ by fixing $\hat{\boldsymbol{\theta}} = \hat{\boldsymbol{\theta}}(\boldsymbol{x}_n)$.

In view of the properties of the K-L information, the larger the expected log-likelihood

$$E_G\left[\log f(Z|\hat{\boldsymbol{\theta}})\right] = \int \log f(z|\hat{\boldsymbol{\theta}})dG(z) \tag{3.70}$$

of the model is, the closer the model is to the true one. Therefore, in the definition of the information criterion, the crucial issue is to obtain a good estimator of the expected log-likelihood. One such estimator is

$$\begin{aligned} E_{\hat{G}}\left[\log f(Z|\hat{\boldsymbol{\theta}})\right] &= \int \log f(z|\hat{\boldsymbol{\theta}})d\hat{G}(z) \\ &= \frac{1}{n}\sum_{\alpha=1}^{n} \log f(x_\alpha|\hat{\boldsymbol{\theta}}), \end{aligned} \tag{3.71}$$

in which the unknown probability distribution G contained in the expected log-likelihood is replaced with an empirical distribution function $\hat{G}$. This is the log-likelihood of the statistical model $f(z|\hat{\boldsymbol{\theta}})$ or the maximum log-likelihood

$$\ell(\hat{\boldsymbol{\theta}}) = \sum_{\alpha=1}^{n} \log f(x_\alpha|\hat{\boldsymbol{\theta}}). \tag{3.72}$$

It is worth noting here that the estimator of the expected log-likelihood $E_G[\log f(Z|\hat{\boldsymbol{\theta}})]$ is $n^{-1}\ell(\hat{\boldsymbol{\theta}})$ and that the log-likelihood $\ell(\hat{\boldsymbol{\theta}})$ is an estimator of $nE_G[\log f(Z|\hat{\boldsymbol{\theta}})]$.

3.4.2 Necessity of Bias Correction for the Log-Likelihood

In practical situations, it is difficult to precisely capture the true structure of given phenomena from a limited number of observed data. For this reason, we construct several candidate statistical models based on the observed data at hand and select the model that most closely approximates the mechanism of the occurrence of the phenomena. In this subsection, we consider the situation in which multiple models $\{f_j(z|\boldsymbol{\theta}_j); j = 1, 2, \ldots, m\}$ exist, and the maximum likelihood estimator $\hat{\boldsymbol{\theta}}_j$ has been obtained for the parameters of the model, $\boldsymbol{\theta}_j$.

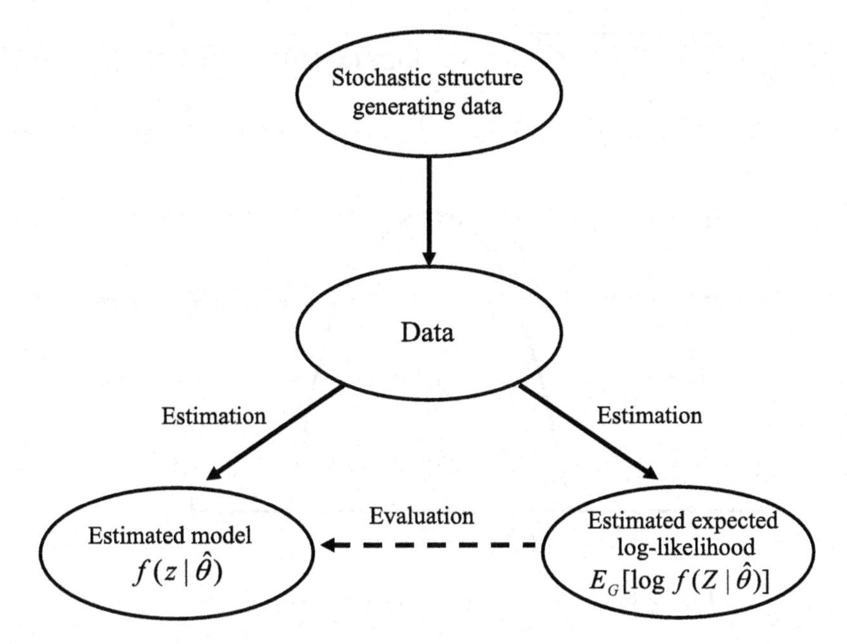

Fig. 3.5. Use of data in the estimations of the parameter of a model and of the expected log-likelihood.

From the foregoing argument, it appears that the goodness of the model specified by $\hat{\boldsymbol{\theta}}_j$, that is, the goodness of the maximum likelihood model $f_j(z|\hat{\boldsymbol{\theta}}_j)$, can be determined by comparing the magnitudes of the maximum log-likelihood $\ell_j(\hat{\boldsymbol{\theta}}_j)$. However, it is known that this approach does not provide a fair comparison of models, since the quantity $\ell_j(\hat{\boldsymbol{\theta}}_j)$ contains a bias as an estimator of the expected log-likelihood $nE_G[\log f_j(z|\hat{\boldsymbol{\theta}}_j)]$, and the magnitude of the bias varies with the dimension of the parameter vector.

This result may seem to contradict the fact that generally $\ell(\boldsymbol{\theta})$ is a good estimator of $nE_G[\log f(Z|\boldsymbol{\theta})]$. However, as is evident from the process by which the log-likelihood in (3.71) was derived, the log-likelihood was obtained by estimating the expected log-likelihood by reusing the data $\boldsymbol{x}_n$ that were initially used to estimate the model in place of the future data (Figure 3.5). The use of the same data twice for estimating the parameters and for estimating the evaluation measure (the expected log-likelihood) of the goodness of the estimated model gives rise to the bias.

Relationship between log-likelihood and expected log-likelihood. Figure 3.6 shows the relationship between the expected log-likelihood function and the log-likelihood function

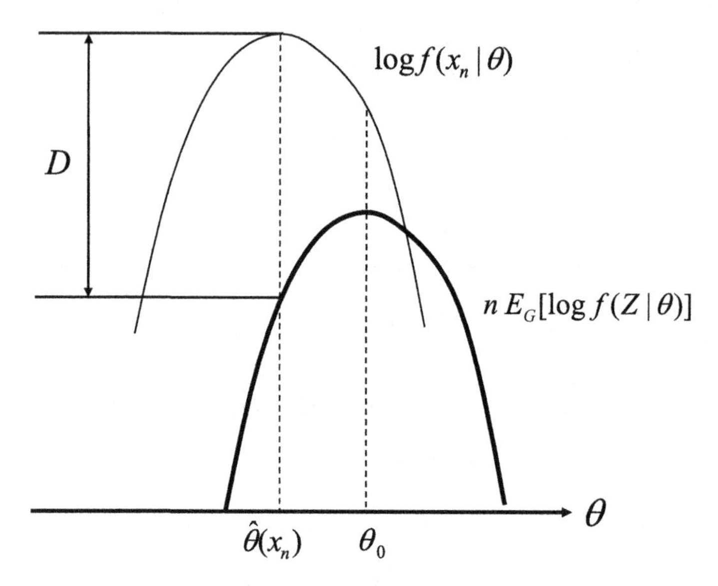

Fig. 3.6. Log-likelihood and expected log-likelihood.

$$n\eta(\theta) = nE_G\left[\log f(Z|\theta)\right], \qquad \ell(\theta) = \sum_{\alpha=1}^{n} \log f(x_\alpha|\theta), \tag{3.73}$$

for a model $f(x|\theta)$ with a one-dimensional parameter θ. The value of θ that maximizes the expected log-likelihood is the true parameter θ_0. On the other hand, the maximum likelihood estimator $\hat{\theta}(\boldsymbol{x}_n)$ is given as the maximizer of the log-likelihood function $\ell(\theta)$. The goodness of the model $f(z|\hat{\theta})$ defined by $\hat{\theta}(\boldsymbol{x}_n)$ should be evaluated in terms of the expected log-likelihood $E_G[\log f(Z|\hat{\theta})]$. However, in actuality, it is evaluated using the log-likelihood $\ell(\hat{\theta})$ that can be calculated from data. In this case, as indicated in Figure 3.6, the true criterion should give $E_G[\log f(Z|\hat{\theta})] \leq E_G[\log f(Z|\theta_0)]$ (see Subsection 3.1.1). However, in the log-likelihood, the relationship $\ell(\hat{\theta}) \geq \ell(\theta_0)$ always holds.

The log-likelihood function fluctuates depending on data, and the geometry between the two functions also varies; however, the above two inequalities always hold. If the two functions have the same form, then the log-likelihood is actually inferior to the extent that it appears to be better than the true model. The objective of the bias evaluation is to compensate for this phenomenon of reversal. Therefore, the prerequisite for a fair comparison of models is evaluation of and correction for the bias. In this subsection, we define an information criterion as a bias-corrected log-likelihood of the model.

Let us assume that n observations $\boldsymbol{x}_n$ generated from the true distribution $G(x)$ or $g(x)$ are realizations of the random variable $\boldsymbol{X}_n = (X_1, X_2, \cdots, X_n)^T$, and let

$$\ell(\hat{\boldsymbol{\theta}}) = \sum_{\alpha=1}^{n} \log f(x_\alpha|\hat{\boldsymbol{\theta}}(\boldsymbol{x}_n)) = \log f(\boldsymbol{x}_n|\hat{\boldsymbol{\theta}}(\boldsymbol{x}_n)) \tag{3.74}$$

represent the log-likelihood of the statistical model $f(z|\hat{\boldsymbol{\theta}}(\boldsymbol{x}_n))$ estimated by the maximum likelihood method. The bias of the log-likelihood as an estimator of the expected log-likelihood given in (3.70) is defined by

$$b(G) = E_{G(\boldsymbol{x}_n)}\left[\log f(\boldsymbol{X}_n|\hat{\boldsymbol{\theta}}(\boldsymbol{X}_n)) - nE_{G(z)}\left[\log f(Z|\hat{\boldsymbol{\theta}}(\boldsymbol{X}_n))\right]\right], \tag{3.75}$$

where the expectation $E_{G(\boldsymbol{x}_n)}$ is taken with respect to the joint distribution, $\prod_{\alpha=1}^{n} G(x_\alpha) = G(\boldsymbol{x}_n)$, of the sample $\boldsymbol{X}_n$, and $E_{G(z)}$ is the expectation on the true distribution $G(z)$. We see that the general form of the information criterion can be constructed by evaluating the bias and correcting for the bias of the log-likelihood as follows:

$$\begin{aligned} \mathrm{IC}(\boldsymbol{X}_n; \hat{G}) &= -2(\text{log-likelihood of statistical model} - \text{bias estimator}) \\ &= -2\sum_{\alpha=1}^{n} \log f(X_\alpha|\hat{\boldsymbol{\theta}}) + 2\left\{\text{estimator for } b(G)\right\}. \end{aligned} \tag{3.76}$$

In general, the bias $b(G)$ can take various forms depending on the relationship between the true distribution generating the data and the specified model and on the method employed to construct a statistical model. In the following, we derive an information criterion for evaluating statistical models constructed by the maximum likelihood method.

3.4.3 Derivation of Bias of the Log-Likelihood

The maximum likelihood estimator $\hat{\boldsymbol{\theta}}$ is given as the p-dimensional parameter $\boldsymbol{\theta}$ that maximizes the log-likelihood function $\ell(\boldsymbol{\theta}) = \sum_{\alpha=1}^{n} \log f(X_\alpha|\boldsymbol{\theta})$ or by solving the likelihood equation

$$\frac{\partial \ell(\boldsymbol{\theta})}{\partial \boldsymbol{\theta}} = \sum_{\alpha=1}^{n} \frac{\partial}{\partial \boldsymbol{\theta}} \log f(X_\alpha|\boldsymbol{\theta}) = \boldsymbol{0}. \tag{3.77}$$

Further, by taking the expectation, we obtain

$$E_{G(\boldsymbol{x}_n)}\left[\sum_{\alpha=1}^{n} \frac{\partial}{\partial \boldsymbol{\theta}} \log f(X_\alpha|\boldsymbol{\theta})\right] = nE_{G(z)}\left[\frac{\partial}{\partial \boldsymbol{\theta}} \log f(Z|\boldsymbol{\theta})\right]. \tag{3.78}$$

Therefore, for a continuous model, if $\boldsymbol{\theta}_0$ is a solution of the equation

$$E_{G(z)}\left[\frac{\partial}{\partial \boldsymbol{\theta}} \log f(Z|\boldsymbol{\theta})\right] = \int g(z) \frac{\partial}{\partial \boldsymbol{\theta}} \log f(z|\boldsymbol{\theta}) dz = \boldsymbol{0}, \tag{3.79}$$

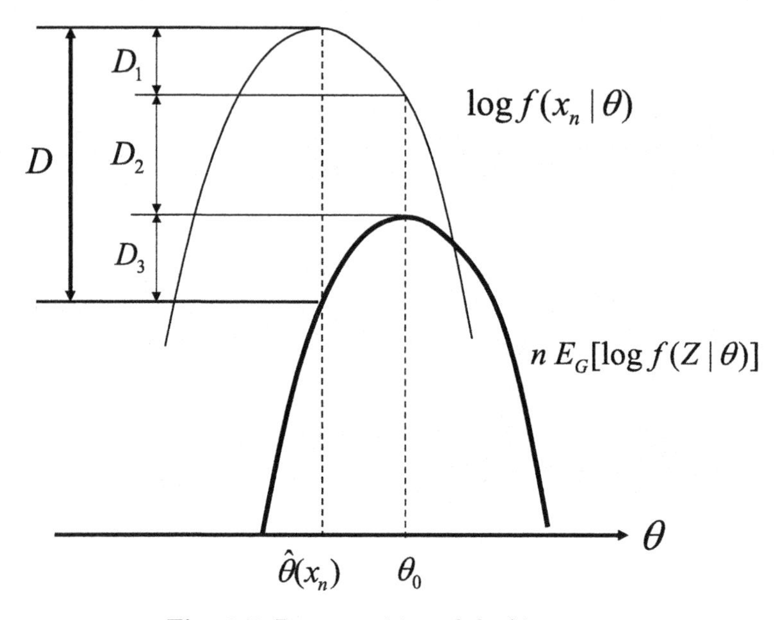

Fig. 3.7. Decomposition of the bias term.

it can be shown that the maximum likelihood estimator $\hat{\boldsymbol{\theta}}$ converges in probability to $\boldsymbol{\theta}_0$ when $n \to +\infty$. For a discrete model, see (3.17).

Using the above results, we now evaluate the bias

$$b(G) = E_{G(\boldsymbol{x}_n)} \left[\log f(\boldsymbol{X}_n | \hat{\boldsymbol{\theta}}(\boldsymbol{X}_n)) - n E_{G(z)} \left[\log f(Z | \hat{\boldsymbol{\theta}}(\boldsymbol{X}_n)) \right] \right] \tag{3.80}$$

when the expected log-likelihood is estimated using the log-likelihood of the statistical model. To this end, we first decompose the bias as follows (Figure 3.7):

$$\begin{aligned} & E_{G(\boldsymbol{x}_n)} \left[\log f(\boldsymbol{X}_n | \hat{\boldsymbol{\theta}}(\boldsymbol{X}_n)) - n E_{G(z)} \left[\log f(Z | \hat{\boldsymbol{\theta}}(\boldsymbol{X}_n)) \right] \right] \\ & \quad = E_{G(\boldsymbol{x}_n)} \left[\log f(\boldsymbol{X}_n | \hat{\boldsymbol{\theta}}(\boldsymbol{X}_n)) - \log f(\boldsymbol{X}_n | \boldsymbol{\theta}_0) \right] \\ & \qquad + E_{G(\boldsymbol{x}_n)} \left[\log f(\boldsymbol{X}_n | \boldsymbol{\theta}_0) - n E_{G(z)} \left[\log f(Z | \boldsymbol{\theta}_0) \right] \right] \\ & \qquad + E_{G(\boldsymbol{x}_n)} \left[n E_{G(z)} \left[\log f(Z | \boldsymbol{\theta}_0) \right] - n E_{G(z)} \left[\log f(Z | \hat{\boldsymbol{\theta}}(\boldsymbol{X}_n)) \right] \right] \\ & \quad = D_1 + D_2 + D_3. \end{aligned} \tag{3.81}$$

Notice that $\hat{\boldsymbol{\theta}} = \hat{\boldsymbol{\theta}}(\boldsymbol{X}_n)$ depends on the sample $\boldsymbol{X}_n$. In the next step, we calculate separately the three expectations D_1, D_2, and D_3.

(1) Calculation of D_2 The easiest case is the evaluation of D_2, which does not contain an estimator. It can easily be seen that

$$\begin{aligned} D_2 &= E_{G(\boldsymbol{x}_n)}\Big[\log f(\boldsymbol{X}_n|\boldsymbol{\theta}_0) - nE_{G(z)}\big[\log f(Z|\boldsymbol{\theta}_0)\big]\Big] \\ &= E_{G(\boldsymbol{x}_n)}\left[\sum_{\alpha=1}^{n} \log f(X_\alpha|\boldsymbol{\theta}_0)\right] - nE_{G(z)}\Big[\log f(Z|\boldsymbol{\theta}_0)\Big] \\ &= 0. \end{aligned} \tag{3.82}$$

This implies that in Figure 3.7, although D_2 varies randomly depending on the data, its expectation becomes 0.

(2) Calculation of D_3 First, we write

$$\eta(\hat{\boldsymbol{\theta}}) := E_{G(z)}\left[\log f(Z|\hat{\boldsymbol{\theta}})\right]. \tag{3.83}$$

By performing a Taylor series expansion of $\eta(\hat{\boldsymbol{\theta}})$ around $\boldsymbol{\theta}_0$ given as a solution to (3.79), we obtain

$$\begin{aligned} \eta(\hat{\boldsymbol{\theta}}) &= \eta(\boldsymbol{\theta}_0) + \sum_{i=1}^{p} (\hat{\theta}_i - \theta_i^{(0)}) \frac{\partial \eta(\boldsymbol{\theta}_0)}{\partial \theta_i} \\ &\quad + \frac{1}{2}\sum_{i=1}^{p}\sum_{j=1}^{p} (\hat{\theta}_i - \theta_i^{(0)})(\hat{\theta}_j - \theta_j^{(0)}) \frac{\partial^2 \eta(\boldsymbol{\theta}_0)}{\partial \theta_i \partial \theta_j} + \cdots, \end{aligned} \tag{3.84}$$

where $\hat{\boldsymbol{\theta}} = (\hat{\theta}_1, \hat{\theta}_2, \ldots, \hat{\theta}_p)^T$ and $\boldsymbol{\theta}_0 = (\theta_1^{(0)}, \theta_2^{(0)}, \ldots, \theta_p^{(0)})^T$. Here, by virtue of the fact that $\boldsymbol{\theta}_0$ is a solution of (3.79), it holds that

$$\frac{\partial \eta(\boldsymbol{\theta}_0)}{\partial \theta_i} = E_{G(z)}\left[\left.\frac{\partial}{\partial \theta_i} \log f(Z|\boldsymbol{\theta})\right|_{\boldsymbol{\theta}_0}\right] = 0, \quad i = 1, 2, \ldots, p, \tag{3.85}$$

where $|_{\boldsymbol{\theta}_0}$ is the value of the partial derivative at the point $\boldsymbol{\theta} = \boldsymbol{\theta}_0$.

Therefore, (3.84) can be approximated as

$$\eta(\hat{\boldsymbol{\theta}}) = \eta(\boldsymbol{\theta}_0) - \frac{1}{2}(\hat{\boldsymbol{\theta}} - \boldsymbol{\theta}_0)^T J(\boldsymbol{\theta}_0)(\hat{\boldsymbol{\theta}} - \boldsymbol{\theta}_0), \tag{3.86}$$

where $J(\boldsymbol{\theta}_0)$ is the $p \times p$ matrix given by

$$J(\boldsymbol{\theta}_0) = -E_{G(z)}\left[\left.\frac{\partial^2 \log f(Z|\boldsymbol{\theta})}{\partial \boldsymbol{\theta} \partial \boldsymbol{\theta}^T}\right|_{\boldsymbol{\theta}_0}\right] = -\int g(z) \left.\frac{\partial^2 \log f(z|\boldsymbol{\theta})}{\partial \boldsymbol{\theta} \partial \boldsymbol{\theta}^T}\right|_{\boldsymbol{\theta}_0} dz \tag{3.87}$$

such that its $(a, b)^{th}$ element is given by

$$j_{ab} = -E_{G(z)}\left[\left.\frac{\partial^2 \log f(Z|\boldsymbol{\theta})}{\partial\theta_a\partial\theta_b}\right|_{\boldsymbol{\theta}_0}\right] = -\int g(z)\left.\frac{\partial^2 \log f(z|\boldsymbol{\theta})}{\partial\theta_a\partial\theta_b}\right|_{\boldsymbol{\theta}_0} dz. \quad (3.88)$$

Then, because D_3 is the expectation of $\eta(\boldsymbol{\theta}_0) - \eta(\hat{\boldsymbol{\theta}})$ with respect to $G(\boldsymbol{x}_n)$, we obtain approximately

$$\begin{aligned} D_3 &= E_{G(\boldsymbol{x}_n)}\left[nE_{G(z)}\left[\log f(Z|\boldsymbol{\theta}_0)\right] - nE_{G(z)}\left[\log f(Z|\hat{\boldsymbol{\theta}})\right]\right] \\ &= \frac{n}{2}E_{G(\boldsymbol{x}_n)}\left[(\hat{\boldsymbol{\theta}} - \boldsymbol{\theta}_0)^T J(\boldsymbol{\theta}_0)(\hat{\boldsymbol{\theta}} - \boldsymbol{\theta}_0)\right] \\ &= \frac{n}{2}E_{G(\boldsymbol{x}_n)}\left[\text{tr}\left\{J(\boldsymbol{\theta}_0)(\hat{\boldsymbol{\theta}} - \boldsymbol{\theta}_0)(\hat{\boldsymbol{\theta}} - \boldsymbol{\theta}_0)^T\right\}\right] \\ &= \frac{n}{2}\text{tr}\left\{J(\boldsymbol{\theta}_0)E_{G(\boldsymbol{x}_n)}\left[(\hat{\boldsymbol{\theta}} - \boldsymbol{\theta}_0)(\hat{\boldsymbol{\theta}} - \boldsymbol{\theta}_0)^T\right]\right\}. \end{aligned} \quad (3.89)$$

By substituting the (asymptotic) variance covariance matrix [see (3.58)]

$$E_{G(\boldsymbol{x}_n)}\left[(\hat{\boldsymbol{\theta}} - \boldsymbol{\theta}_0)(\hat{\boldsymbol{\theta}} - \boldsymbol{\theta}_0)^T\right] = \frac{1}{n}J(\boldsymbol{\theta}_0)^{-1}I(\boldsymbol{\theta}_0)J(\boldsymbol{\theta}_0)^{-1} \quad (3.90)$$

of the maximum likelihood estimator $\hat{\boldsymbol{\theta}}$ into (3.89), we have

$$D_3 = \frac{1}{2}\text{tr}\left\{I(\boldsymbol{\theta}_0)J(\boldsymbol{\theta}_0)^{-1}\right\}, \quad (3.91)$$

where $J(\boldsymbol{\theta}_0)$ is given in (3.87) and $I(\boldsymbol{\theta}_0)$ is the $p \times p$ matrix given by

$$\begin{aligned} I(\boldsymbol{\theta}_0) &= E_{G(z)}\left[\left.\frac{\partial \log f(Z|\boldsymbol{\theta})}{\partial\boldsymbol{\theta}}\frac{\partial \log f(Z|\boldsymbol{\theta})}{\partial\boldsymbol{\theta}^T}\right|_{\boldsymbol{\theta}_0}\right] \\ &= \int g(z)\left.\frac{\partial \log f(z|\boldsymbol{\theta})}{\partial\boldsymbol{\theta}}\frac{\partial \log f(z|\boldsymbol{\theta})}{\partial\boldsymbol{\theta}^T}\right|_{\boldsymbol{\theta}_0} dz. \end{aligned} \quad (3.92)$$

All that remains to do be done now is to calculate D_1.

(3) Calculation of D_1 By writing $\ell(\boldsymbol{\theta}) = \log f(\boldsymbol{X}_n|\boldsymbol{\theta})$ and by applying a Taylor series expansion around the maximum likelihood estimator $\hat{\boldsymbol{\theta}}$, we obtain

$$\ell(\boldsymbol{\theta}) = \ell(\hat{\boldsymbol{\theta}}) + (\boldsymbol{\theta} - \hat{\boldsymbol{\theta}})^T\frac{\partial\ell(\hat{\boldsymbol{\theta}})}{\partial\boldsymbol{\theta}} + \frac{1}{2}(\boldsymbol{\theta} - \hat{\boldsymbol{\theta}})^T\frac{\partial^2\ell(\hat{\boldsymbol{\theta}})}{\partial\boldsymbol{\theta}\partial\boldsymbol{\theta}^T}(\boldsymbol{\theta} - \hat{\boldsymbol{\theta}}) + \cdots. \quad (3.93)$$

Here, the quantity $\hat{\boldsymbol{\theta}}$ satisfies the equation $\partial\ell(\hat{\boldsymbol{\theta}})/\partial\boldsymbol{\theta} = \boldsymbol{0}$ by virtue of the maximum likelihood estimator given as a solution of the likelihood equation $\partial\ell(\boldsymbol{\theta})/\partial\boldsymbol{\theta} = \boldsymbol{0}$.

We see that the quantity

$$\frac{1}{n}\frac{\partial^2 \ell(\hat{\boldsymbol{\theta}})}{\partial \boldsymbol{\theta} \partial \boldsymbol{\theta}^T} = \frac{1}{n}\frac{\partial^2 \log f(\boldsymbol{X}_n|\hat{\boldsymbol{\theta}})}{\partial \boldsymbol{\theta} \partial \boldsymbol{\theta}^T} \tag{3.94}$$

converges in probability to $J(\boldsymbol{\theta}_0)$ in (3.87) when n tends to infinity. This can be derived from the fact that the maximum likelihood estimator $\hat{\boldsymbol{\theta}}$ converges to $\boldsymbol{\theta}_0$ and from the result of (3.63), which was obtained based on the law of large numbers. Using these results, we obtain the approximation

$$\ell(\boldsymbol{\theta}_0) - \ell(\hat{\boldsymbol{\theta}}) \approx -\frac{n}{2}(\boldsymbol{\theta}_0 - \hat{\boldsymbol{\theta}})^T J(\boldsymbol{\theta}_0)(\boldsymbol{\theta}_0 - \hat{\boldsymbol{\theta}}) \tag{3.95}$$

for (3.93). Based on this result and the asymptotic variance covariance matrix (3.90) of the maximum likelihood estimator, D_1 can be calculated approximately as follows:

$$\begin{aligned} D_1 &= E_{G(\boldsymbol{x}_n)}\left[\log f(\boldsymbol{X}_n|\hat{\boldsymbol{\theta}}(\boldsymbol{X}_n)) - \log f(\boldsymbol{X}_n|\boldsymbol{\theta}_0)\right] \\ &= \frac{n}{2}E_{G(\boldsymbol{x}_n)}\left[(\boldsymbol{\theta}_0 - \hat{\boldsymbol{\theta}})^T J(\boldsymbol{\theta}_0)(\boldsymbol{\theta}_0 - \hat{\boldsymbol{\theta}})\right] \\ &= \frac{n}{2}E_{G(\boldsymbol{x}_n)}\left[\mathrm{tr}\left\{J(\boldsymbol{\theta}_0)(\boldsymbol{\theta}_0 - \hat{\boldsymbol{\theta}})(\boldsymbol{\theta}_0 - \hat{\boldsymbol{\theta}})^T\right\}\right] \\ &= \frac{n}{2}\mathrm{tr}\left\{J(\boldsymbol{\theta}_0)E_{G(\boldsymbol{x}_n)}[(\hat{\boldsymbol{\theta}} - \boldsymbol{\theta}_0)(\hat{\boldsymbol{\theta}} - \boldsymbol{\theta}_0)^T]\right\} \\ &= \frac{1}{2}\mathrm{tr}\left\{I(\boldsymbol{\theta}_0)J(\boldsymbol{\theta}_0)^{-1}\right\}. \end{aligned} \tag{3.96}$$

Therefore, combining (3.82), (3.91), and (3.96), the bias resulting from the estimation of the expected log-likelihood using the log-likelihood of the model is asymptotically obtained as

$$\begin{aligned} b(G) &= D_1 + D_2 + D_3 \\ &= \frac{1}{2}\mathrm{tr}\left\{I(\boldsymbol{\theta}_0)J(\boldsymbol{\theta}_0)^{-1}\right\} + 0 + \frac{1}{2}\mathrm{tr}\left\{I(\boldsymbol{\theta}_0)J(\boldsymbol{\theta}_0)^{-1}\right\} \\ &= \mathrm{tr}\left\{I(\boldsymbol{\theta}_0)J(\boldsymbol{\theta}_0)^{-1}\right\}, \end{aligned} \tag{3.97}$$

where $I(\boldsymbol{\theta}_0)$ and $J(\boldsymbol{\theta}_0)$ are respectively given in (3.92) and (3.87).

(4) Estimation of bias Because the bias depends on the unknown probability distribution G that generated the data through $I(\boldsymbol{\theta}_0)$ and $J(\boldsymbol{\theta}_0)$, the bias must be estimated based on observed data. Let $\hat{I}$ and $\hat{J}$ be the consistent estimators of $I(\boldsymbol{\theta}_0)$ and $J(\boldsymbol{\theta}_0)$. In this case, we obtain an estimator of the bias $b(G)$ using

$$\hat{b} = \mathrm{tr}(\hat{I}\hat{J}^{-1}). \tag{3.98}$$

Thus, if we determine the asymptotic bias of the log-likelihood as an estimator of the expected log-likelihood of a statistical model, then the information criterion

$$\begin{aligned}\mathrm{TIC} &= -2\left\{\sum_{\alpha=1}^{n} \log f(X_\alpha|\hat{\boldsymbol{\theta}}) - \mathrm{tr}(\hat{I}\hat{J}^{-1})\right\} \\ &= -2\sum_{\alpha=1}^{n} \log f(X_\alpha|\hat{\boldsymbol{\theta}}) + 2\mathrm{tr}(\hat{I}\hat{J}^{-1}) \end{aligned} \tag{3.99}$$

is derived by correcting the bias of the log-likelihood of the model in the form shown in (3.76). This information criterion, which was investigated by Takeuchi (1976) and Stone (1977), is referred to as the "TIC."

Notice that the matrices $I(\boldsymbol{\theta}_0)$ and $J(\boldsymbol{\theta}_0)$ can be estimated by replacing the unknown probability distribution $G(z)$ or $g(z)$ by an empirical distribution function $\hat{G}(z)$ or $\hat{g}(z)$ based on the observed data as follows:

$$I(\hat{\boldsymbol{\theta}}) = \frac{1}{n}\sum_{\alpha=1}^{n} \left.\frac{\partial \log f(x_\alpha|\boldsymbol{\theta})}{\partial \boldsymbol{\theta}} \frac{\partial \log f(x_\alpha|\boldsymbol{\theta})}{\partial \boldsymbol{\theta}^T}\right|_{\hat{\boldsymbol{\theta}}}, \tag{3.100}$$

$$J(\hat{\boldsymbol{\theta}}) = -\frac{1}{n}\sum_{\alpha=1}^{n} \left.\frac{\partial^2 \log f(x_\alpha|\boldsymbol{\theta})}{\partial \boldsymbol{\theta}\partial \boldsymbol{\theta}^T}\right|_{\hat{\boldsymbol{\theta}}}. \tag{3.101}$$

The $(i,j)^{th}$ elements of these matrices are

$$I_{ij}(\hat{G}) = \frac{1}{n}\sum_{\alpha=1}^{n} \left.\frac{\partial \log f(X_\alpha|\boldsymbol{\theta})}{\partial \theta_i} \frac{\partial \log f(X_\alpha|\boldsymbol{\theta})}{\partial \theta_j}\right|_{\hat{\boldsymbol{\theta}}}, \tag{3.102}$$

$$J_{ij}(\hat{G}) = -\frac{1}{n}\sum_{\alpha=1}^{n} \left.\frac{\partial^2 \log f(X_\alpha|\boldsymbol{\theta})}{\partial \theta_i \partial \theta_j}\right|_{\hat{\boldsymbol{\theta}}}, \tag{3.103}$$

respectively.

3.4.4 Akaike Information Criterion (AIC)

The Akaike Information Criterion (AIC) has played a significant role in solving problems in a wide variety of fields as a model selection criterion for analyzing actual data. The AIC is defined by

$$\mathrm{AIC} = -2(\text{maximum log-likelihood}) + 2(\text{number of free parameters}). \tag{3.104}$$

The number of free parameters in a model refers to the dimensions of the parameter vector $\boldsymbol{\theta}$ contained in the specified model $f(x|\boldsymbol{\theta})$.

The AIC is an evaluation criterion for the badness of the model whose parameters are estimated by the maximum likelihood method, and it indicates that the bias of the log-likelihood (3.80) approximately becomes the "number of free parameters contained in the model." The bias is derived under the assumption that the true distribution $g(x)$ is contained in the specified parametric model $\{f(x|\boldsymbol{\theta}); \boldsymbol{\theta} \in \Theta \subset R^p\}$, that is, there exists a $\boldsymbol{\theta}_0 \in \Theta$ such that the equality $g(x) = f(x|\boldsymbol{\theta}_0)$ holds.

Let us now assume that the parametric model is $\{f(x|\boldsymbol{\theta}); \boldsymbol{\theta} \in \Theta \subset R^p\}$ and that the true distribution $g(x)$ can be expressed as $g(x) = f(x|\boldsymbol{\theta}_0)$ for properly specified $\boldsymbol{\theta}_0 \in \Theta$. Under this assumption, the equality $I(\boldsymbol{\theta}_0) = J(\boldsymbol{\theta}_0)$ holds for the $p \times p$ matrix $J(\boldsymbol{\theta}_0)$ given in (3.87) and the $p \times p$ matrix $I(\boldsymbol{\theta}_0)$ given in (3.92), as stated in Remark 2 of Subsection 3.3.5. Therefore, the bias (3.97) of the log-likelihood is asymptotically given by

$$E_{G(\boldsymbol{x}_n)}\left[\sum_{\alpha=1}^{n} \log f(X_\alpha|\hat{\boldsymbol{\theta}}) - nE_{G(z)} \log f(Z|\hat{\boldsymbol{\theta}})\right]$$

$$= \text{tr}\left\{I(\boldsymbol{\theta}_0)J(\boldsymbol{\theta}_0)^{-1}\right\} = \text{tr}(I_p) = p, \qquad (3.105)$$

where I_p is the identity matrix of dimension p. Hence, the AIC

$$\text{AIC} = -2\sum_{\alpha=1}^{n} \log f(X_\alpha \mid \hat{\boldsymbol{\theta}}) + 2p \qquad (3.106)$$

can be obtained by correcting the asymptotic bias p of the log-likelihood.

The AIC does not require any analytical derivation of the bias correction terms for individual problems and does not depend on the unknown probability distribution G, which removes fluctuations due to the estimation of the bias. Further, Akaike (1974) states that if the true distribution that generated the data exists near the specified parametric model, the bias associated with the log-likelihood of the model based on the maximum likelihood method can be approximated by the number of parameters. These attributes make the AIC a highly flexible technique from a practical standpoint.

Findley and Wei (2002) provided a derivation of AIC and its asymptotic properties for the case of vector time series regression model [see also Findley (1985), Bhansali (1986)]. Burnham and Anderson (2002) provided a nice review and explanation of the use of AIC in the model selection and evaluation problems [see also Linhart and Zucchini (1986), Sakamoto et al. (1986), Bozdogan (1987), Kitagawa and Gersch (1996), Akaike and Kitagawa (1998), McQuarrie and Tsai (1998), and Konishi (1999, 2002)]. Burnham and Anderson (2002) also discussed modeling philosophy and perspectives on model selection from an information-theoretic point of view, focusing on the AIC.

Example 10 (TIC for normal model) We assume a normal distribution for the model

$$f(x|\mu, \sigma^2) = \frac{1}{\sqrt{2\pi\sigma^2}} \exp\left\{-\frac{(x-\mu)^2}{2\sigma^2}\right\}. \qquad (3.107)$$

We start by deriving TIC in (3.99) for any $g(x)$. Given n observations $\{x_1, x_2, \ldots, x_n\}$ that are generated from the true distribution $g(x)$, the statistical model is given by

$$f(x|\hat{\mu}, \hat{\sigma}^2) = \frac{1}{\sqrt{2\pi\hat{\sigma}^2}} \exp\left\{-\frac{(x-\hat{\mu})^2}{2\hat{\sigma}^2}\right\}, \tag{3.108}$$

with the maximum likelihood estimators $\hat{\mu} = n^{-1}\sum_{\alpha=1}^{n} x_\alpha$ and $\hat{\sigma}^2 = n^{-1}\sum_{\alpha=1}^{n} (x_\alpha - \hat{\mu})^2$. Therefore, the bias associated with the estimation of the expected log-likelihood using the log-likelihood of the model,

$$E_G\left[\frac{1}{n}\sum_{\alpha=1}^{n} \log f(X_\alpha|\hat{\mu}, \hat{\sigma}^2) - \int g(z) \log f(z|\hat{\mu}, \hat{\sigma}^2) dz\right], \tag{3.109}$$

can be calculated using the matrix $I(\boldsymbol{\theta})$ of (3.92) and the matrix $J(\boldsymbol{\theta})$ of (3.87). This involves performing the following calculations:

For the log-likelihood function

$$\log f(x|\boldsymbol{\theta}) = -\frac{1}{2}\log(2\pi\sigma^2) - \frac{(x-\mu)^2}{2\sigma^2},$$

the expected value is obtained by

$$E_G[\log f(x|\theta)] = -\frac{1}{2}\log(2\pi\sigma^2) - \sigma^2(G) + \frac{(\mu - \mu(G))^2}{\sigma^2},$$

where $\mu(G)$ and $\sigma^2(G)$ are the mean and the variance of the true distribution $g(x)$, respectively. Therefore, the "true" parameters of the model are given by $\theta_0 = (\mu(G), \sigma^2(G))$.

The partial derivatives with respect to μ and σ^2 are

$$\begin{aligned}
&\frac{\partial}{\partial\mu}\log f(x|\boldsymbol{\theta}) = \frac{x-\mu}{\sigma^2}, \quad \frac{\partial}{\partial\sigma^2}\log f(x|\boldsymbol{\theta}) = -\frac{1}{2\sigma^2} + \frac{(x-\mu)^2}{2\sigma^4},\\
&\frac{\partial^2}{\partial\mu^2}\log f(x|\boldsymbol{\theta}) = -\frac{1}{\sigma^2}, \quad \frac{\partial^2}{\partial\mu\partial\sigma^2}\log f(x|\boldsymbol{\theta}) = -\frac{x-\mu}{\sigma^4},\\
&\frac{\partial^2}{(\partial\sigma^2)^2}\log f(x|\boldsymbol{\theta}) = \frac{1}{2\sigma^4} - \frac{(x-\mu)^2}{\sigma^6}.
\end{aligned}$$

Then the 2×2 matrices $I(\boldsymbol{\theta}_0)$ and $J(\boldsymbol{\theta}_0)$ are given by

$$\begin{aligned}
J(\boldsymbol{\theta}) &= -\begin{bmatrix} E_G\left[\frac{\partial^2}{\partial\mu^2}\log f(X|\boldsymbol{\theta})\right] & E_G\left[\frac{\partial^2}{\partial\sigma^2\partial\mu}\log f(X|\boldsymbol{\theta})\right] \\ E_G\left[\frac{\partial^2}{\partial\mu\partial\sigma^2}\log f(X|\boldsymbol{\theta})\right] & E_G\left[\frac{\partial^2}{(\partial\sigma^2)^2}\log f(X|\boldsymbol{\theta})\right] \end{bmatrix}\\
&= \begin{bmatrix} \frac{1}{\sigma^2} & \frac{E_G[X-\mu]}{\sigma^4}\ \frac{E_G(X-\mu)^2}{\sigma^6} \\ \frac{E_G[X-\mu]}{\sigma^4} & \frac{E_G[(X-\mu)^2]}{\sigma^6} - \frac{1}{2\sigma^4} \end{bmatrix} = \begin{bmatrix} \frac{1}{\sigma^2} & 0 \\ 0 & \frac{1}{2\sigma^4} \end{bmatrix},
\end{aligned}$$

$$I(\boldsymbol{\theta}) = E_G\left[\begin{pmatrix} \dfrac{X-\mu}{\sigma^2} \\ -\dfrac{1}{2\sigma^2} + \dfrac{(X-\mu)^2}{2\sigma^4} \end{pmatrix} \left(\frac{X-\mu}{\sigma^2}, -\frac{1}{2\sigma^2} + \frac{(X-\mu)^2}{2\sigma^4} \right)\right]$$

$$= E_G\begin{bmatrix} \dfrac{(X-\mu)^2}{\sigma^4} & -\dfrac{X-\mu}{2\sigma^4} + \dfrac{(X-\mu)^3}{2\sigma^6} \\ -\dfrac{X-\mu}{2\sigma^4} + \dfrac{(X-\mu)^3}{2\sigma^6} & \dfrac{1}{4\sigma^4} - \dfrac{(X-\mu)^2}{4\sigma^6} + \dfrac{(X-\mu)^4}{4\sigma^8} \end{bmatrix}$$

$$= \begin{bmatrix} \dfrac{1}{\sigma^2} & \dfrac{\mu_3}{2\sigma^6} \\ \dfrac{\mu_3}{2\sigma^6} & \dfrac{\mu_4}{4\sigma^8} - \dfrac{1}{4\sigma^4} \end{bmatrix},$$

where $\mu_j = E_G[(X-\mu)^j]$ $(j = 1, 2, \ldots)$ is the jth-order centralized moment of the true distribution $g(x)$. We note here that, in general, $I(\boldsymbol{\theta}_0) \neq J(\boldsymbol{\theta}_0)$.

From the above preparation, the bias correction term can be calculated as follows:

$$I(\boldsymbol{\theta})J(\boldsymbol{\theta})^{-1} = \begin{bmatrix} \dfrac{1}{\sigma^2} & \dfrac{\mu_3}{2\sigma^6} \\ \dfrac{\mu_3}{2\sigma^6} & \dfrac{\mu_4}{4\sigma^8} - \dfrac{1}{4\sigma^4} \end{bmatrix} \begin{bmatrix} \sigma^2 & 0 \\ 0 & 2\sigma^4 \end{bmatrix}$$

$$= \begin{bmatrix} 1 & \dfrac{\mu_3}{\sigma^2} \\ \dfrac{\mu_3}{2\sigma^4} & \dfrac{\mu_4}{2\sigma^4} - \dfrac{1}{2} \end{bmatrix}.$$

Therefore,

$$\mathrm{tr}\left\{I(\boldsymbol{\theta})J(\boldsymbol{\theta})^{-1}\right\} = 1 + \frac{\mu_4}{2\sigma^4} - \frac{1}{2} = \frac{1}{2}\left(1 + \frac{\mu_4}{\sigma^4}\right).$$

This result is generally not equal to the number of parameters, i.e. two in this case. However, if there exists a $\boldsymbol{\theta}_0$ that satisfies $f(x|\boldsymbol{\theta}_0) = g(x)$, then $g(x)$ is a normal distribution, and we have $\mu_3 = 0$ and $\mu_4 = 3\sigma^4$. Hence, it follows that

$$\frac{1}{2} + \frac{\mu_4}{2\sigma^4} = \frac{1}{2} + \frac{3\sigma^4}{2\sigma^4} = \frac{1}{2} + \frac{3}{2} = 2.$$

Given the data, the estimator for the bias is obtained using

$$\frac{1}{n}\mathrm{tr}(\hat{I}\hat{J}^{-1}) = \frac{1}{n}\left\{\frac{1}{2} + \frac{\hat{\mu}_4}{2\hat{\sigma}^4}\right\}, \tag{3.110}$$

where $\hat{\sigma}^2 = n^{-1}\sum_{\alpha=1}^{n}(x_\alpha - \overline{x})^2$ and $\hat{\mu}_4 = n^{-1}\sum_{\alpha=1}^{n}(x_\alpha - \overline{x})^4$. Consequently, the information criteria TIC and AIC are given by the following formulas, respectively:

$$\mathrm{TIC} = -2\sum_{\alpha=1}^{n} \log f(x_\alpha|\hat{\mu},\hat{\sigma}^2) + 2\left(\frac{1}{2} + \frac{\hat{\mu}_4}{2\hat{\sigma}^4}\right), \tag{3.111}$$

$$\mathrm{AIC} = -2\sum_{\alpha=1}^{n} \log f(x_\alpha|\hat{\mu},\hat{\sigma}^2) + 2 \times 2, \tag{3.112}$$

where the maximum log-likelihood is given by

$$\sum_{\alpha=1}^{n} \log f(x_\alpha|\hat{\mu},\hat{\sigma}^2) = -\frac{n}{2}\log(2\pi\hat{\sigma}^2) - \frac{n}{2}.$$

Table 3.3. Change of the bias correction term $\frac{1}{2}(1+\hat{\mu}_4/\hat{\sigma}^4)$ of the TIC when the true distribution is assumed to be a mixed normal distribution ($\xi_1 = \xi_2 = 0, \sigma_1^2 = 1, \sigma_2^2 = 3$); ε denotes the mixing ratio and n is the number of observations. The mean and standard deviation of the estimated bias correction term for each value of ε and n are shown.

ε	$n = 25$	$n = 100$	$n = 400$	$n = 1600$
0.00	1.89 (0.37)	1.97 (0.23)	1.99 (0.12)	2.00 (0.06)
0.01	2.03 (0.71)	2.40 (1.25)	2.67 (1.11)	2.78 (0.71)
0.02	2.14 (0.83)	2.73 (1.53)	3.18 (1.38)	3.33 (0.81)
0.05	2.44 (1.13)	3.45 (1.78)	4.02 (1.35)	4.24 (0.80)
0.10	2.74 (1.24)	3.87 (1.56)	4.42 (1.09)	4.60 (0.60)
0.15	2.87 (1.18)	3.96 (1.34)	4.38 (0.89)	4.49 (0.46)
0.20	2.91 (1.09)	3.84 (1.12)	4.16 (0.69)	4.24 (0.37)
0.30	2.85 (0.94)	3.48 (0.82)	3.67 (0.48)	3.73 (0.25)
0.40	2.68 (0.80)	3.14 (0.65)	3.26 (0.37)	3.29 (0.19)
0.50	2.52 (0.69)	2.84 (0.50)	2.92 (0.28)	2.95 (0.15)
0.60	2.37 (0.60)	2.61 (0.44)	2.67 (0.24)	2.68 (0.12)
0.70	2.22 (0.53)	2.40 (0.36)	2.45 (0.20)	2.46 (0.10)
0.80	2.10 (0.47)	2.23 (0.30)	2.27 (0.16)	2.28 (0.08)
0.90	1.98 (0.41)	2.09 (0.26)	2.12 (0.14)	2.12 (0.07)
1.00	1.88 (0.36)	1.97 (0.23)	1.99 (0.12)	2.00 (0.06)

Example 11 (TIC for normal model versus mixture of two normal distributions) Let us assume that the true distribution generating data is a mixture of two normal distributions

$$g(x) = (1-\varepsilon)\phi(x|\xi_1,\sigma_1^2) + \varepsilon\phi(x|\xi_2,\sigma_2^2) \qquad (0 \le \varepsilon \le 1), \tag{3.113}$$

where $\phi(x|\xi_i,\sigma_i^2)$ $(i = 1,2)$ is the probability density function of the normal distribution with mean ξ_i and variance σ_i^2. We assume the normal model

$N(\mu, \sigma^2)$ for the model. Table 3.3 shows the mean and the standard deviation of 10,000 simulation runs of the TIC bias correction term $\frac{1}{2}(1 + \hat{\mu}_4/\hat{\sigma}^4)$ in (3.111), which were obtained by varying the mixing ratio and the number of observations in a mixed normal distribution. When n is small and ε is equal to either 0 or 1, the result is smaller than the bias correction term 2 of the AIC. The bias correction term is maximized when the value of ε is in the neighborhood of 0.1 to 0.2. Notice that in the region in which the correction term in the TIC is large, the standard deviation is also large.

Table 3.4. Estimated bias correction terms of TIC and their standard deviations when normal distribution models are fitted to simulated data from the t-distribution.

df	$n = 25$	$n = 100$	$n = 400$	$n = 1,600$
∞	1.89 (0.37)	1.98 (0.23)	2.00 (0.12)	2.00 (0.06)
9	2.12 (0.62)	2.42 (0.69)	2.54 (0.52)	2.58 (0.34)
8	2.17 (0.66)	2.51 (0.82)	2.67 (0.86)	2.73 (0.63)
7	2.21 (0.72)	2.64 (0.99)	2.85 (1.05)	2.95 (0.91)
6	2.29 (0.81)	2.85 (1.43)	3.20 (1.81)	3.36 (1.46)
5	2.43 (1.00)	3.21 (1.96)	3.87 (3.21)	4.28 (4.12)
4	2.67 (1.23)	3.94 (3.01)	5.49 (6.37)	7.46 (15.96)
3	3.06 (1.62)	5.72 (5.38)	10.45 (14.71)	19.79 (41.12)
2	4.01 (2.32)	10.54 (9.39)	30.88 (35.67)	101.32 (138.74)
1	6.64 (3.17)	25.27 (13.94)	100.14 (56.91)	404.12 (232.06)

Example 12 (TIC for normal model versus t-distribution) Table 3.4 shows the means and the standard deviations of the estimated bias correction term of the TIC, $\frac{1}{2}(1 + \hat{\mu}_4/\hat{\sigma}^4)$ in (3.111), when it is assumed that the true distribution is the t-distribution with degrees of freedom *df*,

$$g(x|df) = \frac{\Gamma\left(\frac{df+1}{2}\right)}{\sqrt{df\pi}\,\Gamma\left(\frac{df}{2}\right)}\left(1 + \frac{x^2}{df}\right)^{-\frac{1}{2}(df+1)}, \tag{3.114}$$

which were obtained by repeating 10,000 simulation runs. Four data lengths (n= 25, 100, 400, and 1,600) and 10 different values for the degrees of freedom [1 to 9 and the normal distribution ($df = \infty$)] were examined.

When the degrees of freedom *df* is small and the number of observations is large, the results differ significantly from the correction term 2 of the AIC. Notice that in this case, the standard deviation is also extremely large, exceeding the value of the bias in some cases.

Example 13 (Polynomial regression models) Assume that the following 20 observations, (x, y), are observed in experiments (Figure 3.8):

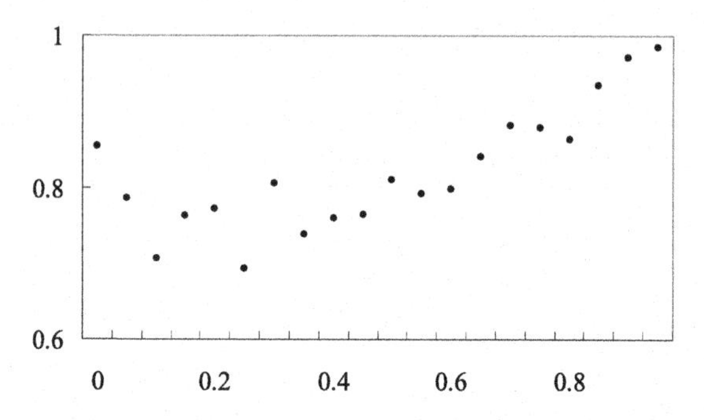

Fig. 3.8. Twenty observations used for polynomial regression models.

(0.00, 0.854), (0.05, 0.786), (0.10, 0.706), (0.15, 0.763), (0.20, 0.772),
(0.25, 0.693), (0.30, 0.805), (0.35, 0.739), (0.40, 0.760), (0.45, 0.764),
(0.50, 0.810), (0.55, 0.791), (0.60, 0.798), (0.65, 0.841), (0.70, 0.882),
(0.75, 0.879), (0.80, 0.863), (0.85, 0.934), (0.90, 0.971), (0.95, 0.985).

A polynomial regression model is then fitted to these 20 observations; specifically, to the following model:

$$y = \beta_0 + \beta_1 x + \beta_2 x^2 + \cdots + \beta_p x^p + \varepsilon, \qquad \varepsilon \sim N(0, \sigma^2). \tag{3.115}$$

Here we write $\boldsymbol{\theta} = (\beta_0, \beta_1, \ldots, \beta_p, \sigma^2)^T$ and when data $\{(y_\alpha, x_\alpha), \alpha = 1, \ldots, n\}$ are given, the log-likelihood function can be written as

$$\ell(\boldsymbol{\theta}) = -\frac{n}{2}\log(2\pi\sigma^2) - \frac{1}{2\sigma^2}\sum_{\alpha=1}^{n}\Big(y_\alpha - \sum_{j=0}^{p}\beta_j x_\alpha^j\Big)^2. \tag{3.116}$$

Therefore, the maximum likelihood estimators $\hat{\beta}_0, \hat{\beta}_1, \ldots, \hat{\beta}_p$ for the coefficients can be obtained by minimizing the following term:

$$\sum_{\alpha=1}^{n}\Big(y_\alpha - \sum_{j=0}^{p}\beta_j x_\alpha^j\Big)^2. \tag{3.117}$$

In addition, the maximum likelihood estimator of the error variance is given by

$$\hat{\sigma}^2 = \frac{1}{n}\sum_{\alpha=1}^{n}\Big(y_\alpha - \sum_{j=0}^{p}\hat{\beta}_j x_\alpha^j\Big)^2. \tag{3.118}$$

By substituting this expression into (3.116), we obtain the maximum log-likelihood

$$\ell(\hat{\boldsymbol{\theta}}) = -\frac{n}{2}\log\big(2\pi\hat{\sigma}^2\big) - \frac{n}{2}. \tag{3.119}$$

Further, because the number of parameters contained in this model is $p + 2$, that is, for $\beta_0, \beta_1, \ldots, \beta_p$ and σ^2, the AIC for evaluating the pth order polynomial regression model is given by

Table 3.5. Results of estimating polynomial regression models.

Order	$\hat{\sigma}^2$	Log-Likelihood	AIC	AIC Difference
—	0.678301	−24.50	50.99	126.49
0	0.006229	22.41	−40.81	34.68
1	0.002587	31.19	−56.38	19.11
2	0.000922	41.51	−75.03	0.47
3	0.000833	42.52	−75.04	0.46
4	0.000737	43.75	−75.50	—
5	0.000688	44.44	−74.89	0.61
6	0.000650	45.00	−74.00	1.49
7	0.000622	45.45	−72.89	2.61
8	0.000607	45.69	−71.38	4.12
9	0.000599	45.83	−69.66	5.84

$$\text{AIC}_p = n(\log 2\pi + 1) + n \log \hat{\sigma}^2 + 2(p+2). \tag{3.120}$$

Table 3.5 summarizes the results obtained by fitting polynomials up to order nine to this set of data. As the order increases, the residual variance reduces, and the log-likelihood increases monotonically. The AIC attains a minimum at $p = 4$, and the model

$$\begin{aligned} y_j &= 0.835 - 1.068x_j + 3.716x_j^2 - 4.573x_j^3 + 2.141x_j^4 + \varepsilon_j, \\ \varepsilon_j &\sim N(0, 0.737 \times 10^{-3}), \end{aligned} \tag{3.121}$$

is selected as the best model.

In order to demonstrate the importance of order selection in a regression model, Figure 3.9 shows the results of running Monte Carlo experiments. Using different random numbers, 20 observations were generated according to (3.115), and using the data, 2nd-, 4th-, and 9th-order polynomials were estimated. Figure 3.9 shows the 10 regression curves that were obtained by repeating these operations 10 times, along with the "true" regression polynomial that was used for generating the data. In the case of the 2nd-order polynomial regression model, while the width of the fluctuations is small, the low order of the polynomial results in a large bias in the regression curves. For the 4th-order polynomial, the 10 estimated values cover the true regression polynomial. By contrast, for the 9th-order polynomial, although the true regression polynomial is covered, the large fluctuations indicate that the estimated values are highly unstable.

Example 14 (Factor analysis model) Suppose that $\boldsymbol{x} = (x_1, \ldots, x_p)^T$ is an observable random vector with mean vector $\boldsymbol{\mu}$ and variance covariance matrix Σ. The factor analysis model is

$$\boldsymbol{x} = \boldsymbol{\mu} + L\boldsymbol{f} + \boldsymbol{\varepsilon}, \tag{3.122}$$

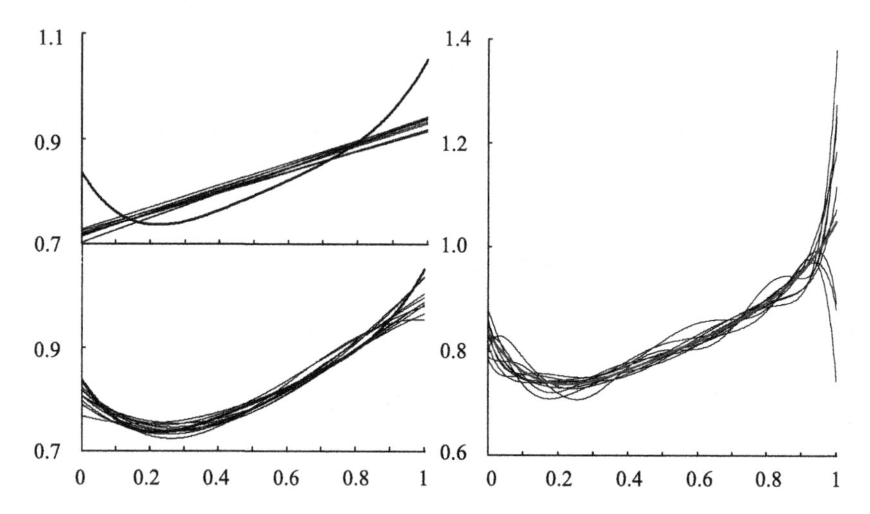

Fig. 3.9. Fluctuations in estimated polynomials for (3.115). Upper left: $p = 2$; lower-left: $p = 4$; right: $p = 9$.

where L is a $p \times m$ matrix of factor loadings, and $\boldsymbol{f} = (f_1, \ldots, f_m)^T$ and $\boldsymbol{\varepsilon} = (\varepsilon_1, \ldots, \varepsilon_p)^T$ are unobservable random vectors. The elements of $\boldsymbol{f}$ are called common factors while the elements of $\boldsymbol{\varepsilon}$ are referred to as specific or unique factors. It is assumed that

$$\begin{aligned} E[\boldsymbol{f}] &= \mathbf{0}, \quad \mathrm{Cov}(\boldsymbol{f}) = E[\boldsymbol{f}\boldsymbol{f}^T] = I_m, \\ E[\boldsymbol{\varepsilon}] &= \mathbf{0}, \quad \mathrm{Cov}(\boldsymbol{\varepsilon}) = E[\boldsymbol{\varepsilon}\boldsymbol{\varepsilon}^T] = \Psi = \mathrm{diag}[\psi_1, \cdots, \psi_p], \\ \mathrm{Cov}(\boldsymbol{f}, \boldsymbol{\varepsilon}) &= E[\boldsymbol{f}\boldsymbol{\varepsilon}^T] = 0, \end{aligned} \tag{3.123}$$

where I_m is the identity matrix of order m and Ψ is a $p \times p$ diagonal matrix with i^{th} diagonal element ψ_i (> 0). It then follows from (3.122) and (3.123) that Σ can be expressed as

$$\Sigma = LL^T + \Psi. \tag{3.124}$$

Assume that the common factors $\boldsymbol{f}$ and the specific factors $\boldsymbol{\varepsilon}$ are normally distributed. Let $\overline{\boldsymbol{x}}$ and S be, respectively, the sample mean vector and sample covariance matrix based on a set of n observations $\{\boldsymbol{x}_1, \ldots, \boldsymbol{x}_n\}$ on $\boldsymbol{x}$. It is known [see, for example, Lawley and Maxwell (1971) and Anderson (2003)] that the maximum likelihood estimates, $\hat{L}$ and $\hat{\Psi}$, of the matrix L of factor loadings and the covariance matrix Ψ of specific factors are obtained by minimizing the discrepancy function

$$Q(L, \Psi) = \log|\Sigma| - \log|S| + \mathrm{tr}\left(\Sigma^{-1}S\right) - p, \tag{3.125}$$

subject to the condition that $L^T\Psi^{-1}L$ is a diagonal matrix. Then, the AIC is defined by

$$\mathrm{AIC} = n\left\{p\log(2\pi) + \log|\hat{\Sigma}| + \mathrm{tr}\left(\hat{\Sigma}^{-1}S\right)\right\} + 2\left\{p(m+1) - \frac{1}{2}m(m-1)\right\}, \tag{3.126}$$

where $\hat{\Sigma} = \hat{L}\hat{L}^T + \hat{\Psi}$.

The use of the AIC in the factor analysis model was considered by Akaike (1973, 1987). Ichikawa and Konishi (1999) derived the TIC for a covariance structure analysis model and investigated the performance of three information criteria, namely the AIC, the TIC, and the bootstrap information criteria (introduced in Chapter 8). The use of AIC-type criteria for selecting variables in principal component, canonical correlation, and discriminant analyses was discussed, in relation to the likelihood ratio tests, by Fujikoshi (1985) and Siotani et al. (1985, Chapter 13).

3.5 Properties of MAICE

The estimators and models selected by minimizing the AIC are referred to as MAICE (minimum AIC estimators). In this section, we discuss several topics related to the properties of MAICE.

3.5.1 Finite Correction of the Information Criterion

In Section 3.4, we derived the AIC for general statistical models estimated using the maximum likelihood method. In contrast, information criterion for particular models such as normal distribution models can be derived directly and analytically by calculating the bias, without having to resort to asymptotic theories such as the Taylor series expansion or the asymptotic normality. Let us first consider a simple normal distribution model, $N(\mu, \sigma^2)$.

Since the logarithm of the probability density function is

$$\log f(x|\mu, \sigma^2) = -\frac{1}{2}\log(2\pi\sigma^2) - \frac{(x-\mu)^2}{2\sigma^2},$$

the log-likelihood of the model based on the data, $\boldsymbol{x}_n = \{x_1, x_2, \ldots, x_n\}$, is given by

$$\ell(\mu, \sigma^2) = -\frac{n}{2}\log(2\pi\sigma^2) - \frac{1}{2\sigma^2}\sum_{\alpha=1}^{n}(x_\alpha - \mu)^2.$$

By substituting the maximum likelihood estimators

$$\hat{\mu} = \frac{1}{n}\sum_{\alpha=1}^{n} x_\alpha, \quad \hat{\sigma}^2 = \frac{1}{n}\sum_{\alpha=1}^{n}(x_\alpha - \hat{\mu})^2,$$

into this expression, we obtain the maximum log-likelihood

$$\ell(\hat{\mu}, \hat{\sigma}^2) = -\frac{n}{2}\log(2\pi\hat{\sigma}^2) - \frac{n}{2}.$$

If the data set is obtained from the same normal distribution $N(\mu, \sigma^2)$, then the expected log-likelihood is given by

$$E_G\left[\log f(Z|\hat{\mu}, \hat{\sigma}^2)\right] = -\frac{1}{2}\log(2\pi\hat{\sigma}^2) - \frac{1}{2\hat{\sigma}^2}\left\{\sigma^2 + (\mu - \hat{\mu})^2\right\},$$

where $G(z)$ is the distribution function of the normal distribution $N(\mu, \sigma^2)$. Therefore, the difference between the two quantities is

$$\ell(\hat{\mu}, \hat{\sigma^2}) - nE_G\left[\log f(Z|\hat{\mu}, \hat{\sigma^2})\right] = \frac{n}{2\hat{\sigma}^2}\left\{\sigma^2 + (\mu - \hat{\mu})^2\right\} - \frac{n}{2}.$$

By taking the expectation with respect to the joint distribution of n observations distributed as the normal distribution $N(\mu, \sigma^2)$, and using

$$E_G\left[\frac{\sigma^2}{\hat{\sigma}^2(\boldsymbol{x}_n)}\right] = \frac{n}{n-3}, \quad E_G\left[\{\mu - \hat{\mu}(\boldsymbol{x}_n)\}^2\right] = \frac{\sigma^2}{n},$$

we obtain the bias correction term for a finite sample as

$$b(G) = \frac{n}{2}\frac{n}{(n-3)\sigma^2}\left(\sigma^2 + \frac{\sigma^2}{n}\right) - \frac{n}{2} = \frac{2n}{n-3}. \tag{3.127}$$

Here we used the fact that for a χ^2 random variable with degrees of freedom r, χ^2_r, we have $E[1/\chi^2_r] = 1/(r-2)$. Therefore, the information criterion (IC) for the normal distribution model is given by

$$\mathrm{IC} = -2\ell(\hat{\mu}, \hat{\sigma}^2) + \frac{4n}{n-3}. \tag{3.128}$$

Table 3.6 shows changes in this bias term $b(G)$ with respect to several values of n. This table shows that $b(G)$ approaches the correction term 2 of the AIC as the number of observations increases.

Table 3.6. Changes of the bias $b(G)$ for normal distribution model as the number of the observations increases.

n	4	6	8	12	18	25	50	100
$b(G)$	8.0	4.0	3.2	2.7	2.4	2.3	2.1	2.1

The topic of a finite correction of the AIC for more general Gaussian linear regression models will be discussed in Subsection 7.2.2.

3.5.2 Distribution of Orders Selected by AIC

Let us consider the problem of order selection in an autoregressive model

$$y_n = \sum_{j=1}^{m} a_j y_{n-j} + \varepsilon_n, \quad \varepsilon_n \sim N(0, \sigma^2). \tag{3.129}$$

In this case, an asymptotic distribution of the number of orders is obtained when the number of orders is selected using the AIC minimization method [Shibata (1976)]. We now define p_j and q_j $(j = 1, \ldots, M)$ by setting $\alpha_i = \Pr(\chi_i^2 > 2i)$, $p_0 = q_0 = 1$, with respect to the χ^2-variate with i degrees of freedom according to the following equations:

$$p_j = \sum \left\{ \prod_{i=1}^{j} \frac{1}{r_i!} \left(\frac{\alpha_i}{i} \right)^{r_i} \right\}, \tag{3.130}$$

$$q_j = \sum \left\{ \prod_{i=1}^{j} \frac{1}{r_i!} \left(\frac{1-\alpha_i}{i} \right)^{r_i} \right\}, \tag{3.131}$$

where $\sum$ is the sum of all combinations of $(r_1, \ldots, r_j)$ that satisfy the equation $r_1 + 2r_2 + \cdots + nr_j = j$. In this case, according to Shibata (1976), if the AR model with order m_0 is the true model, and if the order $0 \leq m \leq M$ of the AR model is selected using the AIC, then the asymptotic distribution of $\hat{m}$ can be obtained as

$$\lim_{n \to +\infty} \Pr(\hat{m} = m) = \begin{cases} p_{m-m_0}\, q_{M-m} & \text{for } m_0 \leq m \leq M, \\ 0 & \text{for } \quad m < m_0. \end{cases} \tag{3.132}$$

This result shows that the probability of selecting the true order using the minimum AIC procedure is not unity even as $n \to +\infty$. In other words, the order selection using the AIC is not consistent. At the same time, since the distribution of the selected order has an asymptotic distribution, the result indicates that it will not spread as n increases.

In general, under the assumptions that the true model is of finite dimension and it is included in the class of candidate models, a criterion that identifies the correct model asymptotically with probability one is said to be consistent. The consistency has been investigated by Shibata (1976, 1981), Nishii (1984), Findley (1985), etc. A review of consistency on model selection criteria was provided by Rao and Wu (2001) and Burnham and Anderson (2002, Section 6.3).

Example 15 (Order selection in linear regression models) Figure 3.10 shows the distribution of the number of explanatory variables that are selected using the AIC for the case of an ordinary regression model

$$y_i = a_1 x_{i1} + \cdots + a_k x_{ik} + \varepsilon_i, \quad \varepsilon_i \sim N(0, \sigma^2).$$

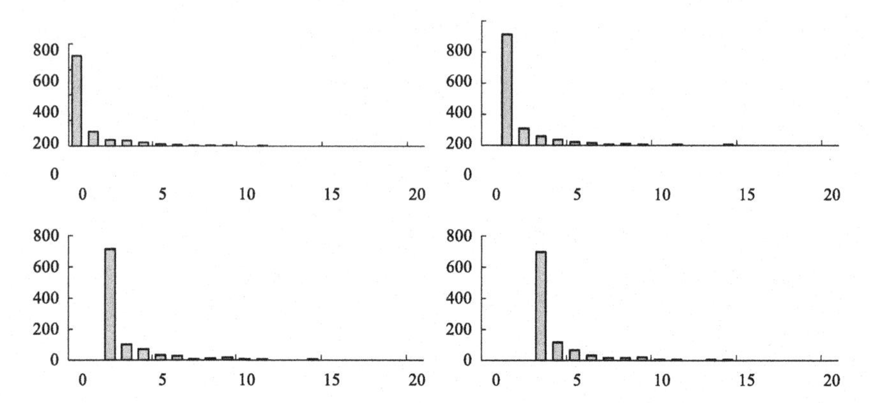

Fig. 3.10. Distributions of orders selected by AIC. The upper left, upper right, lower left, and lower right plots represent the cases in which the true order is 0, 1, 2, and 3, respectively.

It will be demonstrated by simulations that even for the ordinary regression case, we can obtain results that are qualitatively similar to those for the autoregression case.

For simplicity, we assume that x_{ij} ($j = 1, \ldots, 20,\ i = 1, \ldots, n$) are orthonormal variables. We also assume that the true model that generates data is given by $\sigma^2 = 0.01$ and

$$a_j^* = \begin{cases} 0.7^j & \text{for } j = 1, \ldots, k^*, \\ 0 & \text{for } j = k^* + 1, \ldots, 20\,. \end{cases} \tag{3.133}$$

Figure 3.10 shows the distributions of orders obtained by generating data with $n = 400$ and by repeating 1,000 times the process of selecting orders using the AIC. The upper left plot represents the case in which the true order is defined as $k^* = 0$. Similarly, the upper right, lower left, and lower right plots represent the cases for which $k^* = 1, 2, 3$, respectively. These results also indicate that when the number of observations involved is relatively large (for example, $n = 400$) for both the regression model and autoregressive models, the probability with which the true order is obtained is approximately 0.7, which means that the order is overestimated with a probability of 0.3. In this distribution, varying the true order k^* only shifts the location of the maximum probability to the right, while only slightly modifying the shape of the distribution.

Figure 3.11 shows the results of examining changes in distribution as a function of the number of observations for the case $k^* = 1$. The graph on the left shows the case when $n = 100$, while the graph on the right shows the case when $n = 1,600$. The results suggest that when the true order is a finite number, the distribution of orders converges to a certain distribution when the size of n becomes large. Figure 3.12 shows the case for $k^* = 20$, in which

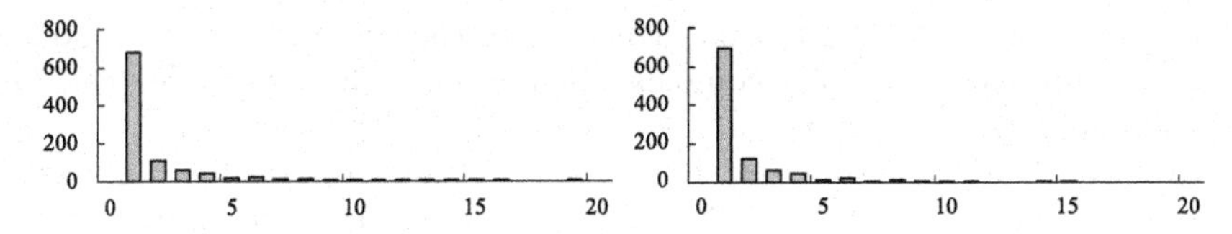

Fig. 3.11. Change in distribution of order selected by the AIC, for different number of observations. Left graph: $n = 100$; right graph: $n = 1,600$.

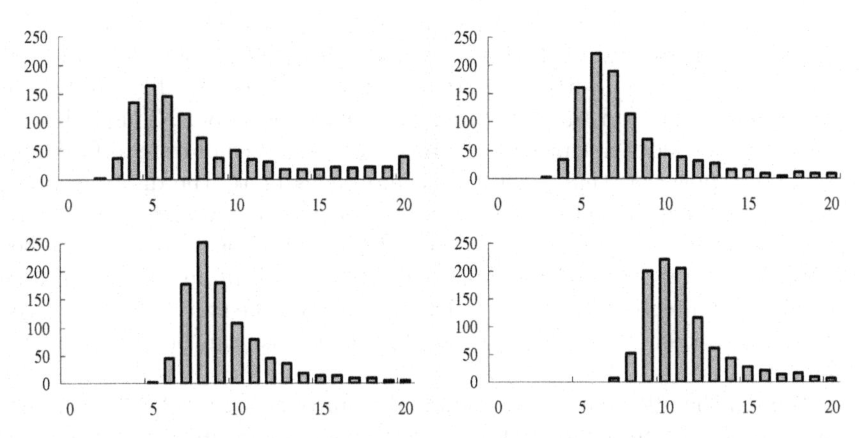

Fig. 3.12. Distributions of orders selected by AIC when the true coefficient decays with the order. The upper left, upper right, lower left, and lower right graphs represent the cases in which the number of observations is 50, 100, 400, and 1,600, respectively.

all of the coefficients are nonzero. The results indicate that the distribution's mode shifts to the right as the number of observations, n, increases and that when complex phenomena are approximated using a relatively simple model, the order selected by the AIC increases with the number of observations.

3.5.3 Discussion

Here we summarize several points regarding the selection of a model using the AIC. The AIC has been criticized because it does not yield a consistent estimator with respect to the selection of orders. Such an argument is frequently misunderstood, and we attempt to clarify these misunderstandings in the following.

(1) First, the objective of our modeling is to obtain a "good" model, rather than a "true" model. If one recalls that statistical models are approximations of complex systems toward certain objectives, the task of estimating the true order is obviously not an appropriate goal. A true model or order can be defined explicitly only in a limited number of situations, such as

when running simulation experiments. From the standpoint that a model is an approximation of a complex phenomenon, the true order can be infinitely large.

(2) Even if a true finite order exists, the order of a good model is not necessarily equal to the true order. In situations where there are only a small number of observations, considering the instability of the parameters being estimated, the AIC reveals the possibility that a higher prediction accuracy can be obtained using models having lower orders.

(3) Shibata's (1976) results described in the previous section indicate that if the true order is assumed, the asymptotic distribution of orders selected by the AIC can be a fixed distribution that is determined solely by the maximum order and the true order of a family of models. This indicates that the AIC does not provide a consistent estimator of orders. It should be noted, however, that when the true order is finite, the distribution of orders that is selected does not vary when the number of observations is increased. It should also be noted that in this case, even if a higher order is selected, when the number of observations is large, each coefficient estimate of a regressor with an order greater than the true order converges to the true value 0 and that a consistent estimator can be obtained as a model.

(4) Although the information criterion makes automatic model selection possible, it should be noted that the model evaluation criterion is a relative evaluation criterion. This means that selecting a model using an information criterion is only a selection from a family of models that we have specified. Therefore, the critical task for us is to set up more appropriate models by making use of knowledge regarding that object.

4

Statistical Modeling by AIC

The majority of the problems in statistical inference can be considered to be problems related to statistical modeling. They are typically formulated as comparisons of several statistical models. In this chapter, we consider using the AIC for various statistical inference problems such as checking the equality of distributions, determining the bin size of a histogram, selecting the order for regression models, detecting structural changes, determining the shape of a distribution, and selecting the Box-Cox transformation.

4.1 Checking the Equality of Two Discrete Distributions

Assume that we have two sets of data each having k categories and that the number of observations in each category is given as follows [Sakamoto et al. (1986)]:

Category	1	2	$\cdots$	k
Data set 1	n_1	n_2	$\cdots$	n_k
Data set 2	m_1	m_2	$\cdots$	m_k

where the total numbers of observations are $n_1 + \cdots + n_k = n$ and $m_1 + \cdots + m_k = m$, respectively. We further assume that these data sets follow the multinomial distributions with k categories

$$p(n_1, \ldots, n_k | p_1, \ldots, p_k) = \frac{n!}{n_1! \ldots n_k!} p_1^{n_1} \cdots p_k^{n_k}, \tag{4.1}$$

$$p(m_1, \ldots, m_k | q_1, \ldots, q_k) = \frac{m!}{m_1! \ldots m_k!} q_1^{m_1} \cdots q_k^{m_k}, \tag{4.2}$$

where p_j and q_j denote the probabilities that each event in Data set 1 and Data set 2 results in the category j, and $\boldsymbol{p} = (p_1, \ldots, p_k)$ and $\boldsymbol{q} = (q_1, \ldots, q_k)$ satisfy $p_i > 0$ and $q_i > 0$ for all i.

The log-likelihood of the model consisting of two individual models for Data set 1 and Data set 2 is defined as

$$\ell_2(p_1,\ldots,p_k,q_1,\ldots,q_k) = \log n! - \sum_{j=1}^{k}\log n_j! + \sum_{j=1}^{k} n_j \log p_j + \log m! - \sum_{j=1}^{k}\log m_j! + \sum_{j=1}^{k} m_j \log q_j. \quad (4.3)$$

Therefore, the maximum likelihood estimates of p_j and q_j are given by

$$\hat{p}_j = \frac{n_j}{n}, \quad \hat{q}_j = \frac{m_j}{m}, \quad (4.4)$$

and the maximum log-likelihood of the model is

$$\ell_2(\hat{p}_1,\ldots,\hat{p}_k,\hat{q}_1,\ldots,\hat{q}_k) = C + \sum_{j=1}^{k} n_j \log\left(\frac{n_j}{n}\right) + \sum_{j=1}^{k} m_j \log\left(\frac{m_j}{m}\right), \quad (4.5)$$

where $C = \log n! + \log m! + \sum_{j=1}^{k}(\log n_j! + \log m_j!)$ is a constant term that is independent of the parameters. Since the number of free parameters of the model is $2(k-1)$, the AIC is given by

$$\begin{aligned} \text{AIC} &= -2\ell_2(\hat{p}_1,\ldots,\hat{p}_k,\hat{q}_1,\ldots,\hat{q}_k) + 2\times 2(k-1) \quad (4.6)\\ &= -2\left\{C + \sum_{j=1}^{k} n_j \log\left(\frac{n_j}{n}\right) + \sum_{j=1}^{k} m_j \log\left(\frac{m_j}{m}\right)\right\} + 4(k-1). \end{aligned}$$

On the other hand, if we assume the two distributions are equal, it holds that $p_j = q_j \equiv r_j$, and the log-likelihood can be expressed as

$$\ell_1(r_1,\ldots,r_k) = C + \sum_{j=1}^{k}(n_j + m_j)\log r_j. \quad (4.7)$$

Then we have the maximum likelihood estimates of r_j as

$$\hat{r}_j = \frac{n_j + m_j}{n + m}, \quad (4.8)$$

and the maximum log-likelihood of the model is given by

$$\ell_1(\hat{p}_1,\ldots,\hat{p}_k) = C + \sum_{j=1}^{k}(n_j + m_j)\log\left(\frac{n_j + m_j}{n + m}\right). \quad (4.9)$$

Since the number of free parameters of the model is $k-1$, the AIC of this model is obtained as

$$\begin{aligned}\text{AIC} &= -2\ell_1(\hat{p}_1,\ldots,\hat{p}_k) + 2(k-1)\\ &= -2\left\{C + \sum_{j=1}^{k}(n_j+m_j)\log\left(\frac{n_j+m_j}{n+m}\right)\right\} + 2(k-1). \qquad (4.10)\end{aligned}$$

Example 1 (Equality of two multinomial distributions) The following table shows two sets of survey data each having five categories.

Category	C_1	C_2	C_3	C_4	C_5
First survey	304	800	400	57	323
Second survey	174	509	362	80	214

From this table we can obtain the maximum likelihood estimates of the parameters of the multinomial distribution; $\hat{p}_j$ and $\hat{q}_j$ indicate the estimated parameters of each model, while $\hat{r}_j$ expresses the estimated parameters obtained by assuming that the two distributions are equal.

Category	C_1	C_2	C_3	C_4	C_5
$\hat{p}_j$	0.16	0.42	0.21	0.03	0.17
$\hat{q}_j$	0.13	0.38	0.27	0.06	0.16
$\hat{r}_j$	0.148	0.406	0.236	0.043	0.167

From this table, ignoring the common constant C, the maximum log-likelihoods of the models are obtained as

$$\begin{aligned}\text{Model 1}: \quad & \sum_{j=1}^{k} n_j \log\left(\frac{n_j}{n}\right) + \sum_{j=1}^{k} m_j \log\left(\frac{m_j}{m}\right) = -2628.644 - 1938.721\\ & \qquad = -4567.365,\\ \text{Model 2}: \quad & \sum_{j=1}^{k}(n_j+m_j)\log\left(\frac{n_j+m_j}{n+m}\right) = -4585.612. \qquad (4.11)\end{aligned}$$

Since the number of free parameters of the models is $2(k-1) = 8$ in Model 1 and $k-1 = 4$ in Model 2, by ignoring the common constant C, the AICs of the models are given as 9,150.731 and 9,179.223, respectively. Namely, the AIC indicates that the two data sets were obtained from different distributions.

4.2 Determining the Bin Size of a Histogram

Histograms are used for representing the properties of a set of observations obtained from either a discrete distribution or a continuous distribution. Assume that we have a histogram $\{n_1, n_2, \ldots, n_k\}$; here k is referred to as the bin size. It is well known that if the bin size is too large, the corresponding histogram may become too sensitive and it is difficult to capture the characteristics of the true distribution. In such a case, we may consider using a histogram having a smaller bin size. However, it is obvious that if we use a bin size that is too small, the histogram cannot capture the shape of the true distribution. Therefore, the selection of an appropriate bin size is an important problem.

A histogram with k bins can be considered as a model specified by a multinomial distribution with k parameters:

$$P(n_1,\ldots,n_k|p_1,\ldots,p_k) = \frac{n!}{n_1!\cdots n_k!}p_1^{n_1}\cdots p_k^{n_k}, \tag{4.12}$$

where $n_1+\cdots+n_k = n$ and $p_1+\cdots+p_k = 1$ [Sakamoto et al. (1986)]. Then the log-likelihood of the model can be written as

$$\ell(p_1,\ldots,p_k) = C + \sum_{j=1}^{k} n_j \log p_j, \tag{4.13}$$

where $C = \log n! - \sum_{j=1}^{k} \log n_j!$ is a constant term that is independent of the values of the parameters p_j. Therefore, the maximum likelihood estimate of p_j is

$$\hat{p}_j = \frac{n_j}{n}. \tag{4.14}$$

Since the number of free parameters is $k-1$, the AIC is given by

$$\mathrm{AIC} = (-2)\left\{C + \sum_{j=1}^{k} n_j \log\left(\frac{n_j}{n}\right)\right\} + 2(k-1). \tag{4.15}$$

To compare this histogram model with a simpler one, we may consider the model obtained by assuming the restriction $p_{2j-1} = p_{2j}$ for $j = 1,\ldots,m$. Here, for simplicity, we assume that $k = 2m$. The maximum likelihood estimates of this restricted model are

$$\hat{p}_{2j-1} = \hat{p}_{2j} = \frac{n_{2j-1}+n_{2j}}{2n}, \tag{4.16}$$

and the AIC is given by

$$\mathrm{AIC} = (-2)\left\{C + \sum_{j=1}^{m} (n_{2j-1}+n_{2j}) \log\left(\frac{n_{2j-1}+n_{2j}}{2n}\right)\right\} + 2(m-1). \tag{4.17}$$

Similarly, we can compute the AICs for histograms with smaller bin sizes such as $k/4$.

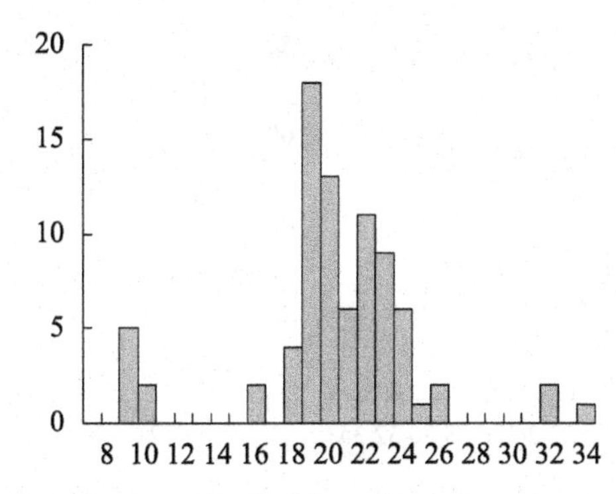

Fig. 4.1. Histogram of galaxy data. Bin size $m = 28$.

Table 4.1. Log-likelihoods and the AICs for three different bin sizes, $k = 28, 14$, and 7.

Bin Size	Log-Likelihood	AIC
28	−189.18942	432.37884
14	−197.71509	421.43018
7	−209.51501	431.03002

Example 2 (Histogram of galaxy data) The following table shows the number of observations in the galaxy data [Roeder (1990)] that fall in the interval $[6+i, 7+i)$, $i = 1, \ldots, 28$. Figure 4.1 shows the original histogram (see Example 9 in Section 2.2).

0	5	2	0	0	0	0	0	2	0	4	18	13	6
11	9	6	1	2	0	0	0	0	0	2	0	1	0

Table 4.1 shows the log-likelihoods and the AICs of the original histogram with bin size $k = 28$ and the ones with $k = 14$ and $k = 7$. The AIC is minimized at $k = 14$, suggesting that the original histogram is too fine and that a histogram with only 7 bins is too coarse. Figure 4.2 shows two histograms for $k = 14$ and 7.

4.3 Equality of the Means and/or the Variances of Normal Distributions

Assume that two sets of data $\{y_1, \ldots, y_n\}$ and $\{y_{n+1}, \ldots, y_{n+m}\}$ are given. To check the equality of these two data sets, we consider the model composed of two normal distributions, $y_1, \ldots, y_n \sim N(\mu_1, \tau_1^2)$ and $y_{n+1}, \ldots, y_{n+m} \sim N(\mu_2, \tau_2^2)$, i.e.,

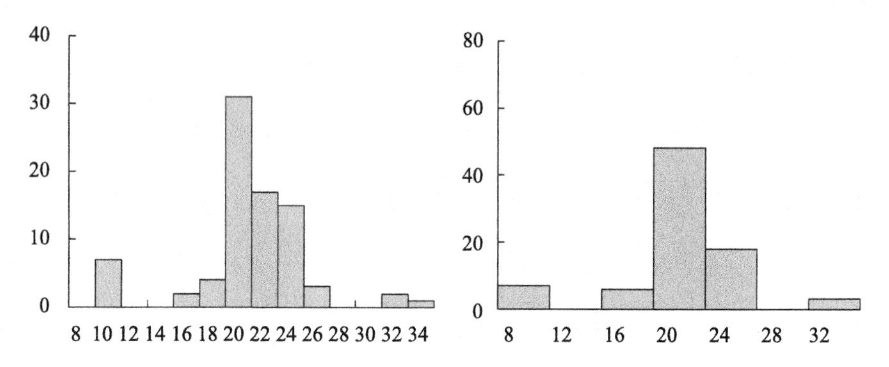

Fig. 4.2. Histogram of galaxy data. Bin sizes $k = 14$ and 7.

$$
\begin{aligned}
f(y_i|\mu_1, \sigma_1^2) &= \frac{1}{\sqrt{2\pi\sigma_1^2}} \exp\left\{-\frac{(y-\mu_1)^2}{2\sigma_1^2}\right\}, \quad i = 1, \ldots, n, \\
f(y_i|\mu_2, \sigma_2^2) &= \frac{1}{\sqrt{2\pi\sigma_2^2}} \exp\left\{-\frac{(y-\mu_2)^2}{2\sigma_2^2}\right\}, \quad i = n+1, \ldots, n+m. \quad (4.18)
\end{aligned}
$$

Given the above data, the log-likelihood of the model is

$$
\begin{aligned}
\ell(\mu_1, \mu_2, \sigma_1^2, \sigma_2^2) = &-\frac{n}{2}\log(2\pi\sigma_1^2) - \frac{1}{2\sigma_1^2}\sum_{j=1}^{n}(y_j - \mu_1)^2 \\
&-\frac{m}{2}\log(2\pi\sigma_2^2) - \frac{1}{2\sigma_2^2}\sum_{j=n+1}^{n+m}(y_j - \mu_2)^2. \quad (4.19)
\end{aligned}
$$

By maximizing the log-likelihood function, we have the maximum likelihood estimates of the models

$$
\begin{aligned}
\hat\mu_1 &= \frac{1}{n}\sum_{j=1}^{n} y_j, \quad \hat\sigma_1^2 = \frac{1}{n}\sum_{j=1}^{n}(y_j - \hat\mu_1)^2, \\
\hat\mu_2 &= \frac{1}{m}\sum_{j=n+1}^{n+m} y_j, \quad \hat\sigma_2^2 = \frac{1}{n}\sum_{j=n+1}^{n+m}(y_j - \hat\mu_2)^2. \quad (4.20)
\end{aligned}
$$

The maximum log-likelihood is

$$
\ell(\hat\mu_1, \hat\mu_2, \hat\sigma_1^2, \hat\sigma_2^2) = -\frac{n}{2}\log(2\pi\hat\sigma_1^2) - \frac{m}{2}\log(2\pi\hat\sigma_2^2) - \frac{n+m}{2}, \quad (4.21)
$$

and since the number of unknown parameters is four, the AIC is given by

$$
\mathrm{AIC} = (n+m)(\log 2\pi + 1) + n\log\hat\sigma_1^2 + m\log\hat\sigma_2^2 + 2\times 4. \quad (4.22)
$$

To check the homogeneity of the two data sets in question, we compare this model with the following three restricted models:

(1) $\mu_1 = \mu_2 = \mu$ and $\sigma_1^2 = \sigma_2^2 = \sigma^2$,
(2) $\sigma_1^2 = \sigma_2^2 = \sigma^2$,
(3) $\mu_1 = \mu_2 = \mu$.

Assumption (1) is equivalent to having $n+m$ observations $y_1, \ldots, y_{n+m}$ from the same normal distribution model. The AIC of this model is given by

$$\mathrm{AIC} = (n+m)\{\log(2\pi\hat{\sigma}^2) + 1\} + 2 \times 2, \tag{4.23}$$

where $\hat{\mu}$ and $\hat{\sigma}^2$ are defined by

$$\hat{\mu} = \frac{1}{n+m}\sum_{j=1}^{n+m} y_j, \quad \hat{\sigma}^2 = \frac{1}{n+m}\sum_{j=1}^{n+m}(y_j - \hat{\mu})^2. \tag{4.24}$$

Under assumption (2), the log-likelihood of the model can be written as

$$\begin{aligned}\ell_2(\mu_1, \mu_2, \sigma^2) = &-\frac{n+m}{2}\log(2\pi\sigma^2) - \frac{1}{2\sigma^2}\sum_{j=1}^{n}(y_j - \mu_1)^2 \\ &- \frac{1}{2\sigma^2}\sum_{j=n+1}^{n+m}(y_j - \mu_2)^2.\end{aligned} \tag{4.25}$$

Therefore, we have the maximum likelihood estimates of the models

$$\begin{aligned}\hat{\mu}_1 &= \frac{1}{n}\sum_{j=1}^{n} y_j, \quad \hat{\mu}_2 = \frac{1}{m}\sum_{j=n+1}^{n+m} y_j, \\ \hat{\sigma}^2 &= \frac{1}{n+m}\left\{\sum_{j=1}^{n}(y_j - \hat{\mu}_1)^2 + \sum_{j=n+1}^{n+m}(y_j - \hat{\mu}_2)^2\right\}.\end{aligned} \tag{4.26}$$

The maximum log-likelihood is then given by

$$\ell_2(\hat{\mu}_1, \hat{\mu}_2, \hat{\sigma}_2^2) = -\frac{n+m}{2}\log(2\pi\hat{\sigma}^2) - \frac{n+m}{2}, \tag{4.27}$$

and since the number of unknown parameters is three, the AIC is given by

$$\mathrm{AIC} = (n+m)\{\log(2\pi\hat{\sigma}^2) + 1\} + 2 \times 3. \tag{4.28}$$

Similarly, under assumption (3), we have the log-likelihood of the model

$$\begin{aligned}\ell_3(\mu, \sigma_1^2, \sigma_2^2) = &-\frac{n}{2}\log(2\pi\sigma_1^2) - \frac{1}{2\sigma_1^2}\sum_{j=1}^{n}(y_j - \mu)^2 \\ &- \frac{m}{2}\log(2\pi\sigma_2^2) - \frac{1}{2\sigma_2^2}\sum_{j=n+1}^{n+m}(y_j - \mu)^2.\end{aligned} \tag{4.29}$$

The maximum likelihood estimates of the models are given as the solutions of the likelihood equations

$$\frac{\partial \ell_3}{\partial \mu} = 0, \quad \frac{\partial \ell_3}{\partial \sigma_1^2} = 0, \quad \frac{\partial \ell_3}{\partial \sigma_2^2} = 0. \tag{4.30}$$

From the equations, the maximum likelihood estimates of the variances are

$$\tilde{\sigma}_1^2 = \frac{1}{n}\sum_{j=1}^{n}(y_j - \mu)^2, \quad \tilde{\sigma}_2^2 = \frac{1}{m}\sum_{j=n+1}^{n+m}(y_j - \mu)^2. \tag{4.31}$$

Therefore, by substituting these into the likelihood equation for the mean, the maximum likelihood estimate of the mean μ is obtained as the solution to the equation

$$\frac{\partial \ell_3}{\partial \mu} = \frac{1}{\tilde{\sigma}_1^2}\sum_{j=1}^{n}(y_j - \mu)^2 + \frac{1}{\tilde{\sigma}_2^2}\sum_{j=n+1}^{n+m}(y_j - \mu)^2 = 0. \tag{4.32}$$

From this, we obtain the equation

$$n\sum_{j=1}^{n}(y_j - \mu)\sum_{j=n+1}^{n+m}(y_j - \mu)^2 + m\sum_{j=n+1}^{n+m}(y_j - \mu)\sum_{j=1}^{n}(y_j - \mu)^2 = 0, \tag{4.33}$$

which can be expressed by the cubic equation

$$\mu^3 + A\mu^2 + B\mu + C = 0. \tag{4.34}$$

Here the coefficients A, B, and C are defined by

$$\begin{aligned} A &= -\{(1 + w_2)\hat{\mu}_1 + (1 + w_1\hat{\mu}_2)\}, \\ B &= 2\hat{\mu}_1\hat{\mu}_2 + w_2 s_1^2 + w_1 s_2^2, \\ C &= -(w_1\hat{\mu}_1 s_2^2 + w_2\hat{\mu}_2 s_1^2), \end{aligned} \tag{4.35}$$

with $w_1 = n/(n+m)$, $w_2 = m/(n+m)$, and

$$s_1^2 = \frac{1}{n}\sum_{j=1}^{n} y_j^2, \quad s_2^2 = \frac{1}{m}\sum_{j=n+1}^{n+m} y_j^2. \tag{4.36}$$

The solution to this cubic equation can be obtained using the Cardano formula shown below. Then the AIC is obtained by

$$\text{AIC} = (n+m)(\log 2\pi + 1) + n\log\tilde{\sigma}_1^2 + m\log\tilde{\sigma}_2^2 + 2\times 3. \tag{4.37}$$

Remark (Cardano's formula) The cubic equation

$$\mu^3 + A\mu^2 + B\mu + C = 0 \tag{4.38}$$

Table 4.2. Comparison of four normal distribution models.

Restriction	ℓ	AIC	$\hat{\mu}_1$	$\hat{\mu}_2$	$\hat{\sigma}_1^2$	$\hat{\sigma}_2^2$
none	−48.411	104.823	0.310	0.857	1.033	3.015
$\sigma_1^2 = \sigma_2^2$	−50.473	106.946	0.310	0.857	1.694	1.694
$\mu_1 = \mu_2$	−48.852	103.703	0.438	0.438	1.049	3.191
$\mu_1 = \mu_2, \sigma_1^2 = \sigma_2^2$	−51.050	106.101	0.492	0.492	1.760	1.760

can be transformed to a reduced form

$$\lambda^3 + 3p\lambda + q = 0 \tag{4.39}$$

by $\lambda = \mu + A/3$, $p = (3B - C^2)/9$, and $q = (2A^3 - 9AB + 27C)/27$. The solutions to this equation are then

$$\lambda = \sqrt[3]{\alpha} + \sqrt[3]{\beta}, \quad \omega\sqrt[3]{\alpha} + \omega^2\sqrt[3]{\beta}, \quad \omega^2\sqrt[3]{\alpha} + \omega\sqrt[3]{\beta}, \tag{4.40}$$

where α, β, and ω are given by

$$\omega = \frac{-1 + \sqrt{3}i}{2},$$
$$\alpha, \beta = \frac{-q \pm \sqrt{q^2 + 4p^3}}{2}. \tag{4.41}$$

Example 3 (Numerical result for the equality of two normal distributions) Consider the two sets of data:

Data set 1

0.26 −1.33 1.07 1.78 −0.16 0.03 −0.79 −1.55 1.27 0.56
−0.95 0.60 0.27 1.67 0.60 −0.42 1.87 0.65 −0.75 1.52

Data set 2

1.70 0.84 1.34 0.11 −0.88 −1.43 3.52 2.69 2.51 −1.83

The sample sizes of Data sets 1 and 2 are $n = 20$ and $m = 10$, respectively. The four models presented above were fitted and the results summarized in Table 4.2. The estimated variance of Data set 2 is about three times larger than that of Data set 1, but the difference in their means is not so large. Therefore, the AIC of the model that assumes equality of the variances is larger than that of the two-normal model without any restrictions. However, the AIC of the model that assumes the equality of the mean values is smaller than the AIC of the no-restriction model. The AIC of the model with the restriction that $\mu_1 = \mu_2, \sigma_1^2 = \sigma_2^2$ is larger than that of the no-restriction model.

4.4 Variable Selection for Regression Model

Suppose we have a response variable y and m explanatory variables $x_1, \ldots, x_m$. The linear regression model is

$$y = a_0 + a_1 x_1 + \cdots + a_m x_m + \varepsilon, \tag{4.42}$$

where the residual term ε is assumed to be a normal random variable with mean zero and variance σ^2.

The conditional distribution of the response variable y given the explanatory variables is

$$p(y|x_1, \ldots, x_m) = (2\pi\sigma^2)^{-\frac{1}{2}} \exp\left\{ -\frac{1}{2\sigma^2}\left(y - a_0 - \sum_{j=1}^{m} a_j x_j \right)^2 \right\}. \tag{4.43}$$

Therefore, given a set of n independent observations $\{(y_i, x_{i1}, \ldots, x_{im}); i = 1, \ldots, n\}$, the likelihood of the regression model is

$$L(a_0, a_1, \ldots, a_m, \sigma^2) = \prod_{i=1}^{n} p(y_i|x_{i1}, \ldots, x_{im}). \tag{4.44}$$

Thus, the log-likelihood is given by

$$\begin{aligned} &\ell(a_0, a_1, \ldots, a_m, \sigma^2) \\ &= -\frac{n}{2}\log(2\pi\sigma^2) - \frac{1}{2\sigma^2}\sum_{i=1}^{n}\left(y_i - a_0 - \sum_{j=1}^{m} a_j x_{ij} \right)^2, \end{aligned} \tag{4.45}$$

and the maximum likelihood estimators $\hat{a}_0, \hat{a}_1, \ldots, \hat{a}_m$ of the regression coefficients $a_0, a_1, \ldots, a_m$ are obtained as the solution to the system of linear equations

$$X^T X \boldsymbol{a} = X^T \boldsymbol{y}, \tag{4.46}$$

where $\boldsymbol{a} = (a_0, a_1, \ldots, a_m)^T$ and the $n \times (m+1)$ matrix X and n-dimensional vector $\boldsymbol{y}$ are defined by

$$X = \begin{bmatrix} 1 & x_{11} & \cdots & x_{1m} \\ 1 & x_{21} & \cdots & x_{2m} \\ \vdots & \vdots & \ddots & \vdots \\ 1 & x_{n1} & \cdots & x_{nm} \end{bmatrix}, \quad \boldsymbol{y} = \begin{bmatrix} y_1 \\ y_2 \\ \vdots \\ y_n \end{bmatrix}. \tag{4.47}$$

The maximum likelihood estimate $\hat{\sigma}^2$ is

$$\hat{\sigma}^2 = \frac{1}{n}\sum_{i=1}^{n}\{y_i - (\hat{a}_0 + \hat{a}_1 x_{i1} + \cdots + \hat{a}_m x_{im})\}^2. \tag{4.48}$$

Substituting this into (4.45) yields the maximum log-likelihood

$$\ell(\hat{a}_0, \hat{a}_1, \ldots, \hat{a}_m, \hat{\sigma}^2) = -\frac{n}{2}\log 2\pi - \frac{n}{2}\log d(x_1, \ldots, x_m) - \frac{n}{2}, \qquad (4.49)$$

where $d(x_1, \ldots, x_m)$ is the estimate of the residual variance σ^2 of the model given by (4.48).

Since the number of free parameters contained in the multiple regression model is $m + 2$, the AIC for this model is

$$\mathrm{AIC} = n(\log 2\pi + 1) + n \log d(x_1, \ldots, x_m) + 2(m + 2). \qquad (4.50)$$

In multiple regression analysis, all of the given explanatory variables may not be necessarily effective for predicting the response variable. An estimated model with an unnecessarily large number of explanatory variables may be unstable. By selecting the model having the minimum AIC for different possible combinations of the explanatory variables, we expect to obtain a reasonable model.

Example 4 (Daily temperature data) Table 4.3 shows the daily minimum temperatures in January averaged from 1971 through 2000, y_i, the latitudes, x_{i1}, longitudes, x_{i2}, and altitudes, x_{i3}, of 25 cities in Japan. A similar data set was analyzed in Sakamoto et al. (1986). To predict the average daily minimum temperature in January, we consider the multiple regression model

$$y_i = a_0 + a_1 x_{i1} + a_2 x_{i2} + a_3 x_{i3} + \varepsilon_i, \qquad (4.51)$$

where the residual ε_i is assumed to be a normal random variable with mean zero and variance σ^2.

Given a set of n (=25) observations $\{(y_i, x_{i1}, x_{i2}, x_{i3}); i = 1, \ldots, n\}$, the likelihood of the multiple regression model is defined by

$$L(a_0, a_1, a_2, a_3, \sigma^2) = \left(\frac{1}{2\pi\sigma^2}\right)^{\frac{n}{2}} \exp\left\{-\frac{1}{2\sigma^2}\sum_{i=1}^{n}\left(y_i - a_0 - \sum_{j=1}^{3} a_j x_{ij}\right)^2\right\}. \qquad (4.52)$$

The log-likelihood is then given by

$$\ell(a_0, a_1, a_2, a_3, \sigma^2) = -\frac{n}{2}\log(2\pi\sigma^2) - \frac{1}{2\sigma^2}\sum_{i=1}^{n}\left(y_i - a_0 - \sum_{j=1}^{3} a_j x_{ij}\right)^2, \qquad (4.53)$$

and the estimators $\hat{a}_0, \hat{a}_1, \ldots, \hat{a}_m$ of the regression coefficients a_0, a_1, a_2, a_3 are obtained by the maximum likelihood or least squares method. Then the maximum likelihood estimate of the residual variance, $\hat{\sigma}^2$, is obtained by

$$\hat{\sigma}^2 = \frac{1}{n}\left\{\sum_{i=1}^{n}\left(y_i - \hat{a}_0 - \sum_{j=1}^{3} \hat{a}_j x_{ij}\right)^2\right\}. \qquad (4.54)$$

Table 4.3. Average daily minimum temperatures (in Celsius) for 25 cities in Japan.

n	Cities	Temp. (y)	Latitude (x_1)	Longitude (x_2)	Altitude (x_3)
1	Wakkanai	−7.6	45.413	141.683	2.8
2	Sapporo	−7.7	43.057	141.332	17.2
3	Kushiro	−11.4	42.983	144.380	4.5
3	Nemuro	−7.4	43.328	145.590	25.2
4	Akita	−2.7	39.715	140.103	6.3
5	Morioka	−5.9	39.695	141.168	155.2
6	Yamagata	−3.6	38.253	140.348	152.5
7	Wajima	0.1	37.390	136.898	5.2
8	Toyama	−0.4	36.707	137.205	8.6
9	Nagano	−4.3	36.660	138.195	418.2
10	Mito	−2.5	36.377	140.470	29.3
11	Karuizawa	−9.0	36.338	138.548	999.1
12	Fukui	0.3	36.053	136.227	8.8
13	Tokyo	2.1	35.687	139.763	6.1
14	Kofu	−2.7	35.663	138.557	272.8
15	Tottori	0.7	35.485	134.240	7.1
16	Nagoya	0.5	35.165	136.968	51.1
17	Kyoto	1.1	35.012	135.735	41.4
18	Shizuoka	1.6	34.972	138.407	14.1
19	Hiroshima	1.7	34.395	132.465	3.6
20	Fukuoka	3.2	33.580	130.377	2.5
21	Kochi	1.3	33.565	133.552	0.5
22	Shionomisaki	4.7	33.448	135.763	73.0
23	Nagasaki	3.6	32.730	129.870	26.9
24	Kagoshima	4.1	31.552	130.552	3.9
25	Naha	14.3	26.203	127.688	28.1

(Source: Chronological Scientific Tables of 2004.)

Substituting this into (4.53), the maximum log-likelihood is given by

$$\ell(\hat{a}_0, \hat{a}_1, \hat{a}_2, \hat{a}_3, \hat{\sigma}^2) = -\frac{n}{2}\log 2\pi - \frac{n}{2}\log\hat{\sigma}^2 - \frac{n}{2}. \tag{4.55}$$

In actual modeling, in addition to this full-order model, we also consider the subset regression models, i.e., the models defined by using a subset of regressors. This is equivalent to assuming that the regression coefficients of excluded variables are zero. Since the number of free parameters contained in the subset regression model is $k+2$, where k is the number of actually used variables or nonzero coefficients, the AIC is defined by

$$\mathrm{AIC}(x_1, \ldots, x_m) = n(\log 2\pi + 1) + n\log\hat{\sigma}^2 + 2(k+2). \tag{4.56}$$

Table 4.4 summarizes the estimated residual variances and coefficients and AICs of various models. It shows that the model having the latitude and the

Table 4.4. Subset regression models: AICs and estimated residual variances and coefficients.

No.	Explanatory variables	Residual variance	k	AIC	Regression coefficients a_0	a_1	a_2	a_3
1	x_1, x_3	1.490	2	88.919	40.490	−1.108	—	−0.010
2	x_1, x_2, x_3	1.484	3	90.812	44.459	−1.071	—	−0.010
3	x_1, x_2	5.108	2	119.715	71.477	−0.835	−0.305	—
4	x_1	5.538	1	119.737	40.069	−1.121	—	—
5	x_2, x_3	5.693	2	122.426	124.127	—	−0.906	−0.007
6	x_2	7.814	1	128.346	131.533	—	−0.965	—
7	x_3	19.959	1	151.879	0.382	—	—	−0.010
8	none	24.474	0	154.887	−0.580	—	—	—

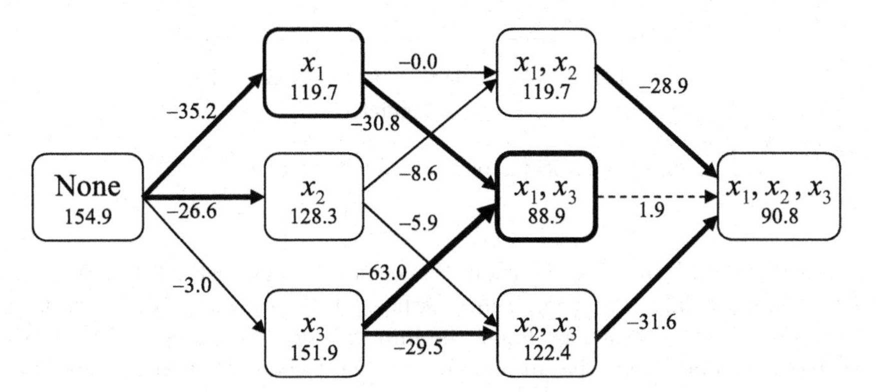

Fig. 4.3. Decrease of AIC values by adding regressors.

altitude as explanatory variables has the smallest value for the AIC. The AIC of the model with all three explanatory variables is larger than that of the model having the lowest value for the AIC. This is because the reduction in the residual variance of the former model is miniscule compared to that of the model having the lowest value of the AIC, and it indicates that knowledge of the longitude x_2 is of little value if we already know the latitude and altitude (x_1 and x_3).

Figure 4.3 shows the change in the AIC value when only one explanatory variable is incorporated in a subset regression model. It is interesting to note that when only one explanatory variable is used, x_3 (altitude) gives the smallest reduction in the AIC value. However, when the models with two explanatory variables are considered, the inclusion of x_3 is very effective in reducing the AIC value, and the AIC best model out of these models had the explanatory variables of x_1 and x_3, i.e., the latitude and the altitude. The AIC of the model with x_1 and x_2 is the same as that of the model with x_1. These suggest that x_1 and x_2 contain similar information, whereas x_3 has independent information. This can be understood from Figure 4.4, which shows

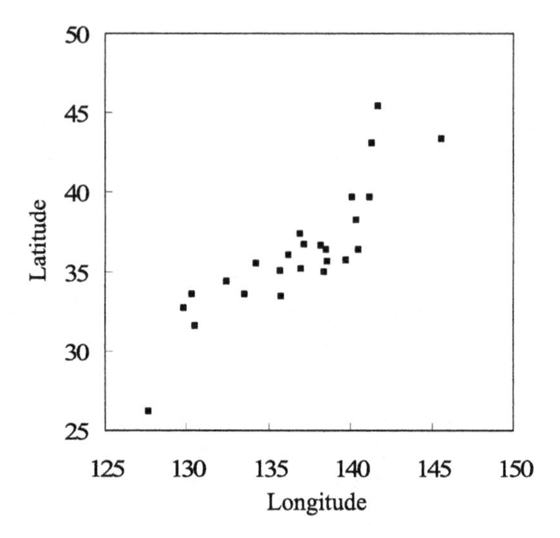

Fig. 4.4. Scatterplot: latitude vs. longitude.

the longitude vs. latitude scatterplot. Since the four main islands of Japan are located along a line that runs from northeast to southwest, x_1 and x_2 have a strong positive correlation. Thus, the information about the longitude of a city has a similar predictive ability for temperature as that of the latitude. However, when the latitude is known, knowledge of the longitude is almost redundant, whereas knowledge of the altitude is very useful and the residual variance becomes less than one third when the altitude is included.

The minimum AIC model is given by

$$y_i = 40.490 - 1.108x_{i1} - 0.010x_{i3} + \varepsilon_i,$$

with $\varepsilon_i \sim N(0, 1.490)$. The regression coefficient for the altitude x_3, -0.010, is about 50% larger than the common knowledge that the temperature should drop by about 6 degrees with a rise in altitude of $1,000$ meters.

Note that when the number of explanatory variables is large, we need to exercise care when comparing subset regression models having a different number of nonzero coefficients. This problem will be considered in section 8.5.2.

4.5 Generalized Linear Models

This section considers various types of regression models in the context of generalized linear models [Nelder and Wedderburn (1972), McCullagh and Nelder (1989)] and introduces a general framework for constructing the AIC.

Suppose that we have n independent observations $y_1, \ldots, y_n$ corresponding to $(p+1)$-dimensional design points $\boldsymbol{x}_\alpha = (1, x_{\alpha 1}, \ldots, x_{\alpha p})^T$ for $\alpha = 1, \ldots, n$. Regression models, in general, consist of a random component and a systematic component. The random component specifies the distribution of the response variable Y_α, while the systematic component represents the mean structure $E[Y_\alpha|\boldsymbol{x}_\alpha] = \mu_\alpha$, $\alpha = 1, \ldots, n$. In generalized linear models, the responses Y_α are assumed to be drawn from the exponential family of distributions with densities

$$f(y_\alpha|\boldsymbol{x}_\alpha; \theta_\alpha, \psi) = \exp\left\{\frac{y_\alpha\theta_\alpha - b(\theta_\alpha)}{\psi} + c(y_\alpha, \psi)\right\}, \quad \alpha = 1, \ldots, n, \tag{4.57}$$

where $b(\cdot)$ and $c(\cdot, \cdot)$ are specific functions and ψ is a scale parameter. The conditional expectation μ_α is related to the predictor η_α by $h(\mu_\alpha) = \eta_\alpha$, where $h(\cdot)$ is a monotone differentiable function called a *link function*. The linear predictor is given by $\eta_\alpha = \boldsymbol{x}_\alpha^T\boldsymbol{\beta}$, where $\boldsymbol{\beta}$ is a $(p+1)$-dimensional vector of unknown parameters.

Let $\ell(\theta_\alpha, \psi)$ be the log-likelihood function

$$\begin{aligned}\ell(\theta_\alpha, \psi) &= \log f(y_\alpha|\boldsymbol{x}_\alpha; \theta_\alpha, \psi) \\ &= \frac{y_\alpha\theta_\alpha - b(\theta_\alpha)}{\psi} + c(y_\alpha, \psi).\end{aligned} \tag{4.58}$$

From the well-known properties

$$E\left[\frac{\partial\ell(\theta_\alpha, \psi)}{\partial\theta_\alpha}\right] = 0, \quad E\left[\left\{\frac{\partial\ell(\theta_\alpha, \psi)}{\partial\theta_\alpha}\right\}^2\right] = -E\left[\frac{\partial^2\ell(\theta_\alpha, \psi)}{\partial\theta_\alpha^2}\right], \tag{4.59}$$

it follows that

$$E[Y_\alpha] = \mu_\alpha = b'(\theta_\alpha), \quad \mathrm{var}(Y_\alpha) = b''(\theta_\alpha)\psi = \frac{\partial\mu_\alpha}{\partial\theta_\alpha}\psi. \tag{4.60}$$

Hence, we have

$$\frac{\partial\ell(\theta_\alpha, \psi)}{\partial\theta_\alpha} = \frac{y_\alpha - b'(\theta_\alpha)}{\psi} = \frac{y_\alpha - \mu_\alpha}{\mathrm{var}(Y_\alpha)}\frac{\partial\mu_\alpha}{\partial\theta_\alpha}. \tag{4.61}$$

Since the linear predictor is related by

$$\eta_\alpha = h(\mu_\alpha) = h(b'(\theta_\alpha)) = \boldsymbol{x}_\alpha^T\boldsymbol{\beta}, \tag{4.62}$$

it can be readily seen that

$$\frac{\partial\mu_\alpha}{\partial\eta_\alpha} = \frac{1}{h'(\mu_\alpha)}, \qquad \frac{\partial\eta_\alpha}{\partial\beta_i} = x_{\alpha i}, \quad i = 0, 1, \ldots, p, \tag{4.63}$$

where $x_{\alpha 0} = 1$.

Therefore, it follows from (4.61) and (4.63) that differentiation of the log-likelihood (4.58) with respect to each β_i gives

$$\begin{aligned}\frac{\partial \ell(\theta_\alpha, \psi)}{\partial \beta_i} &= \frac{\partial \ell(\theta_\alpha, \psi)}{\partial \theta_\alpha} \frac{\partial \theta_\alpha}{\partial \mu_\alpha} \frac{\partial \mu_\alpha}{\partial \eta_\alpha} \frac{\partial \eta_\alpha}{\partial \beta_i} \\ &= \frac{y_\alpha - \mu_\alpha}{\mathrm{var}(Y_\alpha)} \frac{\partial \mu_\alpha}{\partial \theta_\alpha} \frac{\partial \theta_\alpha}{\partial \mu_\alpha} \frac{1}{h'(\mu_\alpha)} x_{\alpha i} \\ &= \frac{y_\alpha - \mu_\alpha}{\mathrm{var}(Y_\alpha)} \frac{1}{h'(\mu_\alpha)} x_{\alpha i}. \end{aligned} \tag{4.64}$$

Consequently, given the observations $y_1, \ldots, y_n$, the maximum likelihood estimator of $\boldsymbol{\beta}$ is given by the solution of the equations

$$\sum_{\alpha=1}^{n} \frac{\partial \ell(\theta_\alpha, \psi)}{\partial \beta_i} = \sum_{\alpha=1}^{n} \frac{y_\alpha - \mu_\alpha}{\mathrm{var}(Y_\alpha)} \frac{1}{h'(\mu_\alpha)} x_{\alpha i} = 0, \qquad i = 0, 1, \ldots, p. \tag{4.65}$$

If the link function has the form of $h(\cdot) = b'^{-1}(\cdot)$, which is the inverse of $b'(\cdot)$, then it follows from (4.62) that

$$\eta_\alpha = h(\mu_\alpha) = h(b'(\theta_\alpha)) = \theta_\alpha = \boldsymbol{x}_\alpha^T \boldsymbol{\beta}. \tag{4.66}$$

Hence, this special link function, known as the *canonical link function*, relates the parameter θ_α in the exponential family (4.57) directly to the linear predictor and leads to

$$f(y_\alpha | \boldsymbol{x}_\alpha; \boldsymbol{\beta}, \psi) = \exp\left\{ \frac{y_\alpha \boldsymbol{x}_\alpha^T \boldsymbol{\beta} - b(\boldsymbol{x}_\alpha^T \boldsymbol{\beta})}{\psi} + c(y_\alpha, \psi) \right\}, \tag{4.67}$$

for $\alpha = 1, \ldots, n$. By replacing the unknown parameters $\boldsymbol{\beta}$ and ψ with the corresponding maximum likelihood estimates $\hat{\boldsymbol{\beta}}$ and $\hat{\psi}$, we have the statistical model $f(y_\alpha | \boldsymbol{x}_\alpha; \hat{\boldsymbol{\beta}}, \hat{\psi})$. The AIC for evaluating the statistical model is then given by

$$\mathrm{AIC} = -2 \sum_{\alpha=1}^{n} \left\{ \frac{y_\alpha \boldsymbol{x}_\alpha^T \hat{\boldsymbol{\beta}} - b(\boldsymbol{x}_\alpha^T \hat{\boldsymbol{\beta}})}{\hat{\psi}} + c(y_\alpha, \hat{\psi}) \right\} + 2(p+2). \tag{4.68}$$

Example 5 (Gaussian linear regression model) Suppose that the observations y_α are independently and normally distributed with mean μ_α and variance σ^2. Then the density function of y_α can be rewritten as

$$\begin{aligned} f(y_\alpha | \mu_\alpha, \sigma^2) &= \frac{1}{\sqrt{2\pi\sigma^2}} \exp\left\{ -\frac{(y_\alpha - \mu_\alpha)^2}{2\sigma^2} \right\} \\ &= \exp\left\{ \frac{y_\alpha \mu_\alpha - \mu_\alpha^2/2}{\sigma^2} - \frac{y_\alpha^2}{2\sigma^2} - \frac{1}{2}\log(2\pi\sigma^2) \right\}. \end{aligned} \tag{4.69}$$

Comparing this density function with the exponential family of densities in (4.57) yields the relations

$$\theta_\alpha = \mu_\alpha, \quad b(\mu_\alpha) = \frac{\mu_\alpha^2}{2}, \quad \psi = \sigma^2,$$
$$c(y_\alpha, \sigma^2) = -\frac{y_\alpha^2}{2\sigma^2} - \frac{1}{2}\log(2\pi\sigma^2). \tag{4.70}$$

By taking

$$\hat{\mu}_\alpha = \boldsymbol{x}_\alpha^T \hat{\boldsymbol{\beta}}, \quad b(\boldsymbol{x}_\alpha^T \hat{\boldsymbol{\beta}}) = \frac{1}{2}\left(\boldsymbol{x}_\alpha^T \hat{\boldsymbol{\beta}}\right)^2,$$
$$c(y_\alpha, \hat{\sigma}^2) = -\frac{y_\alpha^2}{2\hat{\sigma}^2} - \frac{1}{2}\log(2\pi\hat{\sigma}^2) \tag{4.71}$$

in (4.68), we have that the AIC for a Gaussian linear regression model is given by

$$\text{AIC} = n\log(2\pi\hat{\sigma}^2) + n + 2(p+2), \tag{4.72}$$

where $\hat{\sigma}^2 = \sum_{\alpha=1}^n (y_\alpha - \boldsymbol{x}_\alpha^T \hat{\boldsymbol{\beta}})^2/n$.

Example 6 (Linear logistic regression model) Let $y_1, \ldots, y_n$ be an independent sequence of binary random variables taking values 0 and 1 with conditional probabilities

$$\Pr(Y = 1|\boldsymbol{x}_\alpha) = \pi(\boldsymbol{x}_\alpha) \quad \text{and} \quad \Pr(Y = 0|\boldsymbol{x}_\alpha) = 1 - \pi(\boldsymbol{x}_\alpha), \tag{4.73}$$

where $\boldsymbol{x}_\alpha = (1, x_{\alpha 1}, \ldots, x_{\alpha p})^T$ for p explanatory variables. It is assumed that

$$\pi(\boldsymbol{x}_\alpha) = \frac{\exp(\boldsymbol{x}_\alpha^T \boldsymbol{\beta})}{1 + \exp(\boldsymbol{x}_\alpha^T \boldsymbol{\beta})}. \tag{4.74}$$

The y_α have a Bernoulli distribution with mean $\mu_\alpha = \pi(\boldsymbol{x}_\alpha)$, and its density function is given by

$$\begin{aligned} f(y_\alpha|\pi(\boldsymbol{x}_\alpha)) &= \pi(\boldsymbol{x}_\alpha)^{y_\alpha}(1 - \pi(\boldsymbol{x}_\alpha))^{1-y_\alpha} \qquad (4.75) \\ &= \exp\left\{ y_\alpha \log\frac{\pi(\boldsymbol{x}_\alpha)}{1 - \pi(\boldsymbol{x}_\alpha)} + \log(1 - \pi(\boldsymbol{x}_\alpha)) \right\}, \quad y_\alpha = 0, 1. \end{aligned}$$

By comparing with (4.57), it is easy to see that

$$\theta_\alpha = h(\pi(\boldsymbol{x}_\alpha)) = \log\frac{\pi(\boldsymbol{x}_\alpha)}{1 - \pi(\boldsymbol{x}_\alpha)} = \boldsymbol{x}_\alpha^T \boldsymbol{\beta}, \quad \psi = 1, \quad c(y_\alpha, \psi) = 0. \tag{4.76}$$

Noting that $\pi(\boldsymbol{x}_\alpha) = \exp(\theta_\alpha)/\{1 + \exp(\theta_\alpha)\}$, we have

$$b(\theta_\alpha) = -\log(1 - \pi(\boldsymbol{x}_\alpha)) = \log\{1 + \exp(\theta_\alpha)\}. \tag{4.77}$$

Therefore, taking $b(\boldsymbol{x}_\alpha^T\boldsymbol{\beta}) = \log\{1+\exp(\boldsymbol{x}_\alpha^T\boldsymbol{\beta})\}$ in (4.68) and replacing $\boldsymbol{\beta}$ with the maximum likelihood estimate $\hat{\boldsymbol{\beta}}$, we have the AIC for evaluating the statistical model $f(y_\alpha|\boldsymbol{x}_\alpha;\hat{\boldsymbol{\beta}})$ in the form

$$\mathrm{AIC} = 2\sum_{\alpha=1}^{n}\left[\log\left\{1+\exp(\boldsymbol{x}_\alpha^T\hat{\boldsymbol{\beta}})\right\} - y_\alpha\boldsymbol{x}_\alpha^T\hat{\boldsymbol{\beta}}\right] + 2(p+1). \tag{4.78}$$

4.6 Selection of Order of Autoregressive Model

A sequence of observations of a phenomenon that fluctuates with time is called a *time series.* The most fundamental model in time series analysis is the autoregressive (AR) model. For simplicity, we consider here a univariate time series y_t, $t = 1,\ldots,n$. The AR model expresses the present value of a time series as a linear combination of past values and a random component,

$$y_t = \sum_{i=1}^{m} a_i y_{t-i} + \varepsilon_t, \tag{4.79}$$

where m is called the *order* of the AR model, and the a_i are called the *AR coefficients.* The random variable ε_t is assumed to be a normal random variable with mean 0 and variance σ^2. In other words, given the past values, $y_{t-m},\ldots,y_{t-1}$, the y_t are distributed with a normal distribution with mean $a_1y_{t-1}+\cdots+a_my_{t-m}$ and variance σ^2.

For simplicity, assuming that $y_{1-m},\ldots,y_0$ are known, the likelihood of the model given data $y_1,\ldots,y_n$ is obtained by

$$\begin{aligned} L(a_1,\ldots,a_m,\sigma^2) &= f(y_1,\ldots,y_n|y_{1-m},\ldots,y_0) \\ &= \prod_{i=1}^{n} f(y_t|y_{t-m},\ldots,y_{t-1}). \end{aligned} \tag{4.80}$$

Here $f(y_t|y_{t-m},\ldots,y_{t-1})$ is the conditional density of y_t given $y_{t-m},\ldots,y_{t-1}$ and is a normal density with mean $a_1y_{t-1}+\cdots+a_my_{n-m}$ and variance σ^2, i.e.,

$$f(y_t|y_{t-m},\ldots,y_{t-1}) = \frac{1}{\sqrt{2\pi\sigma^2}}\exp\left\{-\frac{1}{2\sigma^2}\left(y_t - \sum_{i=1}^{m} a_i y_{t-i}\right)^2\right\}. \tag{4.81}$$

Thus, assuming that $y_{1-m},\ldots,y_0$ are known, the likelihood of the AR model with order m can be written as

$$L(a_1,\ldots,a_m,\sigma^2) = \left(\frac{1}{2\pi\sigma^2}\right)^{n/2}\exp\left\{-\frac{1}{2\sigma^2}\sum_{i=1}^{n}\left(y_t - \sum_{i=1}^{m} a_i y_{t-i}\right)^2\right\}. \tag{4.82}$$

By taking logarithms of both sides, the log-likelihood of the model can be expressed as

$$\ell(a_1,\ldots,a_m,\sigma^2) = -\frac{n}{2}\log(2\pi\sigma^2) - \frac{1}{2\sigma^2}\sum_{t=1}^{n}\left(y_t - \sum_{i=1}^{m} a_i y_{t-i}\right)^2. \quad (4.83)$$

The maximum likelihood estimators of $a_1,\ldots,a_m$ and σ^2 are obtained by solving the system of equations

$$\begin{aligned}
\frac{\partial \ell}{\partial a_1} &= \frac{1}{\sigma^2}\sum_{t=1}^{n} y_{t-1}\left(y_t - \sum_{i=1}^{m} a_i y_{t-i}\right) = 0,\\
&\vdots \\
\frac{\partial \ell}{\partial a_m} &= \frac{1}{\sigma^2}\sum_{t=1}^{n} y_{t-m}\left(y_t - \sum_{i=1}^{m} a_i y_{t-i}\right) = 0,\\
\frac{\partial \ell}{\partial \sigma^2} &= -\frac{n}{2\sigma^2} + \frac{1}{2\sigma^4}\sum_{t=1}^{n}\left(y_t - \sum_{i=1}^{m} a_i y_{t-i}\right)^2 = 0.
\end{aligned} \quad (4.84)$$

Thus, like other regression models, the maximum likelihood estimators $\hat{a}_1,\ldots,\hat{a}_m$ are obtained as the solution to the normal equation

$$\begin{bmatrix} C(1,1) & \cdots & C(1,m) \\ \vdots & \ddots & \vdots \\ C(m,1) & \cdots & C(m,m) \end{bmatrix} \begin{bmatrix} a_1 \\ \vdots \\ a_m \end{bmatrix} = \begin{bmatrix} C(1,0) \\ \vdots \\ C(m,0) \end{bmatrix}, \quad (4.85)$$

where $C(i,j) = \sum_{t=1}^{n} y_{t-i}y_{t-j}$. The maximum likelihood estimator σ^2 is

$$\hat{\sigma}^2 = \frac{1}{n}\sum_{t=1}^{n}\left(y_t - \sum_{i=1}^{m}\hat{a}_i y_{t-i}\right)^2 = \frac{1}{n}\left(C(0,0) - \sum_{i=1}^{m}\hat{a}_i C(i,0)\right). \quad (4.86)$$

Substitution of this result into (4.83) yields the maximum log-likelihood

$$\ell(\hat{a}_1,\ldots,\hat{a}_m,\hat{\sigma}^2) = -\frac{n}{2}\log(2\pi\hat{\sigma}^2) - \frac{n}{2}. \quad (4.87)$$

Since the autoregressive model with order m has $m+1$ free parameters, the AIC is given by

$$\begin{aligned}
\mathrm{AIC}(m) &= -2\ell(\hat{a}_1,\ldots,\hat{a}_m,\hat{\sigma}^2) + 2(m+1) \\
&= n(\log 2\pi + 1) + n\log\hat{\sigma}^2 + 2(m+1).
\end{aligned} \quad (4.88)$$

Example 7 (Canadian lynx data) The logarithms of the annual numbers of Canadian lynx trapped from 1821 to 1934 recorded by the Hudson Bay Company are shown next [Kitagawa and Gersch (1996)]. The number of observations is $N = 114$.

2.430 2.506 2.767 2.940 3.169 3.450 3.594 3.774 3.695 3.411
2.718 1.991 2.265 2.446 2.612 3.359 3.429 3.533 3.261 2.612
2.179 1.653 1.832 2.328 2.737 3.014 3.328 3.404 2.981 2.557
2.576 2.352 2.556 2.864 3.214 3.435 3.458 3.326 2.835 2.476
2.373 2.389 2.742 3.210 3.520 3.828 3.628 2.837 2.406 2.675
2.554 2.894 3.202 3.224 3.352 3.154 2.878 2.476 2.303 2.360
2.671 2.867 3.310 3.449 3.646 3.400 2.590 1.863 1.581 1.690
1.771 2.274 2.576 3.111 3.605 3.543 2.769 2.021 2.185 2.588
2.880 3.115 3.540 3.845 3.800 3.579 3.264 2.538 2.582 2.907
3.142 3.433 3.580 3.490 3.475 3.579 2.829 1.909 1.903 2.033
2.360 2.601 3.054 3.386 3.553 3.468 3.187 2.723 2.686 2.821
3.000 3.201 3.424 3.531

We considered the AR models up to order 20. To apply the least squares method, the first 20 observations are treated as given in (4.80) and (4.81). Table 4.5 shows the innovation variances and the AIC of the AR models up to order 20. The model with $m = 0$ is the white noise model. The AIC attained is smallest at $m = 11$. Figure 4.5 shows the power spectra obtained using

$$p(f) = \hat{\sigma}^2 \left| 1 - \sum_{j=1}^{m} \hat{a}_j e^{-2\pi i j f} \right|^{-2}, \quad 0 \le f \le 0.5. \tag{4.89}$$

The left plot shows the spectrum obtained from the AR model having the lowest AIC value, $m = 11$, while the right plot shows the spectra obtained from the AR models with orders 0 to 20. The spectrum of the AR model with $m = 11$ is shown using a bold curve. It can be seen that depending on the order of the AR model, the estimated spectrum may become too smooth or too erratic, demonstrating the importance of selecting an appropriate order.

Table 4.5. AR models fitted to Canadian lynx data. m is the order of the AR model, and σ_m^2 is the estimated innovation variance of the AR model with order m.

m	σ_m^2	AIC	m	σ_m^2	AIC
0	0.31607	−106.268	11	0.03319	−296.130
1	0.11482	−199.453	12	0.03255	−295.943
2	0.04847	−278.512	13	0.03248	−294.157
3	0.04828	−276.886	14	0.03237	−292.467
4	0.04657	−278.289	15	0.03235	−290.533
5	0.04616	−277.112	16	0.03187	−289.920
6	0.04512	−277.254	17	0.03183	−288.042
7	0.04312	−279.505	18	0.03127	−287.721
8	0.04201	−279.963	19	0.03088	−286.902
9	0.04128	−279.613	20	0.02998	−287.679
10	0.03829	−284.677			

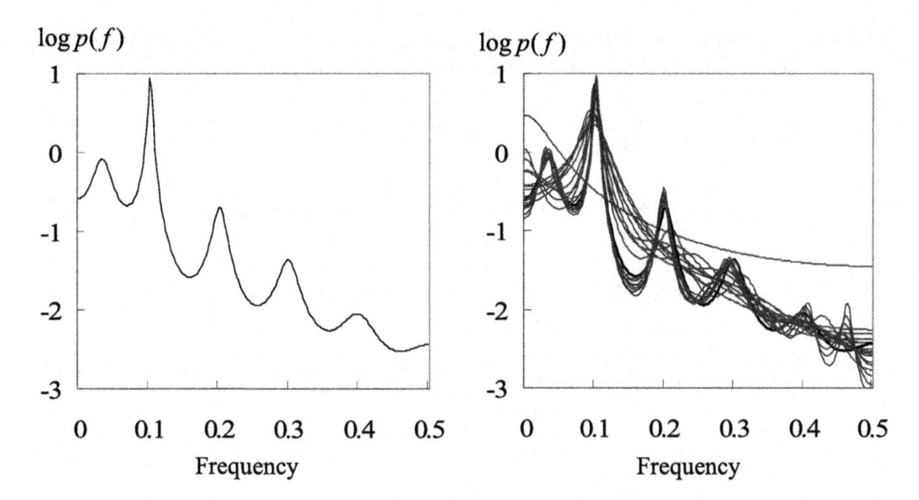

Fig. 4.5. Power spectrum estimates from AR models. Horizontal axis: frequency f, $0 \le f \le 0.5$. Vertical axis: logarithm of power spectrum, $\log p(f)$. Left plot: using the AR model with lowest AIC value, $m = 11$. Right plot: spectra obtained by the AR model with orders up to 20.

In the analysis done so far, the least squares method was used to estimate the AR model. This is a computationally efficient method and has several advantages. However, it uses the initial portion of the data, $y_{1-m}, \cdots, y_0$, only for initialization, that may result in poor estimation for very limited amounts of data. We note here, the exact maximum likelihood estimates of the AR model can be obtained by using the state-space representation with the Kalman filter.

We define the $m \times m$ matrix F and the m-dimensional vectors G, x_n and H by

$$F = \begin{bmatrix} a_1 & a_2 & \cdots & a_m \\ 1 & & & \\ & \ddots & & \\ & & 1 & \end{bmatrix}, \quad G = \begin{bmatrix} 1 \\ 0 \\ \vdots \\ 0 \end{bmatrix}, \quad \boldsymbol{x}_n = \begin{bmatrix} y_n \\ y_{n-1} \\ \vdots \\ y_{n-m+1} \end{bmatrix}, \tag{4.90}$$
$$H = [\, 1 \;\; 0 \;\; \cdots \;\; 0 \,].$$

Then the AR model can be expressed in a state-space model without observation noise as

$$\begin{aligned} \boldsymbol{x}_n &= F\boldsymbol{x}_{n-1} + G\varepsilon_n, \\ y_n &= H\boldsymbol{x}_n. \end{aligned} \tag{4.91}$$

As shown in Subsection 3.3.3, the likelihood of the state-space model can be obtained by using the output of the Kalman filter. The estimates of the

Table 4.6. AR models estimated by the exact maximum likelihood method. m is the order of the AR model, and σ_m^2 is the estimated innovation variance of the AR model with order m.

m	σ_m^2	AIC	m	σ_m^2	AIC
0	0.30888	−115.479	8	0.43017	−296.616
1	0.11695	−210.602	9	0.42580	−295.638
2	0.05121	−291.185	10	0.39878	−300.192
3	0.50575	−290.430	11	0.34580	−312.448
4	0.48525	−292.568	12	0.34081	−311.903
5	0.47426	−292.858	13	0.34009	−310.114
6	0.46962	−291.840	14	0.34002	−308.135
7	0.44137	−296.046			

unknown parameters in the AR model, $\hat{a}_1, \ldots, \hat{a}_m$ and $\hat{\sigma_m^2}$, are obtained by numerically maximizing the log-likelihood function by applying the quasi-Newton method. Table 4.6 shows the exact maximum likelihood estimates of σ_m^2 and the AIC for various orders. Again, order 11 is found to give the lowest AIC order.

4.7 Detection of Structural Changes

In statistical data analysis, we sometimes encounter the situation in which the stochastic structure of the data changes at a certain time or location. We consider here estimation of this change point by the statistical modeling based on the AIC. Hereafter we shall consider the comparatively simple problem of estimating the time point of a level shift of the normal distribution and a more realistic problem of estimating the arrival time of a seismic signal.

4.7.1 Detection of Level Shift

Consider a normal distribution model, $y_n \sim N(\mu_n, \sigma^2)$, or, equivalently,

$$p(y_n|\mu_n, \sigma^2) = (2\pi\sigma^2)^{-\frac{1}{2}} \exp\left\{-\frac{(y_n - \mu_n)^2}{2\sigma^2}\right\}. \tag{4.92}$$

We assume that for some unknown change point k, $\mu_n = \theta_1$ for $n < k$ and $\mu_n = \theta_2$ for $n \geq k$. The integer k is called the *change point*. Given data $y_1, \ldots, y_N$, the likelihood of the model is expressed as

$$L(\theta_1, \theta_2, \sigma_k^2) = \prod_{n=1}^{k-1} p(y_n|\theta_1, \sigma_k^2) \prod_{n=k}^{N} p(y_n|\theta_2, \sigma_k^2). \tag{4.93}$$

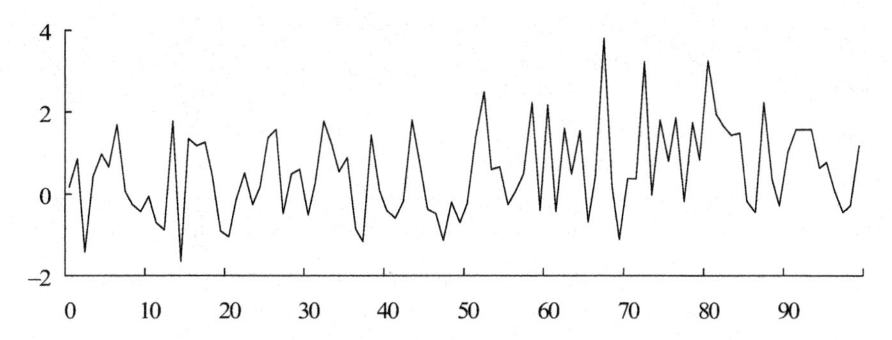

Fig. 4.6. Artificially generated data. Mean value was increased by one at $n = 50$.

Therefore, the log-likelihood is defined by

$$\ell(\theta_1, \theta_2, \sigma_k^2) = -\frac{N}{2}\log(2\pi\sigma_k^2) - \frac{1}{2\sigma_k^2}\left\{\sum_{n=1}^{k-1}(y_n - \theta_1)^2 + \sum_{n=k}^{N}(y_n - \theta_2)^2\right\}. \tag{4.94}$$

It is easy to see that the maximum likelihood estimates are given by

$$\hat{\theta}_1 = \frac{1}{k-1}\sum_{n=1}^{k-1} y_n, \quad \hat{\theta}_2 = \frac{k}{N-k+1}\sum_{n=k}^{N} y_n,$$
$$\hat{\sigma}_k^2 = \frac{1}{N}\left\{\sum_{n=1}^{k-1}(y_n - \hat{\theta}_1)^2 + \sum_{n=k}^{N}(y_n - \hat{\theta}_2)^2\right\}. \tag{4.95}$$

The maximum log-likelihood is

$$\ell(\hat{\theta}_1, \hat{\theta}_2, \hat{\sigma}_k^2) = -\frac{N}{2}\log(2\pi\hat{\sigma}_k^2) - \frac{N}{2},$$

and then the AIC is given by

$$\mathrm{AIC}_k = N\log(2\pi\hat{\sigma}_k^2) + N + 2 \times 3. \tag{4.96}$$

The change point k can be automatically determined by finding the value of k that gives the smallest AIC_k. Note that in the change point problem, the number of parameters does not vary with k; however, the concept of the AIC provides the foundation for estimating the change point by using the likelihood.

Example 8 (Estimating a change point) Figure 4.6 shows a set of data artificially generated using a normal random variable with variance 1. The

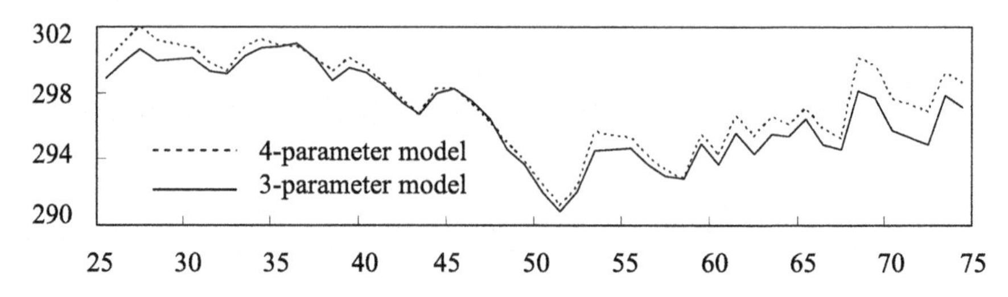

Fig. 4.7. AIC of the level shift model. The black curve shows the AIC of the three-parameter model and the gray curve that of the four-parameter model.

Table 4.7. Results of fitting level shift models.

k	μ_1	μ_2	σ_1^2	σ_2^2	σ^2	AIC'	AIC
40	0.276	0.713	0.877	1.247	1.103	300.127	299.562
41	0.271	0.723	0.856	1.261	1.099	299.474	299.226
42	0.255	0.742	0.846	1.260	1.090	298.577	298.446
43	0.234	0.766	0.843	1.250	1.079	297.564	297.405
44	0.225	0.782	0.827	1.257	1.072	296.641	296.730
45	0.261	0.764	0.863	1.260	1.086	298.272	297.999
46	0.273	0.763	0.851	1.283	1.089	298.247	298.289
47	0.260	0.783	0.841	1.283	1.080	297.295	297.469
48	0.244	0.807	0.835	1.276	1.069	296.237	296.441
49	0.215	0.845	0.857	1.226	1.049	294.972	294.564
50	0.207	0.865	0.842	1.229	1.040	293.902	293.670
51	0.188	0.897	0.843	1.202	1.022	292.423	291.993
52	0.180	0.920	0.829	1.200	1.011	291.183	290.881
53	0.203	0.910	0.841	1.220	1.023	292.320	292.054
54	0.246	0.877	0.920	1.193	1.049	295.681	294.527
55	0.252	0.883	0.905	1.218	1.049	295.470	294.567
56	0.260	0.888	0.892	1.244	1.050	295.309	294.687
57	0.250	0.914	0.881	1.242	1.039	294.190	293.657
58	0.247	0.934	0.866	1.254	1.032	293.286	292.985
59	0.252	0.944	0.852	1.279	1.031	292.817	292.858
60	0.285	0.913	0.902	1.269	1.053	295.479	294.913

mean is 0 for $n = 1, \ldots, 50$ and 1 for $n = 51, \ldots, 100$. Figure 4.7 shows the assumed change point k versus AIC values. Only $26 \leq k \leq 75$ were compared. The solid curve indicates the AIC of the above level shift model with three unknown parameters. On the other hand, the dotted curve shows the AIC of the four-parameter model, which is simply obtained by summing the AICs of two normal distribution models fitted to two data segments. Both AICs have minima at $k = 52$, which is one point away from the true change point.

Table 4.7 shows the estimated mean values μ_1 and μ_2, individual variances σ_1^2 and σ_2^2, and the common variance σ^2 and AICs of the four-parameter model (denoted as AIC') and the three-parameter (level-shift) model (AIC) for $40 \leq k \leq 60$. In most cases, the AIC of the three-parameter model is less than that of the four-parameter model. This reflects the fact that in generating the data, the variance of the series was set to one for the entire interval. Actually, the estimates of the variance by the three-parameter model were closer to the true value.

4.7.2 Arrival Time of a Signal

The location of the epicenter of an earthquake can be estimated based on the arrival times of the seismic signals at several different locations. To utilize the information from the seismic signals to minimize the damage caused by a tsunami or to shut down dangerous industrial plants or to reduce the speed of rapid modes of public transportation, it is necessary to determine the arrival time very quickly. Therefore, development of computationally efficient procedures for automatic estimation of the arrival time of seismic signal is a very important problem.

When an earthquake signal arrives, the characteristics of the time series, such as its variance and spectrum, change abruptly. To estimate the arrival time of a seismic signal, it is assumed that each of the time series before and after the arrival of the seismic signal is stationary and can be expressed by using an autoregressive model as follows [Takanami and Kitagawa (1991)];

Background Noise Model

$$y_n = \sum_{i=1}^{m} a_i y_{n-i} + v_n, \quad v_n \sim N(0, \tau^2), \quad n = 1, \ldots, k, \tag{4.97}$$

Seismic Signal Model

$$y_n = \sum_{i=1}^{\ell} b_i y_{n-i} + w_n, \quad w_n \sim N(0, \sigma^2), \quad n = k+1, \ldots, N, \tag{4.98}$$

where the change point k (precisely $k+1$), the autoregressive orders m and ℓ, the autoregressive coefficients $a_1, \ldots, a_m$, $b_1, \ldots, b_\ell$, and the innovation variances τ^2 and σ^2 are all unknown parameters. Given m and ℓ, the vector consisting of the unknown parameters is denoted by $\boldsymbol{\theta}_{m\ell} = (a_1, \ldots, a_m, \tau^2, b_1, \ldots, b_\ell, \sigma^2)^T$. These two models constitute a simple version of a locally stationary AR model [Ozaki and Tong (1975), Kitagawa and Akaike (1978)].

For simplicity, we assume that the "initial data" $y_{1-M}, \ldots, y_0$ are given, where M is the highest possible AR order. Then given the observations, $y_{1-M}, \ldots, y_N$, the likelihood of the model with respect to the observations $y_1, \ldots, y_N$ is defined by

$$
\begin{aligned}
L(\boldsymbol{\theta}_{m\ell}) &= p(y_1,\ldots,y_N|\boldsymbol{\theta}_{m\ell},y_{1-M},\ldots,y_0)\\
&= p(y_1,\ldots,y_k|y_{1-M},\ldots,y_0,\boldsymbol{\theta}_{m\ell})p(y_{k+1},\ldots,y_N|y_1,\ldots,y_k,\boldsymbol{\theta}_{m\ell})\\
&= \prod_{n=1}^{k} p(y_n|y_{n-1},\ldots,y_{n-m},\boldsymbol{\theta}_{m\ell}) \prod_{n=k+1}^{N} p(y_n|y_{n-1},\ldots,y_{n-\ell},\boldsymbol{\theta}_{m\ell}).
\end{aligned} \tag{4.99}
$$

Therefore, under the assumption of normality of the innovations v_n and w_n, the log-likelihood can be expressed as

$$
\begin{aligned}
\ell(k,m,\ell,\boldsymbol{\theta}_{m\ell}) &= \ell_B(k,m,a_1,\ldots,a_m,\tau^2) + \ell_S(k,\ell,b_1,\ldots,b_\ell,\sigma^2)\\
&= -\frac{k}{2}\log(2\pi\tau^2) - \frac{1}{2\tau^2}\sum_{n=1}^{k}\Big(y_n - \sum_{j=1}^{m} a_j y_{n-j}\Big)^2 \qquad (4.100)\\
&\quad - \frac{N-k}{2}\log(2\pi\sigma^2) - \frac{1}{2\sigma^2}\sum_{n=k+1}^{N}\Big(y_n - \sum_{j=1}^{\ell} b_j y_{n-j}\Big)^2,
\end{aligned}
$$

where ℓ_B and ℓ_S denote the log-likelihoods of the background noise model and the seismic signal model, respectively.

The maximum likelihood estimators $\hat{\theta}_{m\ell} = (\hat{a}_1,\ldots,\hat{a}_m,\hat{b}_1,\ldots,\hat{b}_\ell,\hat{\tau}^2,\hat{\sigma}^2)^T$ are obtained by maximizing this log-likelihood function. In actual computations, the parameters of the background model, $a_1,\ldots,a_m$ and τ^2, and those of the signal model, $b_1,\ldots,b_\ell$ and σ^2, can be estimated independently by maximizing ℓ_B and ℓ_S, respectively.

For a given value of k, the AIC of the current model is given by

$$
\mathrm{AIC}_k = \min_m \mathrm{AIC}_k^B(m) + \min_\ell \mathrm{AIC}_k^S(\ell), \tag{4.101}
$$

where $\mathrm{AIC}_k^B(m)$ and $\mathrm{AIC}_k^S(\ell)$ are the AICs of the background noise model with order m and the seismic signal model with order ℓ, respectively. They are defined by

$$
\begin{aligned}
\mathrm{AIC}_k^B(m) &= k\log(2\pi\hat{\tau}_m^2) + 2(m+1),\\
\mathrm{AIC}_k^S(\ell) &= (N-k)\log(2\pi\hat{\sigma}_\ell^2) + 2(\ell+1),
\end{aligned} \tag{4.102}
$$

where $\hat{\tau}_m^2$ and $\hat{\sigma}_\ell^2$ are the maximum likelihood estimates of the innovation variances of the background noise model with order m and the seismic signal model with order ℓ, respectively. The arrival time of the seismic signal can be estimated by finding the minimum of the AIC_k on a specified interval, say $k \in \{L,\ldots,L+K\}$.

In order to determine the arrival time by the minimum AIC procedure, we have to fit and compare $(K+1)(M+1)^2$ models. Kitagawa and Akaike (1978) developed a very computationally efficient least squares method based on the Householder transformation [Golub (1965)]. The number of necessary computations of this method is only a few times greater than that of fitting

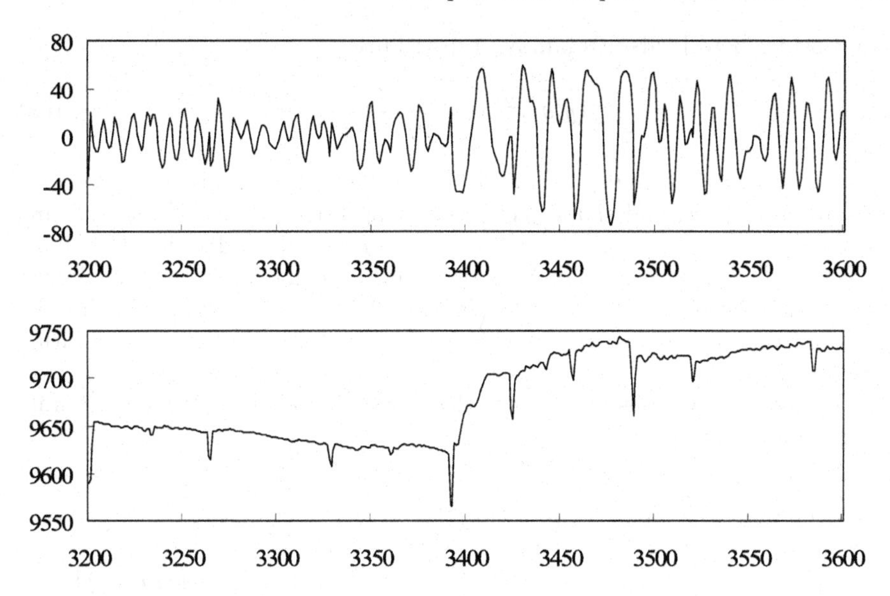

Fig. 4.8. Seismogram and changes of the AIC of the model for estimating the arrival time of a seismic signal. Top plot: east-west component of a seismogram. S wave signal arrives at the middle of the series. Bottom plot: plot of AIC value vs. arrival time.

a single AR model of order M to the entire time series. Namely, the number of necessary computations of this method is reduced to the order of NM^2. Note that if $M = 10$ and $K = 1,000$, the number of necessary computations is reduced to about $1/10,000$ that of the simplistic method.

Example 9 (Detection of a micro earthquake) The top plot of Figure 4.8 shows a portion of the east-west component of a seismogram [Takanami and Kitagawa (1991)] observed at Hokkaido, Japan, y_k, $k = 3200, \ldots, 3600$, where the S wave arrived in the middle of the series. The sampling interval is $\Delta T = 0.01$ second. The bottom plot shows the change of AIC_k for $k = 3200, \ldots, 3600$ when arrival time models are fitted to the data y_j, $j = 2800, \ldots, 4200$. From this figure, it can be seen that the AIC has a minimum at $k = 3393$. There are eight other local minima. However, the variation in the AIC is quite large.

4.8 Comparison of Shapes of Distributions

Assume that we have the 20 observations shown below.

$$
\begin{array}{cccccccccc}
-7.99 & -4.01 & -1.56 & -0.99 & -0.93 & -0.80 & -0.77 & -0.71 & -0.42 & -0.02 \\
0.65 & 0.78 & 0.80 & 1.14 & 1.15 & 1.24 & 1.29 & 2.81 & 4.84 & 6.82
\end{array}
$$

We consider here Pearson's family of distributions

$$f(y|\mu,\tau^2,b) = \frac{C}{(y^2+\tau^2)^b}, \tag{4.103}$$

where $1/2 < b \le \infty$ and μ, τ^2, and b are called the central parameter, dispersion parameter, and shape parameter, respectively. C is the normalizing constant given by $C = \tau^{2b-1}\Gamma(b)/\Gamma\left(b-\frac{1}{2}\right)\Gamma\left(\frac{1}{2}\right)$. By adjusting the shape parameter b, the Pearson's family of distributions can express a broad class of distributions, including Cauchy distribution ($b = 1$), t-distribution with k degrees of freedom [where $b = (k+1)/2$] and normal distribution in its limiting case ($b = \infty$).

Given n observations, $y_1, \ldots, y_N$, the log-likelihood of the Pearson's family of distributions is given by

$$\begin{aligned}
\ell(\mu,\tau^2,b) &= \sum_{n=1}^{N} \log f(y_n|\mu,\tau^2,b) \\
&= N\left\{\left(b-\tfrac{1}{2}\right)\log\tau^2 + \log\Gamma(b) - \log\Gamma(b-\tfrac{1}{2}) - \log\Gamma\left(\tfrac{1}{2}\right)\right\} \\
&\quad -b\sum_{n=1}^{N}\log\left\{(y_n-\mu)^2+\tau^2\right\}.
\end{aligned} \tag{4.104}$$

It is possible to obtain the maximum likelihood estimate of the shape parameter b by using the quasi-Newton method. However, for simplicity, here we shall consider only seven candidates $b = 0.6, 0.75, 1, 1.5, 2, 2.5, 3$, and ∞. Note that $b = 1, 1.5, 2, 2.5, 3$, and ∞ correspond to the Cauchy distribution, the t-distribution with the degrees of freedom 2, 3, 4, 5, and a normal distribution, respectively. Given a value of b, the first derivative of $\ell(\mu,\tau^2,b)$ with respect to μ and τ^2 is, respectively,

$$\begin{aligned}
\frac{\partial \ell}{\partial \mu} &= 2b\sum_{n=1}^{N}\frac{y_n-\mu}{(y_n-\mu)^2+\tau^2}, \\
\frac{\partial \ell}{\partial \tau^2} &= \frac{N(b-1/2)}{\tau^2} - b\sum_{n=1}^{N}\frac{1}{(y_n-\mu)^2+\tau^2}.
\end{aligned} \tag{4.105}$$

For fixed b, the maximum likelihood estimates of μ and τ^2 can be easily obtained using the quasi-Newton method. Table 4.8 shows the maximum likelihood estimates of μ, τ^2, the maximum log-likelihood, and the AIC for each b. Note that for $b = \infty$, the distribution becomes normal and the estimate of the variance, $\hat{\sigma}^2$, is shown instead of the dispersion parameter. As shown in Example 5 of Chapter 3, for the normal distribution model, the mean and the variance are estimated as

$$\hat{\mu} = \frac{1}{N}\sum_{n=1}^{N} y_n = 0.166, \qquad \hat{\sigma}^2 = \frac{1}{N}\sum_{n=1}^{N}(y_n-\hat{\mu})^2 = 8.545, \tag{4.106}$$

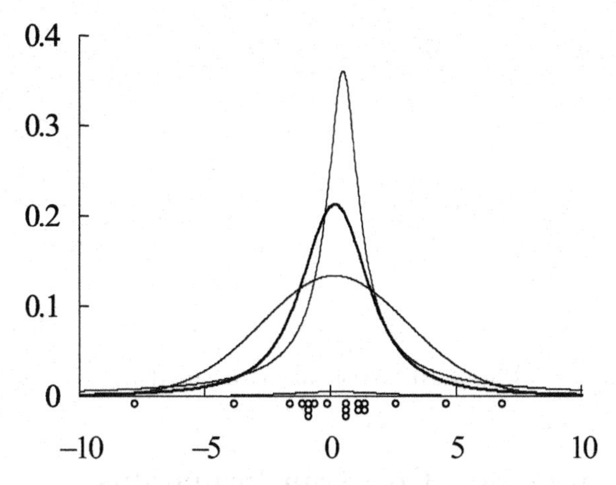

Fig. 4.9. Estimated Pearson's family of distributions for $b = 0.75, 1.5, 2.5$ and the normal distribution. The bold curve indicates the optimal shape parameter ($b = 1.5$). The circles below the x-axis indicate the 20 observations.

with $N = 20$, and the maximum log-likelihood is given by

$$\ell(\hat{\mu}, \hat{\sigma}^2) = -\frac{N}{2}\log(2\pi\hat{\sigma}^2) - \frac{N}{2} = -49.832. \qquad (4.107)$$

It can be seen that the AIC selects $b = 1.5$ as the optimum shape parameter.

Table 4.8. Seven different distributions of Pearson's family of distributions. The maximum likelihood estimates of the central and dispersion parameters, the maximum log-likelihoods, and the AICs are shown. $b = \infty$ shows the normal distribution model.

b	$\hat{\mu}_b$	$\hat{\tau}_b^2$	ℓ	AIC
0.60	0.8012	0.0298	−58.843	121.685
0.75	0.5061	0.4314	−51.397	106.793
1.00	0.1889	1.3801	−47.865	99.730
1.50	0.1853	4.1517	−47.069	98.137
2.00	0.2008	8.3953	−47.428	98.856
2.50	0.2140	13.8696	−47.816	99.633
3.00	0.2224	20.2048	−48.124	100.248
∞	0.1660	8.5445	-49.832	103.663

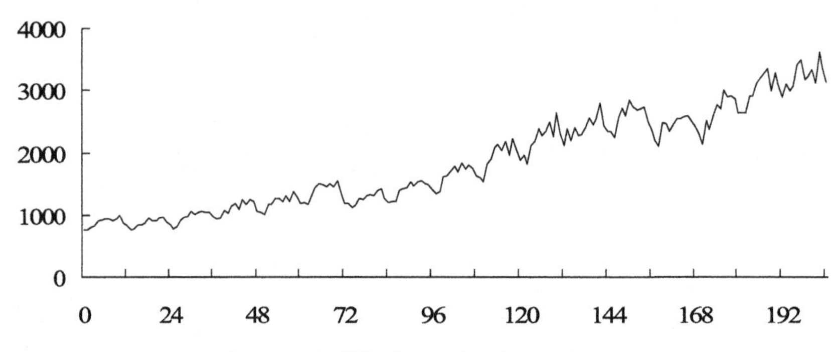

Fig. 4.10. Wholesale hardware data.

4.9 Selection of Box–Cox Transformations

The observations obtained by counting the number of occurrences of a certain event, the number of peoples, or the amount of sales take positive values. These data sets usually have a common feature that the variance increases as the mean value increases. For such data sets, standard statistical models may not fit well because some characteristics of the distribution change depending on the location or the distribution may deviate considerably from the normal distribution.

Figure 4.10 shows the monthly wholesale hardware data published by the U.S. Census Bureau. The annual seasonal variation obviously increases with an increase in the level. For such counted time series, additive seasonal models are usually fit after taking the logarithmic transformation. Here we consider selecting the optimal parameter of the Box–Cox transformation using the AIC.

The Box–Cox transformation [Box and Cox (1964)] is defined by

$$z_n = \begin{cases} \lambda^{-1}(y_n^\lambda - 1), & \text{for } \lambda \neq 0, \\ \log y_n, & \text{for } \lambda = 0. \end{cases} \tag{4.108}$$

It can express various data transformations such as logarithmic transformation and square root transformation by appropriate selection of the value of λ. Except for an additive constant, the Box–Cox transformation becomes the logarithm for $\lambda = 0$, the inverse for $\lambda = -1$, and the square root for $\lambda = 0.5$; it leaves the original data unchanged for $\lambda = 1.0$.

Obviously, the log-likelihood and the AIC values of the transformed data cannot be compared with each other. However, by appropriately compensating the effect of the transformation, we can define the AIC of the model at the original data space. By using this corrected AIC, we can select the optimal value of the transformation parameter λ.

Assume that the data $z_n = h_\lambda(y_n)$ obtained by the Box–Cox transformation follows the probability density function $f(z)$, the probability density

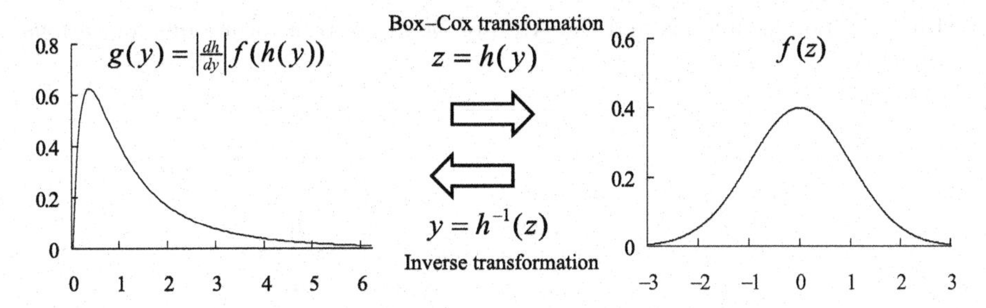

Fig. 4.11. Transformation of the probability density function by a Box–Cox transformation.

function for the original data y_n is given by

$$g(y) = \left|\frac{dh_\lambda}{dy}\right| f(h(y)). \tag{4.109}$$

Here $|dh_\lambda/dy|$ is referred to as the Jacobian of the transformation. Equation (4.109) indicates that the model of the transformed data automatically specifies a model of the original data.

Thus, if, for example, the AICs of the normal distribution models obtained for the original data y_n and the transformed data z_n are denoted as AIC_y and AIC_z, respectively, then by comparing the value of

$$\mathrm{AIC}'_z = \mathrm{AIC}_z - 2\log\left|\frac{dh_\lambda}{dy}\right| \tag{4.110}$$

with AIC_y, we can determine which of the original data or the transformed data can be approximated well by the normal distribution model. Specifically, if $\mathrm{AIC}_y < \mathrm{AIC}'_z$ holds, it is concluded that the original data are better expressed by the normal distribution. On the other hand, if $\mathrm{AIC}_y > \mathrm{AIC}'_z$, then the transformed data are considered to be better. Further, by finding the minimum of AIC'_z, we can determine the best value of λ for the Box–Cox transformation. Note that in the actual statistical modeling, it is necessary to make this correction of the AIC of the fitted model by using the log Jacobian of the Box–Cox transformation.

Table 4.9 shows the values of the log-likelihoods, the AICs, and the transformed AICs for various values of λ. The log-likelihood is a decreasing function of the transformation parameter λ. Since the number of the parameters in the transformed distribution is the same, the AIC takes its maximum at the minimum of the λ, i.e., at $\lambda = -1$. However, the AIC', the corrected AIC obtained by adding the correction term for the data transformation, attains its minimum at $\lambda = 0.1$. This indicates that for the current data set, the best transformation is obtained by $y_n = x_n^{1/10}$. Figure 4.12 shows the Box–Cox transformation of the monthly wholesale hardware data with this AIC best

Table 4.9. Log-likelihoods and the AICs of Box–Cox transformations for various values of λ.

λ	Log-Likelihood	AIC	AIC$'$
1.0	−1645.73	3295.45	3295.45
0.9	−1492.01	2988.02	3290.76
0.8	−1338.56	2681.13	3286.62
0.7	−1185.39	2374.78	3283.01
0.6	−1032.49	2068.99	3279.96
0.5	−879.88	1763.75	3277.47
0.4	−727.54	1459.08	3275.54
0.3	−575.49	1154.98	3274.18
0.2	−423.72	851.44	3273.40
0.1	−272.24	548.49	3273.19
0.0	−121.06	246.11	3273.55
−0.1	29.84	−55.68	3274.50
−0.2	180.45	−356.90	3276.03
−0.3	330.76	−657.53	3278.15
−0.4	480.79	−957.57	3280.85
−0.5	630.52	−1257.04	3284.13
−0.6	779.96	−1555.92	3287.99
−0.7	929.11	−1854.22	3292.43
−0.8	1077.98	−2151.95	3297.44
−0.9	1226.55	−2449.11	3303.03
−1.0	1374.85	−2745.70	3309.19

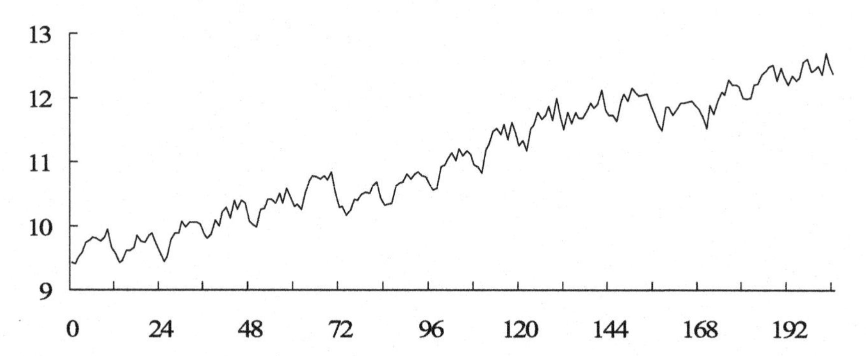

Fig. 4.12. Box–Cox transformation of the wholesale hardware data. The transformation parameter λ selected by the AIC is 0.1.

parameter $\lambda = 0.1$. From this Box–Cox transformation, it can be seen that the variance of the time series becomes almost homogeneous.

5

Generalized Information Criterion (GIC)

We have so far considered the evaluation of statistical models estimated using the maximum likelihood method, for which the AIC is a useful tool for evaluating the estimated models. However, statistical models are constructed to obtain information from observed data in a variety of ways. So if models are developed that employ estimation procedures other than the method of maximum likelihood, how should we construct an information criterion for evaluating such statistical models? With the development of other modeling techniques, it has been necessary to construct information criteria that relax the assumptions imposed on the AIC.

In this chapter, we describe a general framework for constructing information criteria in the context of functional statistics and introduce a generalized information criterion, GIC [Konishi and Kitagawa (1996)]. The GIC can be applied to evaluate statistical models constructed by various types of estimation procedures including the robust estimation procedure and the maximum penalized likelihood procedure. Section 5.1 describes the fundamentals of a functional approach using a probability model having one parameter. In Section 5.2 and subsequent sections, we introduce the generalized information criterion for evaluating statistical models constructed in various ways. We also discuss the relationship among the AIC, TIC, and GIC. Various applications of the GIC to statistical modeling are shown in Chapter 6. Chapter 7 gives the derivation of information criteria and investigates their asymptotic properties with theoretical and numerical improvements.

5.1 Approach Based on Statistical Functionals

5.1.1 Estimators Defined in Terms of Statistical Functionals

The process of statistical inference generally involves building a model that expresses the population distribution or making an inference on the parameters of a specific population distribution, such as a normal distribution. In practice, however, it is difficult to precisely represent the probabilistic mechanism of data generation based on a finite number of observations. Hence, one usually selects an approximating parametric family of probability distributions $\{f(x|\theta);\, \theta \in \Theta \subset R\}$ to the true distribution $G(x)$ [or a density function, $g(x)$] that generates the data. This requires making the assumption that a specified parametric family of probability distributions either does or does not contain the true distribution. A model parameter is, therefore, estimated based on data from the true distribution $G(x)$, but not from $f(x|\theta)$.

From this point of view, we assume that the parameter θ is expressed in the form of a real-valued function of the distribution G, that is, the functional $T(G)$, where $T(G)$ is a real-valued function defined on the set of all distributions on the sample space and does not depend on the sample size n. Then, given data $\{x_1, \ldots, x_n\}$, the estimator $\hat{\theta}$ for θ is given by

$$\hat{\theta} = \hat{\theta}(x_1, \ldots, x_n) = T(\hat{G}) \tag{5.1}$$

in which G is replaced with the empirical distribution function $\hat{G}$, by inserting probability n^{-1} at each observation (see Remark 1). This equation indicates that the estimator depends on data only through the empirical distribution function $\hat{G}$. Such a functional is referred to as a *statistical functional.*

Since various types of estimators, including the maximum likelihood estimator, can be defined in terms of a statistical functional, an information-theoretic approach can provide a unified basis for treating the problem of evaluating statistical models.

Example 1 (Sample mean) If the functional can be written in the form of $T(G) = \int u(x)dG(x)$, then the corresponding estimator is given as

$$T(\hat{G}) = \int u(x)d\hat{G}(x) = \sum_{\alpha=1}^{n} \hat{g}(x_\alpha)u(x_\alpha) = \frac{1}{n}\sum_{\alpha=1}^{n} u(x_\alpha), \tag{5.2}$$

by replacing the unknown probability distribution G with the empirical distribution function $\hat{G}$ and its probability function $\hat{g}(x_\alpha) = n^{-1}$ at each of the observations $\{x_1, \ldots, x_n\}$ [for the notation $dG(x)$, see (3.5) in Chapter 3].

In particular, the mean μ of a probability distribution function $G(x)$ can be expressed as

$$\mu = \int x dG(x) \equiv T_\mu(G). \tag{5.3}$$

By replacing the distribution function G with the empirical distribution function $\hat{G}$, we obtain the estimator for the mean μ:

$$T_\mu(\hat{G}) = \int x d\hat{G}(x) = \frac{1}{n}\sum_{\alpha=1}^{n} x_\alpha = \overline{x}, \tag{5.4}$$

thus obtaining the sample mean.

Example 2 (Sample variance) The functional that defines the variance is given by

$$\begin{aligned} T_{\sigma^2}(G) &= \int (x - T_\mu(G))^2 \, dG(x) \\ &= \int \left(x - \int y dG(y) \right)^2 dG(x) \\ &= \frac{1}{2} \int \int (x-y)^2 dG(x) dG(y), \end{aligned} \tag{5.5}$$

where T_μ is the functional that defines the mean. In this case, by replacing the distribution function G with the empirical distribution function $\hat{G}$ in the first expression of (5.5), the sample variance can be obtained in a natural form as follows:

$$T_{\sigma^2}(\hat{G}) = \int \left(x - T_\mu(\hat{G}) \right)^2 d\hat{G}(x) = \frac{1}{n} \sum_{\alpha=1}^{n} (x_\alpha - \overline{x})^2. \tag{5.6}$$

In addition, from the third expression of (5.5), the well-known formula for the sample variance can be obtained:

$$\begin{aligned} T_{\sigma^2}(\hat{G}) &= \frac{1}{2} \left\{ \int x^2 dG(x) - 2 \int \int xy dG(x) dG(y) + \int y^2 dG(y) \right\} \\ &= \int x^2 d\hat{G}(x) - \left(\int x d\hat{G}(x) \right)^2 \\ &= \frac{1}{n} \sum_{\alpha=1}^{n} x_\alpha^2 - \left(\frac{1}{n} \sum_{\alpha=1}^{n} x_\alpha \right)^2. \end{aligned} \tag{5.7}$$

Example 3 (Maximum likelihood estimator) Consider a probability distribution $f(x|\theta)$ $(\theta \in \Theta \subset R)$ as a candidate model. The unknown parameter θ is then estimated based on the n observations generated from an unknown true distribution $G(x)$. The maximum likelihood estimator, $\hat{\theta}_{\mathrm{ML}}$, is given as the solution of the likelihood equation

$$\sum_{\alpha=1}^{n} \left. \frac{\partial \log f(X_\alpha|\theta)}{\partial \theta} \right|_{\theta=\hat{\theta}_{\mathrm{ML}}} = 0. \tag{5.8}$$

The solution $\hat{\theta}_{\mathrm{ML}}$ can be written as $\hat{\theta}_{\mathrm{ML}} = T_{\mathrm{ML}}(\hat{G})$, where T_{ML} is the functional implicitly defined by

$$\int \left. \frac{\partial \log f(z|\theta)}{\partial \theta} \right|_{\theta=T_{\mathrm{ML}}(G)} dG(z) = 0. \tag{5.9}$$

Example 4 (M-estimator) Huber (1964) generalized the maximum likelihood estimator to a more general estimator, $\hat{\theta}_M$, defined as the solution of the equation

$$\sum_{\alpha=1}^{n} \psi(X_\alpha, \hat{\theta}_M) = 0 \tag{5.10}$$

with ψ being some function on $\mathcal{X} \times \Theta$ ($\Theta \subset R$), where $\mathcal{X}$ is the sample space. The estimator given as a solution of this implicit equation is referred to as the *M-estimator* [Huber (1981), Hampel et al. (1986)]. The maximum likelihood estimator can be considered as a special case of an M-estimator, corresponding to

$$\psi(x, \theta) = \frac{\partial}{\partial \theta} \log f(x|\theta). \tag{5.11}$$

The M-estimator $\hat{\theta}_M$ can be expressed as $\hat{\theta}_M = T_M(\hat{G})$ for the functional $T_M(G)$ given by

$$\int \psi(z, T_M(G)) dG(z) = 0, \tag{5.12}$$

corresponding to the functional $T_{\mathrm{ML}}(G)$ in (5.9) for the maximum likelihood estimator.

We see that Eqs. (5.8) and (5.10) can be respectively obtained by replacing G in (5.9) and (5.12) by the empirical distribution function $\hat{G}$.

Remark 1 (Empirical distribution function) For any real value a, a function $I(x; a)$ defined as follows is referred to as an *indicator function* (Figure 5.1):

$$I(x; a) = \begin{cases} 1 & \text{if} \quad x \geq a, \\ 0 & \text{if} \quad x < a\,. \end{cases} \tag{5.13}$$

Given n observations $\{x_1, x_2, \ldots, x_n\}$, $\hat{G}(x)$ is defined as

$$\hat{G}(x) = \frac{1}{n} \sum_{\alpha=1}^{n} I(x; x_\alpha), \tag{5.14}$$

and then $\hat{G}(x)$ is a step function that jumps by n^{-1} at each observation x_α. The function $\hat{G}(x)$ is an approximation of $G(x)$ and is referred to as

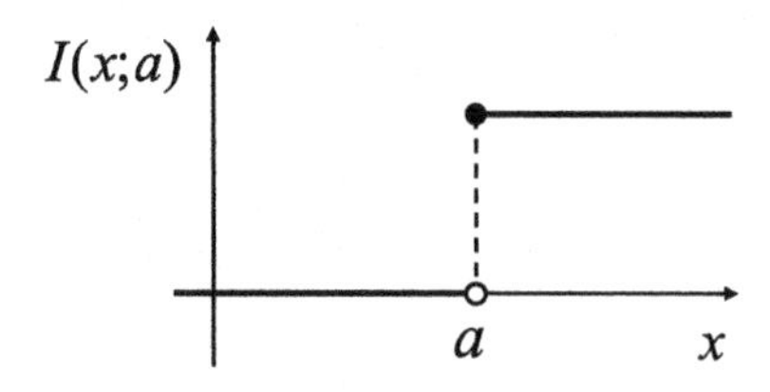

Fig. 5.1. Indicator function.

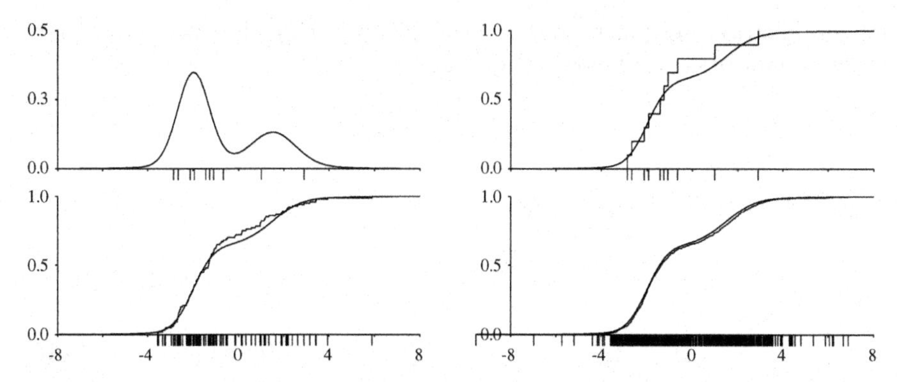

Fig. 5.2. True distribution function and the empirical distribution function. The upper left graph in Fig. 5.2 shows a density function and the 10 observations generated from the distribution. The curve in the upper right graph shows the distribution function that is obtained by integrating the density function in the upper left graph. The stepwise function plotted using a bold line represents an empirical distribution function based on 10 observations. The lower left and lower right graphs show empirical distribution functions obtained from 100 and 1,000 observations, respectively.

an *empirical distribution function*. An empirical distribution function is a distribution function of the probability function $\hat{g}(x_\alpha) = n^{-1}$ $(\alpha = 1, 2, \ldots, n)$, which has an equal probability n^{-1} at each of the n observations.

Figure 5.2 shows that as the number of observations increases, the empirical distribution function approaches the true distribution function and provides a good approximation of the true distribution function that generates data.

In the case of a multivariate distribution function for general p-dimensional random variables $\boldsymbol{X} = (X_1, X_2, \ldots, X_p)^T$, for any $\boldsymbol{a}$ such that $\boldsymbol{a} = (a_1, a_2, \ldots, a_p)^T \in \mathrm{R}^p$, the indicator function in p-dimensional space is defined by

$$I(x; \boldsymbol{a}) = \begin{cases} 1 & \text{if} \quad x_i \geq a_i \quad \text{for all } i, \\ 0 & \text{otherwise.} \end{cases} \tag{5.15}$$

5.1.2 Derivatives of the Functional and the Influence Function

Given the functional $T(G)$, the directional derivative with respect to the distribution function G is defined as a real-valued function $T^{(1)}(x; G)$ that satisfies the equation

$$\begin{aligned} \lim_{\varepsilon \to 0} \frac{T((1-\varepsilon)G + \varepsilon H) - T(G)}{\varepsilon} &= \frac{\partial}{\partial \varepsilon} \{T((1-\varepsilon)G + \varepsilon H)\}\bigg|_{\varepsilon=0} \\ &= \int T^{(1)}(x; G) d\{H(x) - G(x)\} \end{aligned} \tag{5.16}$$

for any distribution function $H(x)$ [von Mises (1947)]. Further, in order to ensure uniqueness, the following equation must hold:

$$\int T^{(1)}(x;G)dG(x) = 0. \tag{5.17}$$

Then, Eq. (5.16) can be written as

$$\begin{aligned}\lim_{\varepsilon\to 0}\frac{T((1-\varepsilon)G+\varepsilon H)-T(G)}{\varepsilon} &= \frac{\partial}{\partial\varepsilon}\{T((1-\varepsilon)G+\varepsilon H)\}\bigg|_{\varepsilon=0}\\ &= \int T^{(1)}(x;G)dH(x).\end{aligned} \tag{5.18}$$

By taking the distribution function H as a delta function δ_x that has a probability of 1 at point x in (5.18), we have

$$\begin{aligned}\lim_{\varepsilon\to 0}\frac{T((1-\varepsilon)G+\varepsilon\delta_x)-T(G)}{\varepsilon} &= \frac{\partial}{\partial\varepsilon}\{T((1-\varepsilon)G+\varepsilon\delta_x)\}\bigg|_{\varepsilon=0}\\ &= \int T^{(1)}(x;G)d\delta_x\\ &= T^{(1)}(x;G).\end{aligned} \tag{5.19}$$

This function, which is called an *influence function*, is used to describe the effect of an infinitesimal contamination at the point x in the robust estimation procedure. The influence function plays a critical role in constructing a generalized information criterion.

Example 5 (Influence function for the sample mean) For the functional that can be represented in the form of $T(G) = \int u(x)dG(x)$, we have

$$\begin{aligned}T((1-\varepsilon)G+\varepsilon\delta_x) &= \int u(y)d\{(1-\varepsilon)G(y)+\varepsilon\delta_x(y)\}\\ &= (1-\varepsilon)T(G)+\varepsilon u(x).\end{aligned} \tag{5.20}$$

Then the influence function can be obtained easily as follows:

$$\begin{aligned}&\lim_{\varepsilon\to 0}\frac{T((1-\varepsilon)G+\varepsilon\delta_x)-T(G)}{\varepsilon}\\ &\quad= \lim_{\varepsilon\to 0}\frac{(1-\varepsilon)T(G)+\varepsilon u(x)-T(G)}{\varepsilon}\\ &\quad= u(x)-T(G).\end{aligned} \tag{5.21}$$

As a direct consequence of this result, the influence function of the functional $T_\mu(G) = \int x dG(x)$ that defines the mean μ is given by

$$T_\mu^{(1)}(x;G) = x - T_\mu(G). \tag{5.22}$$

Example 6 (Influence function for the sample variance) Consider an influence function for the functional $T_{\sigma^2}(G)$ in (5.5) that defines a variance. Noting that

$$\begin{aligned} T_{\sigma^2}(G) &= \int (y - T_\mu(G))^2 \, dG(y) \\ &= \frac{1}{2} \int \int (y - z)^2 dG(y) dG(z), \end{aligned} \tag{5.23}$$

we have

$$\begin{aligned} & T_{\sigma^2}((1-\varepsilon)G + \varepsilon\delta_x) \\ & \quad = (1-\varepsilon)^2 T_{\sigma^2}(G) + \varepsilon(1-\varepsilon) \int (y-x)^2 dG(y). \end{aligned} \tag{5.24}$$

Hence, by using

$$\begin{aligned} \int (y-x)^2 dG(y) &= \int \{(y - T_\mu(G)) + (T_\mu(G) - x)\}^2 dG(y) \\ &= \int (y - T_\mu(G))^2 dG(y) + (T_\mu(G) - x)^2 \\ &= T_{\sigma^2}(G) + (T_\mu(G) - x)^2, \end{aligned} \tag{5.25}$$

we obtain the influence function as follows:

$$\begin{aligned} T_{\sigma^2}^{(1)}(x;G) &= \lim_{\varepsilon \to 0} \frac{T_{\sigma^2}((1-\varepsilon)G + \varepsilon\delta_x) - T_{\sigma^2}(G)}{\varepsilon} \\ &= \lim_{\varepsilon \to 0} \frac{(1-\varepsilon)^2 T_{\sigma^2}(G) + \varepsilon(1-\varepsilon)\{T_{\sigma^2}(G) + (T_\mu(G)-x)^2\} - T_{\sigma^2}(G)}{\varepsilon} \\ &= -2T_{\sigma^2}(G) + T_{\sigma^2}(G) + (x - T_\mu(G))^2 \\ &= (x - T_\mu(G))^2 - T_{\sigma^2}(G). \end{aligned} \tag{5.26}$$

Example 7 (Influence function for the M-estimator) We obtain an influence function for a statistical functional defined by an implicit equation, such as the M-estimator. It is assumed that the functional $T_M(G)$ is given as a solution of the implicit equation

$$\int \psi(x, T_M(G)) dG(x) = 0. \tag{5.27}$$

We directly calculate the derivative

$$\left. \frac{\partial}{\partial \varepsilon} \{T_M((1-\varepsilon)G + \varepsilon\delta_x)\} \right|_{\varepsilon=0} \tag{5.28}$$

for the functional $T_M(G)$.

First, by substituting $(1-\varepsilon)G + \varepsilon\delta_x$ for G in (5.27), we have

$$\int \psi(y, T_M((1-\varepsilon)G + \varepsilon\delta_x)) d\{(1-\varepsilon)G(y) + \varepsilon\delta_x(y)\} = 0. \tag{5.29}$$

Differentiating both sides of the equation with respect to ε and setting $\varepsilon = 0$ yield

$$\begin{aligned} &\int \psi(y, T_M(G)) d\{\delta_x(y) - G(y)\} \\ &\quad + \int \frac{\partial}{\partial \theta}\psi(y,\theta)\bigg|_{\theta=T_M(G)} dG(y) \cdot \frac{\partial}{\partial \varepsilon}\{T_M((1-\varepsilon)G + \varepsilon\delta_x)\}\bigg|_{\varepsilon=0} = 0. \end{aligned} \tag{5.30}$$

Consequently, the influence function, $T_M^{(1)}(x;G)$, of the functional that defines the M-estimator is given by

$$\begin{aligned} &\frac{\partial}{\partial \varepsilon}\left\{T_M\left((1-\varepsilon)G + \varepsilon\delta_x\right)\right\}\bigg|_{\varepsilon=0} \\ &\qquad = -\left\{\int \frac{\partial}{\partial \theta}\psi(y,\theta)\bigg|_{\theta=T_M(G)} dG(y)\right\}^{-1} \psi(x, T_M(G)) \\ &\qquad \equiv T_M^{(1)}(x;G). \end{aligned} \tag{5.31}$$

Example 8 (Influence function for the maximum likelihood estimator) Given a parametric model, $f(x|\theta)$ $(\theta \in \Theta \subset R)$, the functional $T_{\rm ML}(G)$ for the maximum likelihood estimator of θ is given as the solution of the equation

$$\int \frac{\partial \log f(z|\theta)}{\partial \theta}\bigg|_{\theta=T_{\rm ML}(G)} dG(z) = 0, \tag{5.32}$$

corresponding to (5.9). Therefore, by taking

$$\psi(x,\theta) = \frac{\partial}{\partial \theta}\log f(x|\theta) \tag{5.33}$$

in (5.31), it can be readily shown that the influence function, $T_{\rm ML}^{(1)}(x;G)$, of the functional $T_{\rm ML}(G)$ is given by

$$T_{\rm ML}^{(1)}(x;G) = J(G)^{-1}\frac{\partial}{\partial \theta}\log f(x|\theta)\bigg|_{\theta=T_{\rm ML}(G)}, \tag{5.34}$$

where

$$J(G) = -\int \frac{\partial^2}{\partial \theta^2}\log f(x|\theta)\bigg|_{\theta=T_{\rm ML}(G)} dG(x). \tag{5.35}$$

5.1.3 Extension of the Information Criteria AIC and TIC

We have shown that various estimators, including maximum likelihood estimators, can be addressed within the framework of functionals. The following problem arises: How do we construct an information criterion in the context of statistical functional? Before answering this question theoretically, we shall re-examine, using functionals, the information criteria AIC and TIC, which provide criteria for statistical models estimated by the maximum likelihood method.

Let $f(x|\hat{\theta}_{\rm ML})$ be a statistical model fitted to the observed data drawn from the true distribution G by the method of maximum likelihood. The maximum likelihood estimator $\hat{\theta}_{\rm ML}$ can be expressed as $\hat{\theta}_{\rm ML} = T_{\rm ML}(\hat{G})$ for the functional given in (5.32). As discussed in Chapter 3, the essential idea in constructing an information criterion is a bias correction for the log-likelihood of $f(x|\hat{\theta}_{\rm ML})$ in estimating the expected log-likelihood $\mathrm{E}_G\left[\log f(Z|\hat{\theta}_{\rm ML})\right]$, and from (3.97) its bias was given by

$$\begin{aligned} & E_G\left[\sum_{\alpha=1}^n \log f(X_\alpha|\hat{\theta}_{\rm ML}) - n\int \log f(z|\hat{\theta}_{\rm ML})dG(z)\right] \\ & \quad = J(G)^{-1}I(G) + O(n^{-1}), \end{aligned} \tag{5.36}$$

where

$$J(G) = -\int \frac{\partial^2}{\partial\theta^2}\log f(x|\theta)\bigg|_{\theta=T_{\rm ML}(G)} dG(x), \tag{5.37}$$

$$I(G) = \int \left\{\frac{\partial \log f(x|\theta)}{\partial\theta}\right\}^2\bigg|_{\theta=T_{\rm ML}(G)} dG(x). \tag{5.38}$$

Using the influence function for the maximum likelihood estimator given by (5.34), we can rewrite the bias as

$$\begin{aligned} J(G)^{-1}I(G) &= \int J(G)^{-1}\left\{\frac{\partial \log f(x|\theta)}{\partial\theta}\right\}^2\bigg|_{\theta=T_{\rm ML}(G)} dG(x) \\ &= \int J(G)^{-1}\frac{\partial \log f(x|\theta)}{\partial\theta}\frac{\partial \log f(x|\theta)}{\partial\theta}\bigg|_{\theta=T_{\rm ML}(G)} dG(x) \\ &= \int T_{\rm ML}^{(1)}(x;G)\frac{\partial \log f(x|\theta)}{\partial\theta}\bigg|_{\theta=T_{\rm ML}(G)} dG(x). \end{aligned} \tag{5.39}$$

This implies that the (asymptotic) bias can be represented as the integral of the product of the influence function for the maximum likelihood estimator and the score function for the probability model $f(x|\theta)$.

More generally, we consider a statistical model $f(x|\hat{\theta})$ fitted to the data from $G(x)$, where the estimator is given by using the functional $T(G)$ as $\hat{\theta}$

$= T(\hat{G})$. It is then expected that the bias of the log-likelihood for the model $f(x|\hat{\theta})$ in estimating the expected log-likelihood will be

$$E_G\left[\sum_{\alpha=1}^{n}\log f(X_\alpha|\hat{\theta}) - n\int \log f(z|\hat{\theta})dG(z)\right]$$
$$= \int T^{(1)}(x;G)\left.\frac{\partial \log f(x|\theta)}{\partial\theta}\right|_{\theta=T(G)} dG(x) + O(n^{-1}). \quad (5.40)$$

This conjecture is, in fact, correct, as will be shown in Section 7.1. The asymptotic bias of the log-likelihood for the model with the estimator defined by a functional is generally given in the form of the integral of the product of an influence function, $T^{(1)}(x;G)$, of the estimator and the score function, $\partial \log f(x|\theta)/\partial\theta$, of a specified model.

By replacing the unknown distribution G by the empirical distribution $\hat{G}$ in (5.40) and subtracting the asymptotic bias estimate from the log-likelihood, we have an information criterion for the statistical model $f(x|\hat{\theta})$ with functional estimator in the following:

$$\mathrm{GIC} = -2\sum_{\alpha=1}^{n}\log f(x_\alpha|\hat{\theta}) + \frac{2}{n}\sum_{\alpha=1}^{n} T^{(1)}(x_\alpha;\hat{G})\left.\frac{\partial \log f(x_\alpha|\theta)}{\partial\theta}\right|_{\theta=T(\hat{G})}. \quad (5.41)$$

This information criterion is more general than the AIC and TIC, enabling evaluation of the model whose parameter θ is estimated by $\hat{\theta} = T(\hat{G})$ in terms of a statistical functional $T(G)$.

Example 9 (Information criterion for a model estimated by M-estimation) Consider a statistical model $f(x|\hat{\theta}_M)$ estimated using the M-estimation procedure. It follows from (5.31) that the influence function for the M-estimator is given by

$$T_M^{(1)}(x;G) = R(\psi,G)^{-1}\psi(x,T_M(G)), \quad (5.42)$$

where

$$R(\psi,G) = -\int \left.\frac{\partial}{\partial\theta}\psi(x,\theta)\right|_{\theta=T_M(G)} dG(x). \quad (5.43)$$

Substituting the influence function into (5.40) gives the bias of the log-likelihood of $f(x|\hat{\theta}_M)$ as follows:

$$E_G\left[\sum_{\alpha=1}^{n}\log f(X_\alpha|\hat{\theta}_M) - n\int \log f(z|\hat{\theta}_M)dG(z)\right]$$
$$= R(\psi,G)^{-1}Q(\psi,G) + O(n^{-1}), \quad (5.44)$$

where

$$Q(\psi, G) = \int \psi(x, \theta) \left. \frac{\partial \log f(x|\theta)}{\partial \theta} \right|_{\theta = T_M(G)} dG(x). \tag{5.45}$$

By replacing the unknown distribution G by the empirical distribution $\hat{G}$ in (5.44) and subtracting the asymptotic bias estimate from the log-likelihood, we have an information criterion for evaluating a model estimated by the M-estimation procedure as follows:

$$\mathrm{GIC}_M = -2 \sum_{\alpha=1}^{n} \log f(x_\alpha | \hat{\theta}_M) + 2R(\psi, \hat{G})^{-1} Q(\psi, \hat{G}), \tag{5.46}$$

where

$$R(\psi, \hat{G}) = -\frac{1}{n} \sum_{\alpha=1}^{n} \left. \frac{\partial \psi(x_\alpha, \theta)}{\partial \theta} \right|_{\theta = \hat{\theta}_M},$$

$$Q(\psi, \hat{G}) = \frac{1}{n} \sum_{\alpha=1}^{n} \psi(x_\alpha, \theta) \left. \frac{\partial \log f(x_\alpha | \theta)}{\partial \theta} \right|_{\theta = \hat{\theta}_M}. \tag{5.47}$$

Fisher consistency. We now consider the situation that the specified parametric family of probability distributions $\{f(x|\theta);\ \theta \in \Theta \subset R\}$ includes the true density $g(x)$ within the framework of the functional approach. Let $F_\theta(x)$ be the distribution function of the specified model $f(x|\theta)$. Assuming that the functional $T(G)$ that gives the estimator of an unknown parameter θ satisfies the condition $T(F_\theta) = \theta$ at $G = F_\theta$, the estimator $T(\hat{F}_\theta)$ is an asymptotically natural estimator for θ, where $\hat{F}_\theta$ is the empirical distribution function. Generally, if the equation

$$T(F_\theta) = \theta \tag{5.48}$$

holds for any θ in the parameter space Θ, the functional $T(G)$ is said to be *Fisher consistent* [Kallianpur and Rao (1955), Hampel et al. (1986, p. 83)]. For example, for the functional $T_\mu(G) = \int x dG(x)$, we have

$$T_\mu(F_\mu) = \int x dF_\mu(x) = \mu \quad \text{for any} \quad \mu \in \Theta \subset R, \tag{5.49}$$

where F_μ is a normal distribution function with mean μ.

We assume that the functional $T_M(G)$ for an M-estimator is Fisher consistent, so that $T_M(F_\theta) = \theta$ for all $\theta \in \Theta$, where F_θ is the distribution function of $f(x|\theta)$. It then follows from (5.27) that

$$\int \psi(x, \theta) dF_\theta(x) = 0, \tag{5.50}$$

for any θ. Differentiating both sides of this equation with respect to θ yields

$$\int \frac{\partial}{\partial \theta}\psi(x,\theta)dF_\theta(x) + \int \psi(x,\theta)d\left\{\frac{\partial}{\partial \theta}F_\theta(x)\right\} = 0. \tag{5.51}$$

By using

$$\begin{aligned} d\left\{\frac{\partial}{\partial \theta}F_\theta(x)\right\} &= \frac{\partial}{\partial \theta}f(x|\theta)dx \\ &= \frac{\partial}{\partial \theta}\{\log f(x|\theta)\}f(x|\theta)dx \\ &= \frac{\partial}{\partial \theta}\log f(x|\theta)dF_\theta(x), \end{aligned} \tag{5.52}$$

Eq. (5.51) can be rewritten as

$$\int \frac{\partial}{\partial \theta}\psi(x,\theta)dF_\theta(x) = -\int \psi(x,\theta)\frac{\partial}{\partial \theta}\log f(x|\theta)dF_\theta(x). \tag{5.53}$$

Therefore, under the assumption that the true model is contained in the specified parametric model, it follows from (5.31) that the influence function of the functional for the M-estimator can be written as

$$T_M^{(1)}(x;F_\theta) = R(\psi,F_\theta)^{-1}\psi(x,\theta), \tag{5.54}$$

where

$$\begin{aligned} R(\psi,F_\theta) &= -\int \frac{\partial}{\partial \theta}\psi(x,\theta)dF_\theta(x) \\ &= \int \psi(x,\theta)\frac{\partial}{\partial \theta}\log f(x|\theta)dF_\theta(x) \\ &= Q(\psi,F_\theta). \end{aligned} \tag{5.55}$$

By substituting this influence function into (5.40) and noting that $R(\psi,F_\theta) = Q(\psi,F_\theta)$ holds in (5.44) when $G = F_\theta$, we see that the information criterion (5.46) can be reduced to

$$\text{GIC}_M = -2\sum_{\alpha=1}^{n}\log f(x_\alpha|\hat{\theta}_M) + 2\times 1. \tag{5.56}$$

We thus observe that the AIC may be used directly for evaluating statistical models estimated using the M-estimation procedure, since there is only one free parameter in the model $f(x|\theta)$.

5.2 Generalized Information Criterion (GIC)

In the preceding section, we introduced the fundamentals of functional approach by using a probability model with one parameter, and the AIC can be extended naturally to a more general information criterion by relaxing the assumptions that (i) estimation is by maximum likelihood, and that (ii) this is carried out in a parametric family of distributions including the true model.

In this section, we demonstrate that within the framework of statistical functionals, the information criteria for evaluating models estimated by maximum likelihood, by maximum penalized likelihood, and by robust procedures can be derived in a unified manner, and we introduce the generalized information criterion (GIC) that can be used to evaluate a variety of models. Examples are given to illustrate how to construct criteria for models estimated by a variety of estimation procedures including the maximum likelihood and maximum penalized likelihood methods.

5.2.1 Definition of the GIC

Let $G(x)$ be the true distribution function with density $g(x)$ that generated data, and let $\hat{G}(x)$ be the empirical distribution function based on n observations, $\boldsymbol{x}_n = \{x_1, x_2, \ldots, x_n\}$, drawn from $G(x)$. On the basis of the information contained in the observations, we choose a parametric model that consists of a family of probability distributions $\{f(x|\boldsymbol{\theta}); \boldsymbol{\theta} \in \Theta \subset R^p\}$, where $\boldsymbol{\theta} = (\theta_1, \ldots, \theta_p)^T$ is the p-dimensional vector of unknown parameters and Θ is an open subset of R_p. This specified family of probability distributions may or may not contain the true density $g(x)$, but it is expected that its deviation from the parametric model will not be too large. The adopted parametric model is estimated by replacing the unknown parameter vector $\boldsymbol{\theta}$ by some estimate $\hat{\boldsymbol{\theta}}$, for which maximum likelihood, penalized likelihood, or robust procedures may be used for estimating parameters.

In order to construct an information criterion that enables us to evaluate various types of statistical models, we employ a functional estimator that is Fisher consistent. Let us assume that the estimator $\hat{\theta}_i$ for the ith parameter θ_i is given by

$$\hat{\theta}_i = T_i(\hat{G}), \qquad i = 1, 2, \ldots, p, \tag{5.57}$$

for a functional $T_i(\cdot)$. If we write the p-dimensional functional vector with $T_i(G)$ as the ith element by

$$\boldsymbol{T}(G) = (T_1(G), T_2(G), \ldots, T_p(G))^T, \tag{5.58}$$

then the p-dimensional estimator can be expressed as

$$\hat{\boldsymbol{\theta}} = \boldsymbol{T}(\hat{G}) = \left(T_1(\hat{G}), T_2(\hat{G}), \ldots, T_p(\hat{G})\right)^T. \tag{5.59}$$

Given a functional $T_i(G)$ $(i = 1, 2, \ldots, p)$, the influence function, which is the directional derivative of the functional at the distribution G, is defined by

$$T_i^{(1)}(x; G) = \lim_{\epsilon \to 0} \frac{T_i((1-\epsilon)G + \epsilon\, \delta_x) - T_i(G)}{\epsilon}, \tag{5.60}$$

where δ_x is a distribution function having a probability of 1 at point x. As shown in Section 5.1, the influence function plays an essential role in the derivation of an information criterion. We define the p-dimensional vector of influence function having $T_i^{(1)}(x;G)$ as the ith element by

$$\boldsymbol{T}^{(1)}(x;G) = \left(T_1^{(1)}(x;G), T_2^{(1)}(x;G), \ldots, T_p^{(1)}(x;G)\right)^T. \tag{5.61}$$

Then the asymptotic bias in (5.40) for a statistical model with one parameter may be extended to the following:

Bias of the log-likelihood. The bias of the log-likelihood for the model $f(x|\hat{\boldsymbol{\theta}})$ in estimating the expected log-likelihood is given by

$$\begin{aligned} b(G) &= E_G\left[\sum_{\alpha=1}^{n} \log f(X_\alpha|\hat{\boldsymbol{\theta}}) - n\int \log f(z|\hat{\boldsymbol{\theta}})dG(z)\right] \qquad (5.62) \\ &= \mathrm{tr}\left\{\int \boldsymbol{T}^{(1)}(z;G)\left.\frac{\partial \log f(z|\boldsymbol{\theta})}{\partial \boldsymbol{\theta}^T}\right|_{\boldsymbol{\theta}=\boldsymbol{T}(G)} dG(z)\right\} + O(n^{-1}), \end{aligned}$$

where $\partial/\partial\boldsymbol{\theta} = (\partial/\partial\theta_1, \partial/\partial\theta_2, \ldots, \partial/\partial\theta_p)^T$. The integrand function is a $p \times p$ matrix, and the integral of the matrix function is defined as the integral of each element

$$\int T_i^{(1)}(x;G)\left.\frac{\partial \log f(z|\boldsymbol{\theta})}{\partial \theta_j}\right|_{\boldsymbol{\theta}=\boldsymbol{T}(G)} dG(z). \tag{5.63}$$

The asymptotic bias of the log-likelihood can be estimated by replacing the unknown probability distribution G with an empirical distribution function $\hat{G}$ based on the observed data, eliminating the need to determine the integral analytically, and we thus obtain the following result:

Generalized information criterion (GIC). An information criterion for evaluating the statistical model $f(x|\hat{\boldsymbol{\theta}})$ with a p-dimensional functional estimator $\hat{\boldsymbol{\theta}} = \boldsymbol{T}(\hat{G})$ is given by

$$\begin{aligned} \mathrm{GIC} &= -2\sum_{\alpha=1}^{n} \log f(x_\alpha|\hat{\boldsymbol{\theta}}) \\ &\quad + \frac{2}{n}\sum_{\alpha=1}^{n} \mathrm{tr}\left\{\boldsymbol{T}^{(1)}(x_\alpha;\hat{G})\left.\frac{\partial \log f(x_\alpha|\boldsymbol{\theta})}{\partial \boldsymbol{\theta}^T}\right|_{\boldsymbol{\theta}=\hat{\boldsymbol{\theta}}}\right\}, \qquad (5.64) \end{aligned}$$

where $\boldsymbol{T}^{(1)}(x_\alpha;\hat{G}) = (T_1^{(1)}(x_\alpha;\hat{G}), \ldots, T_p^{(1)}(x_\alpha;\hat{G}))^T$ and $T_i^{(1)}(x_\alpha;\hat{G})$ is the empirical influence function defined by

$$T_i^{(1)}(x_\alpha;\hat{G}) = \lim_{\varepsilon\to 0}\frac{T_i((1-\varepsilon)\hat{G}+\varepsilon\delta_{x_\alpha})-T_i(\hat{G})}{\varepsilon}, \tag{5.65}$$

with δ_{x_α} being a point mass at x_α.

When selecting the best model from various different models, we select the model for which the value of the information criterion GIC is smallest.

By rewriting the asymptotic bias in the GIC, we have

$$\sum_{\alpha=1}^{n}\mathrm{tr}\left\{\boldsymbol{T}^{(1)}(x_\alpha;\hat{G})\frac{\partial\log f(x_\alpha|\boldsymbol{\theta})}{\partial\boldsymbol{\theta}^T}\bigg|_{\boldsymbol{\theta}=\hat{\boldsymbol{\theta}}}\right\}$$
$$=\sum_{i=1}^{p}\sum_{\alpha=1}^{n}T_i^{(1)}(x_\alpha;\hat{G})\left.\frac{\partial\log f(x_\alpha|\boldsymbol{\theta})}{\partial\theta_i}\right|_{\boldsymbol{\theta}=\hat{\boldsymbol{\theta}}}. \tag{5.66}$$

This implies that the asymptotic bias is given as the sum of products of the empirical influence function $T_i^{(1)}(x_\alpha;\hat{G})$ of the estimator $\hat{\theta}_i$ and the estimated score function of the model.

The generalized information criterion (GIC) is used to evaluate statistical models constructed by various estimation procedures including the maximum likelihood and maximum penalized likelihood methods, and even the Bayesian approach. Detailed derivations and applications of GIC are given in Konishi and Kitagawa (1996, 2003), and Konishi (1999, 2002).

Example 10 (Normal model) Suppose that n independent observations $\{x_1,\ldots,x_n\}$ are generated from the true distribution $G(x)$ having the density function $g(x)$. Consider, as a candidate model, a parametric family of normal densities

$$f(x|\boldsymbol{\theta}) = \frac{1}{\sigma}\phi\left(\frac{x-\mu}{\sigma}\right)$$
$$= \frac{1}{\sqrt{2\pi\sigma^2}}\exp\left\{-\frac{(x-\mu)^2}{2\sigma^2}\right\},\quad \boldsymbol{\theta}=(\mu,\sigma^2)^T\in\Theta. \tag{5.67}$$

If the parametric model is correctly specified, the family $\{f(x|\boldsymbol{\theta});\boldsymbol{\theta}\in\Theta\subset R^p\}$ contains the true density as an element $g(x)=\sigma_0^{-1}\phi((x-\mu_0)/\sigma_0)$ for some $\boldsymbol{\theta}_0 = (\mu_0,\sigma_0^2)^T\in\Theta$. The statistical model estimated by the method of maximum likelihood is

$$f(x|\hat{\boldsymbol{\theta}}) = \frac{1}{\hat{\sigma}}\phi\left(\frac{x-\overline{x}}{\hat{\sigma}}\right)$$
$$= \frac{1}{\sqrt{2\pi\hat{\sigma}^2}}\exp\left\{-\frac{(x-\overline{x})^2}{2\hat{\sigma}^2}\right\},\quad \hat{\boldsymbol{\theta}}=(\overline{x},\hat{\sigma}^2)^T, \tag{5.68}$$

where $\overline{x}$ and $\hat{\sigma}^2$ are the sample mean and the sample variance, respectively. Then the log-likelihood of the statistical model is given by

$$\sum_{\alpha=1}^{n} \log f(x_\alpha|\hat{\boldsymbol{\theta}}) = -\frac{n}{2}\left\{1 + \log(2\pi) + \log\hat{\sigma}^2\right\}. \tag{5.69}$$

As shown in (5.3) and (5.5) in the preceding section, the sample mean and the sample variance are defined, respectively, by the functionals

$$T_\mu(G) = \int x dG(x) \quad \text{and} \quad T_{\sigma^2}(G) = \int (x - T_\mu(G))^2 dG(x). \tag{5.70}$$

Recall that it was shown in (5.22) and (5.26) that these influence functions are given by

$$\begin{aligned} T_\mu^{(1)}(x;G) &= x - T_\mu(G), \\ T_{\sigma^2}^{(1)}(x;G) &= (x - T_\mu(G))^2 - T_{\sigma^2}(G). \end{aligned} \tag{5.71}$$

On the other hand, the partial derivative of the log-likelihood function is

$$\begin{aligned} \left.\frac{\partial \log f(x|\mu,\sigma^2)}{\partial \mu}\right|_{\boldsymbol{\theta}=\boldsymbol{T}(G)} &= \frac{x - T_\mu(G)}{T_{\sigma^2}(G)}, \\ \left.\frac{\partial \log f(x|\mu,\sigma^2)}{\partial \sigma^2}\right|_{\boldsymbol{\theta}=\boldsymbol{T}(G)} &= -\frac{1}{2T_{\sigma^2}(G)} + \frac{(x - T_\mu(G))^2}{2T_{\sigma^2}(G)^2}, \end{aligned} \tag{5.72}$$

where $\boldsymbol{\theta} = (\mu, \sigma^2)^T$ and $\boldsymbol{T}(G) = (T_\mu(G), T_{\sigma^2}(G))^T$.

By substituting these results into (5.62), the (asymptotic) bias of the log-likelihood can be obtained as

$$\begin{aligned} b(G) &= \int T_\mu^{(1)}(x;G) \left.\frac{\partial \log f(x|\mu,\sigma^2)}{\partial \mu}\right|_{\boldsymbol{\theta}=\boldsymbol{T}(G)} dG(x) \\ &\quad + \int T_{\sigma^2}^{(1)}(x;G) \left.\frac{\partial \log f(x|\mu,\sigma^2)}{\partial \sigma^2}\right|_{\boldsymbol{\theta}=\boldsymbol{T}(G)} dG(x) \\ &= \frac{1}{2}\left\{1 + \frac{\mu_4(G)}{T_{\sigma^2}(G)^2}\right\}, \end{aligned} \tag{5.73}$$

where $\mu_4(G)$ is defined by

$$\mu_4(G) = \int (x - T_\mu(G))^4 dG(x). \tag{5.74}$$

By replacing the unknown distribution G in the bias correction term with the empirical distribution function $\hat{G}$, we have

$$b(\hat{G}) = \frac{1}{2}\left\{1 + \frac{\mu_4(\hat{G})}{\hat{\sigma}^4}\right\}, \tag{5.75}$$

where

$$\mu_4(\hat{G}) = \int (x - T_\mu(\hat{G}))^4 d\hat{G}(x)$$
$$= \frac{1}{n}\sum_{\alpha=1}^{n}(x_\alpha - \overline{x})^4. \tag{5.76}$$

Hence, it follows from (5.64) that the GIC is given by

$$\text{GIC} = n\left\{1 + \log(2\pi) + \log \hat{\sigma}^2\right\} + 2\left\{\frac{1}{2} + \frac{1}{2n\hat{\sigma}^4}\sum_{\alpha=1}^{n}(x_\alpha - \overline{x})^4\right\}. \tag{5.77}$$

In a particular situation where the normal model contains the true density, that is, $g(x) = \sigma_0^{-1}\phi((x - \mu_0)/\sigma_0)$ for some $\boldsymbol{\theta} = (\mu_0, \sigma_0^2)^T \in \Theta$, the fourth central moment μ_4 equals $3\sigma_0^4$, and hence we have

$$b(G) = \frac{1}{2} + \frac{\mu_4}{2\sigma_0^4} = 2, \tag{5.78}$$

the asymptotic bias for the AIC.

Example 11 (Numerical comparison) Suppose that the true density $g(x)$ and the parametric model $f(x|\boldsymbol{\theta})$ are respectively

$$g(x) = (1 - \varepsilon)\frac{1}{\sigma_{01}}\phi\left(\frac{x - \mu_{01}}{\sigma_{01}}\right) + \varepsilon\frac{1}{\sigma_{02}}\phi\left(\frac{x - \mu_{02}}{\sigma_{02}}\right), \quad 0 \le \varepsilon \le 1, \tag{5.79}$$

$$f(x|\boldsymbol{\theta}) = \frac{1}{\sigma}\phi\left(\frac{x - \mu}{\sigma}\right), \qquad \boldsymbol{\theta} = (\mu, \sigma^2)^T, \tag{5.80}$$

where $\phi(x)$ denotes the density function of a standard normal distribution. The statistical model is constructed based on n independent observations from the mixture distribution $g(x)$ and is given by (5.68).

Under this situation, the expected log-likelihood for $f(z|\overline{x}, \hat{\sigma}^2)$ can be written as

$$\begin{aligned}\int g(z)\log f(z|\overline{x}, \hat{\sigma}^2)dz &= -\frac{1}{2}\log(2\pi) - \frac{1}{2}\log\hat{\sigma}^2 - \frac{1}{2\hat{\sigma}^2}\int (z - \overline{x})^2 g(z)dz \\ &= -\frac{1}{2}\log(2\pi) - \frac{1}{2}\log\hat{\sigma}^2 \\ &\quad - \frac{1}{2\hat{\sigma}^2}\left[(1 - \varepsilon)\left\{\sigma_{01}^2 + (\mu_{01} - \overline{x})^2\right\}\right. \\ &\quad \left. + \varepsilon\left\{\sigma_{02}^2 + (\mu_{02} - \overline{x})^2\right\}\right].\end{aligned} \tag{5.81}$$

From the results of Example 10, the bias of the log-likelihood in estimating this expected log-likelihood is approximated by

$$\begin{aligned}
b(G) &= E_G \left[\sum_{\alpha=1}^{n} \log f(X_\alpha | \overline{X}, \hat{\sigma}^2) - n \int g(z) \log f(z|\overline{X}, \hat{\sigma}^2) dz \right] \\
&= E_G \left[-\frac{n}{2} + \frac{n}{2\hat{\sigma}^2} \left\{ (1-\varepsilon) \left(\sigma_{01}^2 + (\mu_{01} - \overline{X})^2 \right) + \varepsilon \left(\sigma_{02}^2 + (\mu_{02} - \overline{X})^2 \right) \right\} \right] \\
&\approx \frac{1}{2} + \frac{\mu_4(G)}{2\sigma^4(G)},
\end{aligned} \tag{5.82}$$

where $\sigma^2(G)$ and $\mu_4(G)$ are the variance and the fourth central moment of the mixture distribution $g(x)$, respectively. Hence, we have the bias estimate

$$b(\hat{G}) \approx \frac{1}{2} + \frac{1}{2n\hat{\sigma}^4} \sum_{\alpha=1}^{n} (x_\alpha - \overline{x})^4. \tag{5.83}$$

A Monte Carlo simulation was performed to examine the accuracy of the asymptotic bias. Repeated random samples were generated from a mixture of normal distributions $g(x)$ in (5.79) for different combinations of parameters, in which we took (i) $(\mu_{01}, \mu_{02}, \sigma_{01}, \sigma_{02}) = (0, 0, 1, 3)$ in the left panels of Figure 5.3 and (ii) $(\mu_{01}, \mu_{02}, \sigma_{01}, \sigma_{02}) = (0, 5, 1, 1)$ in the right panels of Figure 5.3.

Figure 5.3 shows a plot of the true bias $b(G)$ and the asymptotic bias estimate $b(\hat{G})$ given by (5.83) with standard errors for various values of the mixing proportion ε. The quantities are estimated by a Monte Carlo simulation with 100,000 repetitions.

It can be seen from the figure that the log-likelihood of a fitted model has a significant bias as an estimate of the expected log-likelihood and that the bias is considerably larger than 2, the approximation of the AIC, if the values of the mixing proportion ε are around $0.05 \sim 0.1$. In the case that $\varepsilon = 0$ or 1, the true distribution $g(x)$ belongs to the specified parametric model and the bias is approximated well by the number of estimated parameters. We also see that for larger sample sizes, the true bias and the estimated asymptotic bias (5.83) coincide well. On the other hand, for smaller sample sizes such as $n = 25$, the estimated asymptotic bias underestimates the true bias.

5.2.2 Maximum Likelihood Method: Relationship Among AIC, TIC, and GIC

According to the assumptions made for model estimation and the relationship between the specified model and the true model, the GIC in (5.64) takes a different form, and consequently we obtain the AIC and TIC proposed previously.

Let us assume that the maximum likelihood method is used for estimating a specified model $f(x|\boldsymbol{\theta})$ based on the observed data from $G(x)$. The maximum likelihood estimator, $\hat{\boldsymbol{\theta}}_{\mathrm{ML}}$, is defined as a solution of the equation

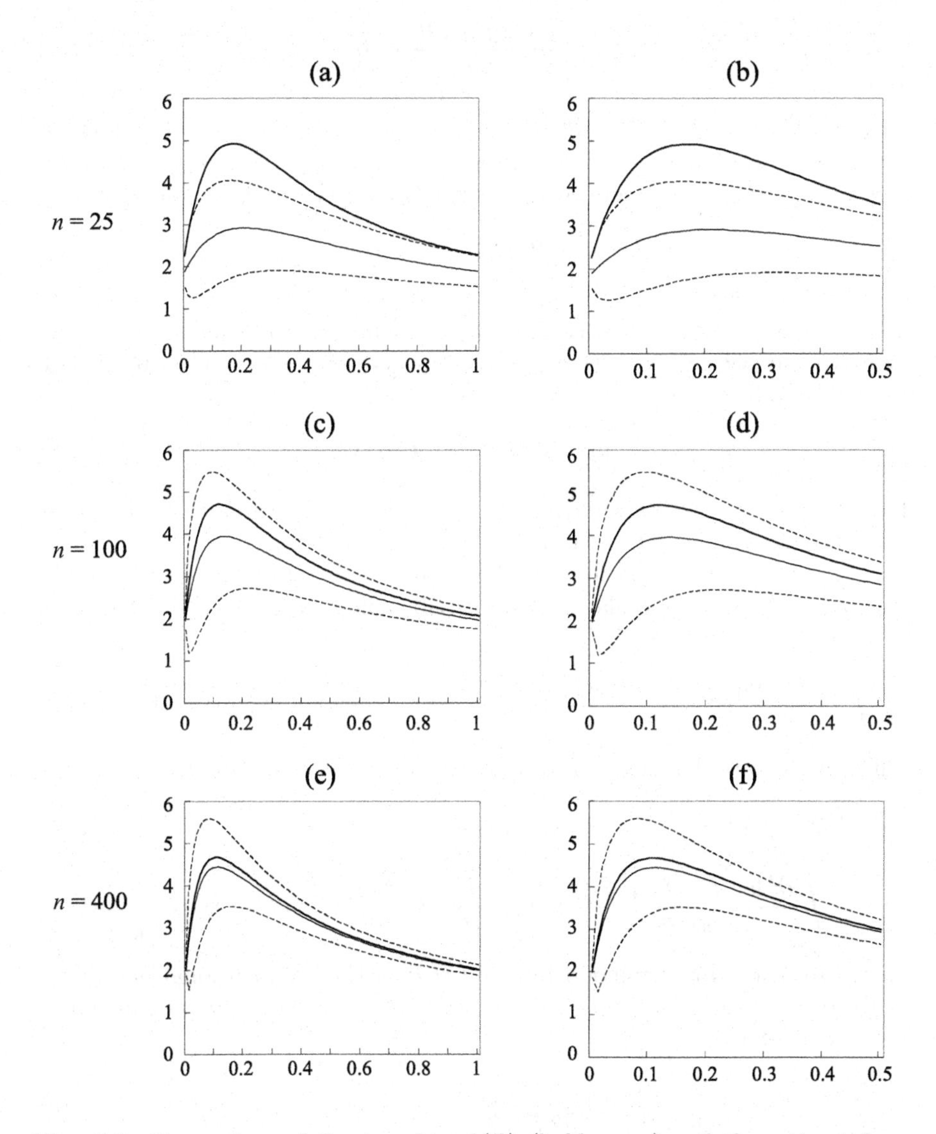

Fig. 5.3. Comparison of the true bias $b(G)$ (bold curve) and the estimated asymptotic bias $b(\hat{G})$ (thin curve) with standard errors $(\cdots\cdots)$ for the sample sizes $n = 25, 100,$ and 200. (a), (c), (e) $(\mu_{01}, \mu_{02}, \sigma_{01}, \sigma_{02}) = (0, 0, 1, 3)$ and (b), (d), (f) $(\mu_{01}, \mu_{02}, \sigma_{01}, \sigma_{02}) = (0, 5, 1, 1)$.

$$\sum_{\alpha=1}^{n} \frac{\partial \log f(x_\alpha|\boldsymbol{\theta})}{\partial \boldsymbol{\theta}} = \mathbf{0}, \tag{5.84}$$

where $\partial/\partial\boldsymbol{\theta} = (\partial/\partial\theta_1, \ldots, \partial/\partial\theta_p)^T$ and $\mathbf{0}$ is the p-dimensional null vector. For any distribution function G, the solution can be expressed as $\hat{\boldsymbol{\theta}}_{\mathrm{ML}} = \boldsymbol{T}_{\mathrm{ML}}(\hat{G})$ with respect to the p-dimensional functional $\boldsymbol{T}_{\mathrm{ML}}(G)$ implicitly defined by

$$\int \left. \frac{\partial \log f(x|\boldsymbol{\theta})}{\partial \boldsymbol{\theta}} \right|_{\boldsymbol{\theta}=\boldsymbol{T}_{\mathrm{ML}}(G)} dG(x) = \mathbf{0}. \tag{5.85}$$

Hence, under certain regularity conditions, the maximum likelihood estimator converges almost surely to the solution $\boldsymbol{T}_{\mathrm{ML}}(G)$ of (5.85) as the sample size tends to infinity, that is,

$$\lim_{n\to+\infty} \boldsymbol{T}_{\mathrm{ML}}(\hat{G}) = \boldsymbol{T}_{\mathrm{ML}}(G). \tag{5.86}$$

This is equivalent to convergence almost surely to the value that minimizes the Kullback–Leibler information.

The influence function for the maximum likelihood estimator can be obtained as follows: By replacing the distribution function G in (5.85) with $(1-\varepsilon)G + \varepsilon\delta_x$, we have

$$\int \frac{\partial \log f(y|\boldsymbol{T}_{\mathrm{ML}}((1-\varepsilon)G + \varepsilon\delta_x))}{\partial \boldsymbol{\theta}} d\left\{(1-\varepsilon)G(y) + \varepsilon\delta_x(y)\right\} = \mathbf{0}. \tag{5.87}$$

Differentiating both sides with respect to ε and setting $\varepsilon = 0$ yield

$$\begin{aligned} &\int \frac{\partial \log f(y|\boldsymbol{T}_{\mathrm{ML}}(G))}{\partial \boldsymbol{\theta}} d\left\{\delta_x(y) - G(y)\right\} \\ &\quad + \int \frac{\partial^2 \log f(y|\boldsymbol{T}_{\mathrm{ML}}(G))}{\partial \boldsymbol{\theta}\partial \boldsymbol{\theta}^T} dG(y) \cdot \left. \frac{\partial}{\partial \varepsilon} \left\{\boldsymbol{T}_{\mathrm{ML}}((1-\varepsilon)G + \varepsilon\delta_x)\right\} \right|_{\varepsilon=0} = \mathbf{0}, \end{aligned} \tag{5.88}$$

where, given the log-likelihood function $\ell(\boldsymbol{\theta})$ of the p-dimensional parameter vector $\boldsymbol{\theta}$, the second-order partial derivative with respect to $\boldsymbol{\theta}$ is defined as a $p \times p$ symmetric matrix

$$\frac{\partial^2 \ell(\boldsymbol{\theta})}{\partial \boldsymbol{\theta}\partial \boldsymbol{\theta}^T} = \left[\frac{\partial^2 \ell(\boldsymbol{\theta})}{\partial \theta_i \partial \theta_j}\right], \qquad i, j = 1, 2, \ldots, p. \tag{5.89}$$

Consequently, by noting that

$$\int \frac{\partial \log f(y|\boldsymbol{T}_{\mathrm{ML}}(G))}{\partial \boldsymbol{\theta}} d\delta_x(y) = \frac{\partial \log f(x|\boldsymbol{T}_{\mathrm{ML}}(G))}{\partial \boldsymbol{\theta}} \tag{5.90}$$

and using (5.85), we obtain the following result:

Influence function for a maximum likelihood estimator. From (5.88), we have the p-dimensional influence function for the maximum likelihood estimator $\hat{\boldsymbol{\theta}}_{\mathrm{ML}} = \boldsymbol{T}_{\mathrm{ML}}(\hat{G})$ in the form

$$\frac{\partial}{\partial \varepsilon}\left\{\boldsymbol{T}_{\mathrm{ML}}((1-\varepsilon)G+\varepsilon\delta_x)\right\}\bigg|_{\varepsilon=0} = J(G)^{-1}\frac{\partial \log f(x|\boldsymbol{\theta})}{\partial \boldsymbol{\theta}}\bigg|_{\boldsymbol{\theta}=\boldsymbol{T}_{\mathrm{ML}}(G)}$$
$$\equiv \boldsymbol{T}_{\mathrm{ML}}^{(1)}(x;G), \tag{5.91}$$

where $J(G)$ is a $p \times p$ matrix given by

$$J(G) = -\int \frac{\partial^2 \log f(x|\boldsymbol{\theta})}{\partial \boldsymbol{\theta} \partial \boldsymbol{\theta}^T}\bigg|_{\boldsymbol{\theta}=\boldsymbol{T}_{\mathrm{ML}}(G)} dG(x). \tag{5.92}$$

By replacing the influence function $\boldsymbol{T}^{(1)}(x;G)$ in (5.62) with the influence function for the maximum likelihood estimator, we obtain the asymptotic bias of the log-likelihood for the estimated model $f(x|\hat{\boldsymbol{\theta}}_{\mathrm{ML}})$:

$$\begin{aligned} b_{\mathrm{ML}}(G) &= \mathrm{tr}\left\{\int \boldsymbol{T}_{\mathrm{ML}}^{(1)}(x;G)\,\frac{\partial \log f(x|\boldsymbol{\theta})}{\partial \boldsymbol{\theta}^T}\bigg|_{\boldsymbol{\theta}=\boldsymbol{T}_{\mathrm{ML}}(G)} dG(x)\right\} \\ &= \mathrm{tr}\left\{J(G)^{-1}\int \frac{\partial \log f(x|\boldsymbol{\theta})}{\partial \boldsymbol{\theta}}\frac{\partial \log f(x|\boldsymbol{\theta})}{\partial \boldsymbol{\theta}^T}\bigg|_{\boldsymbol{\theta}=\boldsymbol{T}_{\mathrm{ML}}(G)} dG(x)\right\} \\ &= \mathrm{tr}\left\{J(G)^{-1}I(G)\right\}, \end{aligned} \tag{5.93}$$

where the $p \times p$ matrix $I(G)$ is given by

$$I(G) = \int \frac{\partial \log f(x|\boldsymbol{\theta})}{\partial \boldsymbol{\theta}}\frac{\partial \log f(x|\boldsymbol{\theta})}{\partial \boldsymbol{\theta}^T}\bigg|_{\boldsymbol{\theta}=\boldsymbol{T}_{\mathrm{ML}}(G)} dG(x). \tag{5.94}$$

Therefore, for the model $f(x|\hat{\boldsymbol{\theta}}_{\mathrm{ML}})$ estimated by the maximum likelihood method, the generalized information criterion in (5.64) is reduced to

$$\mathrm{TIC} = -2\sum_{\alpha=1}^{n} \log f(x_\alpha|\hat{\boldsymbol{\theta}}_{\mathrm{ML}}) + 2\mathrm{tr}\left\{J(\hat{G})^{-1}I(\hat{G})\right\}, \tag{5.95}$$

which agrees with the TIC [Takeuchi (1976)] given by (3.99) in Subsection 3.4.3.

We now consider the case where the true probability distribution $G(x)$ [or the density $g(x)$] is contained in the specified parametric model $\{f(x|\boldsymbol{\theta});\ \boldsymbol{\theta} \in \Theta \subset R^p\}$. Let $f(x|\boldsymbol{\theta})$ and $F_{\boldsymbol{\theta}}$ be, respectively, the true density and its distribution function generating the data. It is assumed that the functional $\boldsymbol{T}_{\mathrm{ML}}(G)$ in (5.85) for the maximum likelihood estimator is Fisher consistent, that is,

$$\boldsymbol{T}_{\mathrm{ML}}(F_{\boldsymbol{\theta}}) = \boldsymbol{\theta} \quad \text{for all } \boldsymbol{\theta} \in \Theta \subset R^p \tag{5.96}$$

[for Fisher consistency, see (5.48) in the preceding section]. Under this assumption, (5.85) can be rewritten as

$$\int \frac{\partial \log f(x|\boldsymbol{\theta})}{\partial \boldsymbol{\theta}} dF_{\boldsymbol{\theta}}(x) = \mathbf{0}. \tag{5.97}$$

Differentiating both sides of this equality with respect to $\boldsymbol{\theta}$ gives

$$\int \frac{\partial^2 \log f(x|\boldsymbol{\theta})}{\partial \boldsymbol{\theta} \partial \boldsymbol{\theta}^T} dF_{\boldsymbol{\theta}}(x) + \int \frac{\partial \log f(x|\boldsymbol{\theta})}{\partial \boldsymbol{\theta}} \frac{\partial \log f(x|\boldsymbol{\theta})}{\partial \boldsymbol{\theta}^T} dF_{\boldsymbol{\theta}}(z) = \mathbf{0}. \tag{5.98}$$

Hence, we have $I(F_{\boldsymbol{\theta}}) = J(F_{\boldsymbol{\theta}})$, called the *Fisher information matrix*, and then the bias of the log-likelihood for $f(x|\hat{\boldsymbol{\theta}}_{\mathrm{ML}})$ in (5.93) is further reduced to

$$b_{\mathrm{ML}}(F_{\boldsymbol{\theta}}) = \mathrm{tr}\left\{ J(F_{\boldsymbol{\theta}})^{-1} I(F_{\boldsymbol{\theta}}) \right\} = p, \tag{5.99}$$

the number of estimated parameters in the specified model $f(x|\boldsymbol{\theta})$. Therefore, we obtain the AIC:

$$\mathrm{AIC} = -2 \sum_{\alpha=1}^{n} \log f(x_\alpha|\hat{\boldsymbol{\theta}}_{\mathrm{ML}}) + 2p. \tag{5.100}$$

Thus, by determining an influence function from the functional that defines a maximum likelihood estimator, it can be shown that the GIC is reduced to the TIC, and by assuming Fisher consistency for the functional, the GIC is further reduced to the AIC.

5.2.3 Robust Estimation

In this subsection, we derive an information criterion for evaluating a statistical model estimated by robust procedures, using the GIC in (5.64).

Suppose that $f(x|\hat{\boldsymbol{\theta}}_M)$ is the estimated model based on data drawn from the true distribution $G(x)$, where $\hat{\boldsymbol{\theta}}_M$ is a p-dimensional M-estimator defined as the solution of the system of implicit equations

$$\sum_{\alpha=1}^{n} \psi_i(x_\alpha, \hat{\boldsymbol{\theta}}_M) = 0, \qquad i = 1, \ldots, p, \tag{5.101}$$

or, in vector notation,

$$\sum_{\alpha=1}^{n} \boldsymbol{\psi}(x_\alpha, \hat{\boldsymbol{\theta}}_M) = \mathbf{0}. \tag{5.102}$$

Here, $\psi_i(x, \boldsymbol{\theta})$ is a real-valued function defined on the product space of the sample and parameter spaces, and $\boldsymbol{\psi} = (\psi_1, \psi_2, \ldots, \psi_p)^T$ is referred to as a $\boldsymbol{\psi}$-function. The M-estimator $\hat{\boldsymbol{\theta}}_M$ is given by $\hat{\boldsymbol{\theta}}_M = \boldsymbol{T}_M(\hat{G})$ for the p-dimensional functional vector $\boldsymbol{T}_M(G)$ defined as the solution of the implicit equations

$$\int \psi_i(x, \boldsymbol{T}_M(G)) dG(x) = 0, \qquad i = 1, \ldots, p, \tag{5.103}$$

or, in vector notation,

$$\int \boldsymbol{\psi}(x, \boldsymbol{T}_M(G)) dG(x) = \mathbf{0}. \tag{5.104}$$

In order to apply the GIC of (5.64), we employ arguments similar to those used in the previous subsection to obtain the influence function for the M-estimator $\hat{\boldsymbol{\theta}}_M$. We first replace the distribution function G with $(1-\varepsilon)G+\varepsilon\delta_x$ in (5.104) as follows:

$$\int \boldsymbol{\psi}(y, \boldsymbol{T}_M((1-\varepsilon)G+\varepsilon\delta_x)\, d\left\{(1-\varepsilon)G(y)+\varepsilon\delta_x(y)\right\} = \mathbf{0}. \tag{5.105}$$

Differentiating both sides of this equation with respect to ε and setting $\varepsilon = 0$, we have

$$\begin{aligned} &\int \boldsymbol{\psi}(y, \boldsymbol{T}_M(G)) d\left\{\delta_x(y) - G(y)\right\} \qquad (5.106) \\ &\quad + \int \frac{\partial \boldsymbol{\psi}(y, \boldsymbol{T}_M(G))^T}{\partial \boldsymbol{\theta}} dG(y) \cdot \frac{\partial}{\partial \varepsilon} \left\{\boldsymbol{T}_M((1-\varepsilon)G+\varepsilon\delta_x)\right\}\Big|_{\varepsilon=0} = \mathbf{0}, \end{aligned}$$

where $\boldsymbol{\psi}(y, \boldsymbol{T}_M(G))^T$ represents a p-dimensional row vector.

Consequently, by making use of (5.104) and

$$\int \boldsymbol{\psi}(y, \boldsymbol{T}_M(G)) d\delta_x(y) = \boldsymbol{\psi}(x, \boldsymbol{T}_M(G)), \tag{5.107}$$

we have the following result:

Influence function for the M-estimator. The p-dimensional influence function, $\boldsymbol{T}_M^{(1)}(x; G)$, for the M-estimator is given by

$$\begin{aligned} \frac{\partial}{\partial \varepsilon} \left\{\boldsymbol{T}_M((1-\varepsilon)G+\varepsilon\delta_x)\right\}_{\varepsilon=0} &= R(\boldsymbol{\psi}, G)^{-1} \boldsymbol{\psi}(x, \boldsymbol{T}_M(G)) \\ &\equiv \boldsymbol{T}_M^{(1)}(x; G), \end{aligned} \tag{5.108}$$

where $R(\boldsymbol{\psi}, G)$ is defined as a $p \times p$ matrix given by

$$R(\boldsymbol{\psi}, G) = -\int \frac{\partial \boldsymbol{\psi}(x, \boldsymbol{\theta})^T}{\partial \boldsymbol{\theta}}\Big|_{\boldsymbol{\theta}=\boldsymbol{T}_M(G)} dG(x), \tag{5.109}$$

with the $(i, j)^{th}$ element

$$-\int \frac{\partial \psi_j(x, \boldsymbol{\theta})}{\partial \theta_i}\Big|_{\boldsymbol{\theta}=\boldsymbol{T}_M(G)} dG(x), \qquad i, j = 1, \ldots, p. \tag{5.110}$$

Substituting this influence function $\boldsymbol{T}_M^{(1)}(x; G)$ into (5.62), we have the asymptotic bias of the log-likelihood of the model $f(x|\hat{\boldsymbol{\theta}}_M)$ in estimating the expected log-likelihood in the form

$$\begin{aligned} b_M(G) &= \operatorname{tr}\left\{ \int \boldsymbol{T}_M^{(1)}(x;G) \left.\frac{\partial \log f(x|\boldsymbol{\theta})}{\partial \boldsymbol{\theta}^T}\right|_{\boldsymbol{\theta}=\boldsymbol{T}_M(G)} dG(x) \right\} \\ &= \operatorname{tr}\left\{ R(\psi,G)^{-1} \int \psi(x,\boldsymbol{T}_M(G)) \left.\frac{\partial \log f(x|\boldsymbol{\theta})}{\partial \boldsymbol{\theta}^T}\right|_{\boldsymbol{\theta}=\boldsymbol{T}_M(G)} dG(x) \right\} \\ &= \operatorname{tr}\left\{ R(\psi,G)^{-1} Q(\psi,G) \right\}, \end{aligned} \tag{5.111}$$

where $Q(\psi, G)$ is a $p \times p$ matrix defined by

$$Q(\psi,G) = \int \psi(x,\boldsymbol{T}_M(G)) \left.\frac{\partial \log f(x|\boldsymbol{\theta})}{\partial \boldsymbol{\theta}^T}\right|_{\boldsymbol{\theta}=\boldsymbol{T}_M(G)} dG(x), \tag{5.112}$$

with the $(i,j)^{th}$ element

$$\int \psi_i(x,\boldsymbol{T}_M(G)) \left.\frac{\partial \log f(x|\boldsymbol{\theta})}{\partial \theta_j}\right|_{\boldsymbol{\theta}=\boldsymbol{T}_M(G)}, \qquad i,j = 1,\ldots,p. \tag{5.113}$$

Then, by using the GIC in (5.64), we have the following result:

Information criterion for a model estimated by a robust procedure. An information criterion for evaluating the statistical model $f(x|\hat{\boldsymbol{\theta}}_M)$ with the M-estimator $\hat{\boldsymbol{\theta}}_M$ is given by

$$\mathrm{GIC}_M = -2\sum_{\alpha=1}^{n} \log f(x_\alpha|\hat{\boldsymbol{\theta}}_M) + 2\operatorname{tr}\left\{ R(\psi,\hat{G})^{-1} Q(\psi,\hat{G}) \right\}, \tag{5.114}$$

where $R(\psi, \hat{G})$ and $Q(\psi, \hat{G})$ are $p \times p$ matrices given by

$$\begin{aligned} R(\psi,\hat{G}) &= -\frac{1}{n}\sum_{\alpha=1}^{n} \left.\frac{\partial \psi(x_\alpha,\boldsymbol{\theta})^T}{\partial \boldsymbol{\theta}}\right|_{\boldsymbol{\theta}=\hat{\boldsymbol{\theta}}}, \\ Q(\psi,\hat{G}) &= \frac{1}{n}\sum_{\alpha=1}^{n} \psi(x_\alpha,\hat{\boldsymbol{\theta}}) \left.\frac{\partial \log f(x_\alpha|\boldsymbol{\theta})}{\partial \boldsymbol{\theta}^T}\right|_{\boldsymbol{\theta}=\hat{\boldsymbol{\theta}}}. \end{aligned} \tag{5.115}$$

The maximum likelihood estimator is an M-estimator, corresponding to $\psi(x|\boldsymbol{\theta}) = \partial \log f(x|\boldsymbol{\theta})/\partial \boldsymbol{\theta}$. By taking this ψ-function in (5.109) and (5.112), we have

$$R(\psi,\hat{G}) = J(G) \quad \text{and} \quad Q(\psi,\hat{G}) = I(G), \tag{5.116}$$

where $J(G)$ and $I(G)$ are respectively given by (5.92) and (5.94). Therefore, we know that the information criterion GIC_M produces in a simple way the TIC given in (5.95).

We now consider the situation in which the parametric family of probability distributions $\{f(x|\boldsymbol{\theta});\ \boldsymbol{\theta} \in \Theta \subset R^p\}$ contains the true distribution $g(x)$ and the functional $\boldsymbol{T}_M$ defined by (5.104) is Fisher consistent, so that $\boldsymbol{T}_M(F_{\boldsymbol{\theta}}) = \boldsymbol{\theta}$ for all $\boldsymbol{\theta} \in \Theta \subset R^p$, where $F_{\boldsymbol{\theta}}(x)$ is the distribution function of $f(x|\boldsymbol{\theta})$. It is then easy to see that (5.104) can be expressed as

$$\int \boldsymbol{\psi}(x, \boldsymbol{\theta}) dF_{\boldsymbol{\theta}}(x) = \boldsymbol{0}. \tag{5.117}$$

By differentiating both sides of the equation with respect to $\boldsymbol{\theta}$, we have

$$\int \frac{\partial \boldsymbol{\psi}(x, \boldsymbol{\theta})^T}{\partial \boldsymbol{\theta}} dF_{\boldsymbol{\theta}}(x) + \int \boldsymbol{\psi}(x, \boldsymbol{\theta}) \frac{\partial \log f(x|\boldsymbol{\theta})}{\partial \boldsymbol{\theta}^T} dF_{\boldsymbol{\theta}}(x) = \boldsymbol{0}. \tag{5.118}$$

[See also the result of (5.98) in the preceding section.] It therefore follows that $Q(\boldsymbol{\psi}, F_{\boldsymbol{\theta}}) = R(\boldsymbol{\psi}, F_{\boldsymbol{\theta}})$, so that the asymptotic bias in (5.111) can be further reduced to

$$b_M(F_{\boldsymbol{\theta}}) = \operatorname{tr}\left\{R(\boldsymbol{\psi}, F_{\boldsymbol{\theta}})^{-1} Q(\boldsymbol{\psi}, F_{\boldsymbol{\theta}})\right\} = p. \tag{5.119}$$

Hence, we have

$$\text{AIC} = -2 \sum_{\alpha=1}^{n} \log f(x_\alpha | \hat{\boldsymbol{\theta}}_M) + 2p. \tag{5.120}$$

This implies that the AIC can be applied directly to evaluate statistical models within the framework of M-estimation.

Example 12 (Normal model estimated by a robust procedure) Consider the parametric model $F_{\boldsymbol{\theta}}(x) = \Phi((x - \mu)/\sigma)$, where Φ is the standard normal distribution function. It is assumed that the parametric family of distributions $\{F_{\boldsymbol{\theta}}(x);\ \boldsymbol{\theta} \in \Theta \subset R^2\}$ $(\boldsymbol{\theta} = (\mu, \sigma)^T)$ contains the true distribution generating the data $\{x_1, \ldots, x_n\}$. The location and scale parameters are respectively estimated by the median, $\hat{\mu}_m$, and the median absolute deviation, $\hat{\sigma}_m$, given by

$$\hat{\mu}_m = \text{med}_i\{x_i\} \quad \text{and} \quad \hat{\sigma}_m = \frac{1}{c}\text{med}_i\{|x_i - \text{med}_j(x_j)|\}, \tag{5.121}$$

where $c = \Phi^{-1}(0.75)$ is chosen to make $\hat{\sigma}_m$ Fisher consistent for Φ. The M-estimators $\hat{\mu}_m$ and $\hat{\sigma}_m$ are defined by the $\boldsymbol{\psi}$-function vector

$$\boldsymbol{\psi}(z; \mu, \sigma) = \left(\text{sign}(z - \mu),\quad c^{-1}\text{sign}(|z - \mu| - c\sigma)\right)^T, \tag{5.122}$$

and their influence functions are

$$\begin{aligned} T_\mu^{(1)}(z; F_{\boldsymbol{\theta}}) &= \frac{\text{sign}(z - \mu)}{2\phi(0)}, \\ T_\sigma^{(1)}(z; F_{\boldsymbol{\theta}}) &= \frac{\text{sign}(|z - \mu| - c\sigma)}{4c\phi(c)}, \end{aligned} \tag{5.123}$$

where ϕ is the standard normal density function [see Huber (1981, p. 137)].

Then, in estimating the expected log-likelihood

$$\int \frac{1}{\hat{\sigma}_m}\phi\left(\frac{x-\hat{\mu}_m}{\hat{\sigma}_m}\right) d\Phi(x), \tag{5.124}$$

the bias correction term (5.111) for the log-likelihood

$$\sum_{\alpha=1}^{n} \log\left\{\frac{1}{\hat{\sigma}_m}\phi\left(\frac{x_\alpha-\hat{\mu}_m}{\hat{\sigma}_m}\right)\right\} \tag{5.125}$$

is [writing $y=(z-\mu)/\sigma$]

$$\int \frac{\text{sign}(y)}{2\phi(0)} y d\Phi(y) + \int \frac{\text{sign}(|y|-c)}{4c\phi(c)}(y^2-1)d\Phi(y) = 2,$$

which is the number of estimated parameters in the normal model and yields the result given in (5.120). We observe that the AIC also holds within the framework of the robust procedure.

Example 13 (M-estimation for linear regression) Let $\{(y_\alpha, \boldsymbol{x}_\alpha);\ \alpha = 1,\ldots,n\}$ ($y_\alpha \in R$, $\boldsymbol{x}_\alpha \in R^p$) be a sample of independent, identically distributed random variables with common distribution $G(y,\boldsymbol{x})$ having density $g(y,\boldsymbol{x})$. Consider the linear model

$$y_\alpha = \boldsymbol{x}_\alpha^T\boldsymbol{\beta} + \varepsilon_\alpha, \qquad \alpha = 1,\ldots,n, \tag{5.126}$$

where $\boldsymbol{\beta}$ is a p-dimensional parameter vector. Let $F(y,\boldsymbol{x}|\boldsymbol{\beta})$ be a model distribution with density $f(y,\boldsymbol{x}|\boldsymbol{\beta}) = f_1(y-\boldsymbol{x}^T\boldsymbol{\beta})f_2(\boldsymbol{x})$, in which the error ε_α is assumed to be independent of $\boldsymbol{x}_\alpha$ and its scale parameter is ignored.

For the linear regression model, we use M-estimates of the regression coefficients $\boldsymbol{\beta}$ given as the solution of the system of equations

$$\sum_{\alpha=1}^{n} \psi(y_\alpha - \boldsymbol{x}_\alpha^T\hat{\boldsymbol{\beta}}_R)\boldsymbol{x}_\alpha = \boldsymbol{0}, \tag{5.127}$$

where $\psi(\cdot)$ is a real-valued function. The influence function of the M-estimator defined by the above equation at the distribution G is

$$\boldsymbol{T}_R^{(1)}(G) = \left\{\int \psi'(y-\boldsymbol{x}^T\boldsymbol{T}_R(G))\boldsymbol{x}\boldsymbol{x}^T dG\right\}^{-1} \psi(y-\boldsymbol{x}^T\boldsymbol{T}_R(G))\boldsymbol{x}, \tag{5.128}$$

where $\psi'(z) = \partial\psi(z)/\partial z$ and $\boldsymbol{T}_R(G)$ is the functional given by

$$\int \psi(y-\boldsymbol{x}^T\boldsymbol{T}_R(G))\boldsymbol{x}dG = \boldsymbol{0}. \tag{5.129}$$

It then follows from (5.111) that the asymptotic bias of the log-likelihood of $f(y, \boldsymbol{x}|\hat{\boldsymbol{\beta}}_R)$ is

$$b_R^{(1)}(G) = \mathrm{tr}\left(\left[\int \psi'\left(y - \boldsymbol{x}^T \boldsymbol{T}_R(G)\right) \boldsymbol{x}\boldsymbol{x}^T dG\right]^{-1} \times \int \psi\left(y - \boldsymbol{x}^T \boldsymbol{T}_R(G)\right) \boldsymbol{x} \frac{\partial \log f(y, \boldsymbol{x}|\boldsymbol{\beta})}{\partial \boldsymbol{\beta}^T}\bigg|_{\boldsymbol{\beta}=\boldsymbol{T}_R(G)} dG\right). \tag{5.130}$$

Suppose that the true density g can be written in the form $g(y, \boldsymbol{x}) = g_1(y - \boldsymbol{x}^T\boldsymbol{\beta})g_2(\boldsymbol{x})$ and that the M-estimator defined by (5.127) is the maximum likelihood estimator for the model $f(y, \boldsymbol{x}|\boldsymbol{\beta})$, that is, $\partial \log f(y, \boldsymbol{x}|\boldsymbol{\beta})/\partial\boldsymbol{\beta} = \psi(y - \boldsymbol{x}^T\boldsymbol{\beta})\boldsymbol{x}$. Then the asymptotic bias $b_R^{(1)}(G)$ in (5.130) can be reduced to $E_{g_1}(\psi')^{-1}E_{g_1}(\psi^2)p$, which agrees with the result given by Ronchetti (1985, p. 23).

Example 14 (Numerical comparison) Consider the normal model $F_{\boldsymbol{\theta}}(x) = \Phi((x - \mu)/\sigma)$ having the density $f(x|\boldsymbol{\theta}) = \sigma^{-1}\phi((x - \mu)/\sigma)$, where $\boldsymbol{\theta} = (\mu, \sigma)^T$. It is assumed that the parametric family of distributions $\{F_{\boldsymbol{\theta}}(x);\ \boldsymbol{\theta} \in \Theta \subset R^2\}$ contains the true distribution that generates the data. The location and scale parameters are respectively estimated by the median, $\hat{\mu}_m = \mathrm{med}_i\{x_i\}$, and the median absolute deviation, $\hat{\sigma}_m = (1/c)\mathrm{med}_i\{|x_i - \mathrm{med}_j(X_j)|\}$, where $c = \Phi^{-1}(0.75)$ is chosen to make $\hat{\sigma}_m$ Fisher consistent for Φ.

Table 5.1. Biases of the log-likelihoods for the M-estimators and the maximum likelihood estimators.

n	25	50	100	200	400	800	1600
M-estimators	3.839	2.569	2.250	2.125	2.056	2.029	2.012
MLE	2.229	2.079	2.047	2.032	2.014	2.002	2.003

Table 5.1 compares the finite-sample biases $b(G)$ of (5.62) of the log-likelihoods for the M-estimator $(\hat{\mu}_m, \hat{\sigma}_m)$ and the maximum likelihood estimator $(\hat{\mu}, \hat{\sigma}^2)$ obtained by averaging over 100,000 repeated Monte Carlo trials. Note that the bias for the maximum likelihood estimator is analytically given by $b(G) = 2n/(n - 3)$ as shown in (3.127).

From the table it may be observed that in the case of the maximum likelihood estimator, the biases are relatively close to 2, which is the asymptotic bias, even when the number of observations involved is small. In contrast, in the case of the M-estimator, the bias is considerably large when $n = 25$. Both of the biases actually converge to the asymptotic bias, 2, as the sample size n becomes large and the convergence of the bias of the robust estimator is slower than that of the maximum likelihood estimator.

5.2.4 Maximum Penalized Likelihood Methods

Nonlinear statistical modeling has received considerable attention in various fields of research such as statistical science, information science, engineering, and artifical intelligence. Nonlinear models are generally characterized by including a large number of parameters. Since maximum likelihood methods yield unstable parameter estimates, the adopted model is usually estimated using the maximum penalized likelihood method or the method of regularization [Good and Gaskins (1971, 1980), Green and Silverman (1994)]. We introduce an information criterion for statistical models constructed by regularization through the case of a regression model and discuss the choice of a smoothing parameter.

Suppose that we have n observations $\{(y_\alpha, \boldsymbol{x}_\alpha);\ \alpha = 1, \cdots, n\}$, where y_α are independent random response variables, $\boldsymbol{x}_\alpha$ are vectors of explanatory variables, and y_α are generated from an unknown true distribution $G(y|\boldsymbol{x})$ having a probability density $g(y|\boldsymbol{x})$. Regression models, in general, consist of a random component and a systematic component. The random component specifies the distribution of the response variable y, while the systematic component represents the mean structure

$$E[Y_\alpha|\boldsymbol{x}_\alpha] = u(\boldsymbol{x}_\alpha), \qquad \alpha = 1, 2, \ldots, n. \tag{5.131}$$

Regression models are used for determining the structure of systems, and such models are generally represented as

$$u(\boldsymbol{x}_\alpha; \boldsymbol{w}), \qquad \alpha = 1, 2, \ldots, n, \tag{5.132}$$

where $\boldsymbol{w}$ is a vector consisting of the unknown parameters contained in each model. The following models are used as regression functions that approximate the mean structure: (i) linear regression, (ii) polynomial regression, (iii) natural cubic splines given by piecewise polynomials [Green and Silverman (1994, p. 12)], (iv) B-splines [de Boor (1978), Imoto (2001), Imoto and Konishi (2003)], (v) kernel functions [Simonoff (1996)], and (vi) neural networks [Bishop (1995), Ripley (1996)].

Let $f(y_\alpha|\boldsymbol{x}_\alpha; \boldsymbol{\theta})$ be a specified parametric model, where $\boldsymbol{\theta}$ is a vector of unknown parameters included in the model. For example, a regression model with Gaussian noise is expressed as

$$f(y_\alpha|\boldsymbol{x}_\alpha; \boldsymbol{\theta}) = \frac{1}{\sqrt{2\pi\sigma^2}} \exp\left[-\frac{\{y_\alpha - u(\boldsymbol{x}_\alpha; \boldsymbol{w})\}^2}{2\sigma^2}\right], \tag{5.133}$$

where $\boldsymbol{\theta} = (\boldsymbol{w}^T, \sigma^2)^T$. The parametric model may be estimated by various procedures including maximum likelihood, robust procedures for handling outliers

[Huber (1981), Hampel et al. (1986)]. Shrinkage estimators provide an alternative estimation method that may be used to advantage when the explanatory variables are highly correlated or when the number of explanatory variables is relatively large compared with the number of observations.

In the estimation of nonlinear regression models for analyzing data with complex structure, the maximum likelihood method often yields unstable parameter estimates and complicated regression curves or surfaces. Instead of maximizing the log-likelihood function, we choose the values of unknown parameters to maximize the penalized log-likelihood function (or the regularized log-likelihood function)

$$\ell_\lambda(\boldsymbol{\theta}) = \sum_{\alpha=1}^{n} \log f(y_\alpha|\boldsymbol{x}_\alpha;\boldsymbol{\theta}) - \frac{n}{2}\lambda H(\boldsymbol{w}). \tag{5.134}$$

This estimation procedure is referred to as the *maximum penalized likelihood method* or the *regularization method.*

The first term in (5.134) is a measure of goodness of fit to the data, while the second term penalizes the roughness of the regression function. The parameter λ (> 0), called a *smoothing parameter* or a *regularization parameter*, performs the function of controlling the trade-off between the smoothness of the function and the goodness of fit to the data. A crucial aspect of model construction is the choice of the smoothing parameter λ. We consider the use of the GIC as a smoothing parameter selector.

The method based on maximizing the penalized log-likelihood function was originally introduced by Good and Gaskins (1971) in the context of density estimation. The Bayesian justification of the method and its relation to shrinkage estimators have been investigated by many authors [Wahba (1978, 1990), Akaike (1980b), Silverman (1985), Shibata (1989), and Kitagawa and Gersch (1996)].

Candidate penalties or regularization terms $H(\boldsymbol{w})$ with an m-dimensional parameter vector $\boldsymbol{w}$ (i) are the discrete approximation of the integration of a second-order derivative that takes the curvature of the function into account, (ii) are finite differences of the unknown parameters, and (iii) sum of squares of w_i are used, depending on the regression functions and data structure under consideration. These are given, respectively, by

$$\begin{aligned} &\text{(i)} \quad H_1(\boldsymbol{w}) = \frac{1}{n}\sum_{\alpha=1}^{n}\sum_{i=1}^{p}\left\{\frac{\partial^2 u(\boldsymbol{x}_\alpha;\boldsymbol{w})}{\partial x_i^2}\right\}^2, \\ &\text{(ii)} \quad H_2(\boldsymbol{w}) = \sum_{i=k+1}^{m} (\Delta^k w_i)^2, \\ &\text{(iii)} \quad H_3(\boldsymbol{w}) = \sum_{i=1}^{m} w_i^2, \end{aligned} \tag{5.135}$$

where Δ represents the difference operator such that $\Delta w_i = w_i - w_{i-1}$.

The regularization term can often be represented as the quadratic function $\boldsymbol{w}^T K \boldsymbol{w}$ of the parameter vector $\boldsymbol{w}$, where K is a known $m \times m$ nonnegative definite matrix. For example, using the $m \times m$ identity matrix I_m, we can write $H_3(\boldsymbol{w})$ as $H_3(\boldsymbol{w}) = \boldsymbol{w}^T I_m \boldsymbol{w}$. Similarly, the regularization term $H_2(\boldsymbol{w})$ based on the difference operator can be represented as

$$H_2(\boldsymbol{w}) = \boldsymbol{w}^T D_k^T D_k \boldsymbol{w} = \boldsymbol{w}^T K \boldsymbol{w}, \tag{5.136}$$

where D_k is an $(m-k) \times m$ matrix given by

$$D_k = \begin{bmatrix} {}_kC_0 & -{}_kC_1 & \cdots & (-1)^k {}_kC_k & 0 & \cdots & 0 \\ 0 & {}_kC_0 & -{}_kC_1 & \cdots & (-1)^k {}_kC_k & \ddots & \vdots \\ \vdots & \ddots & \ddots & \ddots & \ddots & 0 & 0 \\ 0 & \cdots & 0 & {}_kC_0 & -{}_kC_0 & \cdots & (-1)^k {}_kC_k \end{bmatrix} \tag{5.137}$$

with the binomial coefficient ${}_kC_i$. A regularization term frequently used in practice is a second-order difference term given by

$$D_2 = \begin{bmatrix} 1 & -2 & 1 & 0 & \cdots & 0 \\ 0 & 1 & -2 & 1 & \ddots & \vdots \\ \vdots & \ddots & \ddots & \ddots & \ddots & 0 \\ 0 & \cdots & 0 & 1 & -2 & 1 \end{bmatrix}. \tag{5.138}$$

The use of difference penalties has been investigated by Whittaker (1923), Green and Yandell (1985), O'Sullivan et al. (1986), and Kitagawa and Gersch (1996).

We now consider the penalized log-likelihood function expressed as

$$\ell_\lambda(\boldsymbol{\theta}) = \sum_{\alpha=1}^{n} \log f(y_\alpha|\boldsymbol{x}_\alpha; \boldsymbol{\theta}) - \frac{n\lambda}{2} \boldsymbol{w}^T K \boldsymbol{w}. \tag{5.139}$$

Let $\hat{\boldsymbol{\theta}}_P$ be the estimator that maximizes the penalized log-likelihood function (5.139). Then it can be seen that the estimator $\hat{\boldsymbol{\theta}}_P$ is given as the solution of the implicit equation

$$\sum_{\alpha=1}^{n} \boldsymbol{\psi}_P(y_\alpha, \boldsymbol{\theta}) = \boldsymbol{0}, \tag{5.140}$$

where

$$\boldsymbol{\psi}_P(y_\alpha, \boldsymbol{\theta}) = \frac{\partial}{\partial \boldsymbol{\theta}} \left\{ \log f(y_\alpha|\boldsymbol{x}_\alpha; \boldsymbol{\theta}) - \frac{\lambda}{2} \boldsymbol{w}^T K \boldsymbol{w} \right\}. \tag{5.141}$$

Therefore, an information criterion for evaluating the model $f(y|\boldsymbol{x}; \hat{\boldsymbol{\theta}}_P)$ estimated by regularization can be easily obtained within the framework of robust estimation.

In (5.114), by replacing the ψ-function with ψ_P given by (5.141), we obtain the following result:

Information criterion for a model estimated by regularization. An information criterion for the model $f(y|\boldsymbol{x};\hat{\boldsymbol{\theta}}_P)$ with $\hat{\boldsymbol{\theta}}_P$ obtained by maximizing (5.139) is given by

$$\text{GIC}_P = -2\sum_{\alpha=1}^{n} \log f(y_\alpha|\boldsymbol{x}_\alpha;\hat{\boldsymbol{\theta}}_P) + 2\text{tr}\left\{R(\boldsymbol{\psi}_P,\hat{G})^{-1}Q(\boldsymbol{\psi}_P,\hat{G})\right\}, \quad (5.142)$$

where $R(\boldsymbol{\psi}_P,\hat{G})$ and $Q(\boldsymbol{\psi}_P,\hat{G})$ are $(m+1)\times(m+1)$ matrices respectively given by

$$R(\boldsymbol{\psi}_P,\hat{G}) = -\frac{1}{n}\sum_{\alpha=1}^{n} \left.\frac{\partial \boldsymbol{\psi}_P(y_\alpha,\boldsymbol{\theta})^T}{\partial \boldsymbol{\theta}}\right|_{\boldsymbol{\theta}=\hat{\boldsymbol{\theta}}_P},$$

$$Q(\boldsymbol{\psi}_P,\hat{G}) = \frac{1}{n}\sum_{\alpha=1}^{n} \boldsymbol{\psi}_P(y_\alpha,\boldsymbol{\theta})\left.\frac{\partial \log f(y_\alpha|\boldsymbol{x}_\alpha;\boldsymbol{\theta})}{\partial \boldsymbol{\theta}^T}\right|_{\boldsymbol{\theta}=\hat{\boldsymbol{\theta}}_P}. \quad (5.143)$$

Furthermore, by setting $\ell_\alpha(\boldsymbol{\theta}) = \log f(y_\alpha|\boldsymbol{x}_\alpha;\boldsymbol{\theta})$ with $\boldsymbol{\theta} = (\boldsymbol{w}^T,\sigma^2)^T$, these matrices can be expressed as follows:

$$\frac{\partial \boldsymbol{\psi}_P(y_\alpha,\boldsymbol{\theta})^T}{\partial \boldsymbol{\theta}} = \begin{bmatrix} \dfrac{\partial^2 \ell_\alpha(\boldsymbol{\theta})}{\partial \boldsymbol{w}\partial \boldsymbol{w}^T} - \lambda K & \dfrac{\partial^2 \ell_\alpha(\boldsymbol{\theta})}{\partial \boldsymbol{w}\partial \sigma^2} \\ \dfrac{\partial^2 \ell_\alpha(\boldsymbol{\theta})}{\partial \sigma^2\partial \boldsymbol{w}^T} & \dfrac{\partial^2 \ell_\alpha(\boldsymbol{\theta})}{\partial \sigma^2\partial \sigma^2} \end{bmatrix}, \quad (5.144)$$

$$\boldsymbol{\psi}_P(y_\alpha,\boldsymbol{\theta}_P)\frac{\partial \log f(y_\alpha|\boldsymbol{x}_\alpha;\boldsymbol{\theta})}{\partial \boldsymbol{\theta}^T} \quad (5.145)$$

$$= \begin{bmatrix} \dfrac{\partial \ell_\alpha(\boldsymbol{\theta})}{\partial \boldsymbol{w}}\dfrac{\partial \ell_\alpha(\boldsymbol{\theta})}{\partial \boldsymbol{w}^T} - \lambda K\boldsymbol{w}\dfrac{\partial \ell_\alpha(\boldsymbol{\theta})}{\partial \boldsymbol{w}^T} & \dfrac{\partial \ell_\alpha(\boldsymbol{\theta})}{\partial \boldsymbol{w}}\dfrac{\partial \ell_\alpha(\boldsymbol{\theta})}{\partial \sigma^2} - \lambda K\boldsymbol{w}\dfrac{\partial \ell_\alpha(\boldsymbol{\theta})}{\partial \sigma^2} \\ \dfrac{\partial \ell_\alpha(\boldsymbol{\theta})}{\partial \sigma^2}\dfrac{\partial \ell_\alpha(\boldsymbol{\theta})}{\partial \boldsymbol{w}^T} & \left\{\dfrac{\partial \ell_\alpha(\boldsymbol{\theta})}{\partial \sigma^2}\right\}^2 \end{bmatrix}.$$

A crucial issue with nonlinear modeling is the choice of a smoothing parameter, since the estimated model $f(y|\boldsymbol{x};\hat{\boldsymbol{\theta}}_P)$ depends on a smoothing parameter λ. Selection of the smoothing parameter in the modeling process can be viewed as a model selection and evaluation problem. Therefore, an information criterion for evaluating the model $f(y|\boldsymbol{x};\hat{\boldsymbol{\theta}}_P)$ estimated by regularization may be used as a smoothing parameter selector. By evaluating statistical models determined according to the various values of the smoothing parameter, we take the optimal value of the smoothing parameter λ to be that which minimizes the value of GIC_P.

Shibata (1989) introduced an information criterion for evaluating models estimated by regularization and called RIC for regularized information criterion. In neural network models Murata et al. (1994) proposed a network information criterion (NIC) as an estimator of the expected loss for a loss function $-\ell(\boldsymbol{\theta}) + \lambda H(\boldsymbol{\theta})$, where $H(\boldsymbol{\theta})$ is a regularization term.

6

Statistical Modeling by GIC

The current wide availability of fast and inexpensive computers enables us to construct various types of nonlinear models for analyzing data having a complex structure. Crucial issues associated with nonlinear modeling are the choice of adjusted parameters including the smoothing parameter, the number of basis functions in splines and B-splines, and the number of hidden units in neural networks. Selection of these parameters in the modeling process can be viewed as a model selection and evaluation problem. This chapter addresses these issues as a model selection and evaluation problem and provides criteria for evaluating various types of statistical models.

6.1 Nonlinear Regression Modeling via Basis Expansions

In this section, we consider the problem of evaluating nonlinear regression models constructed by the method of regularization. The information criterion GIC is applied to the choice of smoothing parameters and the number of basis functions in the model building process.

Suppose we have n independent observations $\{(y_\alpha, \boldsymbol{x}_\alpha);\ \alpha = 1, 2, \ldots, n\}$, where y_α are random response variables and $\boldsymbol{x}_\alpha$ are p-dimensional vectors of the explanatory variables. In order to extract information from the data, we use the Gaussian nonlinear regression model

$$y_\alpha = u(\boldsymbol{x}_\alpha) + \varepsilon_\alpha, \quad \alpha = 1, 2, \ldots, n, \tag{6.1}$$

where $u(\cdot)$ is an unknown smooth function and the errors ε_α are independently, normally distributed with mean zero and variance σ^2. The problem to be considered is estimating the function $u(\cdot)$ from the observed data, for which we use a regression function expressed as a linear combination of a prescribed set of m basis functions in the following:

$$u(\boldsymbol{x}_\alpha) \approx u(\boldsymbol{x}_\alpha; \boldsymbol{w}) = \sum_{i=1}^{m} w_i b_i(\boldsymbol{x}_\alpha), \tag{6.2}$$

where $b_i(\boldsymbol{x})$ are real-valued functions of a p-dimensional vector of explanatory variables $\boldsymbol{x} = (x_1, x_2, \dots, x_p)^T$.

For example, a linear regression model can be expressed as

$$\sum_{i=0}^{p} w_i b_i(\boldsymbol{x}) = w_0 + w_1 x_1 + w_2 x_2 + \cdots + w_p x_p, \tag{6.3}$$

by putting either $b_1(\boldsymbol{x}) = 1$, $b_i(\boldsymbol{x}) = x_{i-1}$ $(i = 2, 3, \dots, p+1)$, or $b_i(\boldsymbol{x}) = x_i$ $(i = 1, 2, \dots, p)$ and adding a basis function $b_0(\boldsymbol{x}) \equiv 1$ for the intercept w_0. Similarly, the polynomial regression of an explanatory variable x can be expressed as

$$\sum_{i=0}^{m} w_i b_i(x) = w_0 + w_1 x + w_2 x^2 + \cdots + w_m x^m,$$

by adding the basis function $b_0(x) = 1$ for the intercept w_0 and setting $b_i(x) = x^i$.

The Fourier series is the most popular source of basis functions and is defined by $b_0(x) = \sqrt{1/T}$ and

$$b_j(x) = \begin{cases} \sqrt{\dfrac{2}{T}} \sin(w_j x), & w_j = \dfrac{(j+1)\pi}{T} \quad \text{if } j \text{ is odd}, \\ \sqrt{\dfrac{2}{T}} \cos(w_j x), & w_j = \dfrac{j\pi}{T} \quad \text{if } j \text{ is even}, \end{cases} \tag{6.4}$$

for $j = 1, 2, \dots, m$ and the interval $[0, T]$. The Fourier series is useful for basis functions if the observed data are periodic and have sinusoidal features. The natural cubic spline given in Example 17 in Subsection 2.3.1 is also represented by basis functions. Other basis functions, such as the B-spline and radial basis functions, are described in Section 6.2. For basis expansions, we refer to Hastie et al. (2001, Chapter 5).

The regression model based on the basis expansion is represented by

$$\begin{aligned} y_\alpha &= \sum_{i=1}^{m} w_i b_i(\boldsymbol{x}_\alpha) + \varepsilon_\alpha \\ &= \boldsymbol{w}^T \boldsymbol{b}(\boldsymbol{x}_\alpha) + \varepsilon_\alpha, \qquad \alpha = 1, 2, \dots, n, \end{aligned} \tag{6.5}$$

where $\boldsymbol{b}(\boldsymbol{x}) = (b_1(\boldsymbol{x}), b_2(\boldsymbol{x}), \dots, b_m(\boldsymbol{x}))^T$ is an m-dimensional vector of basis functions and $\boldsymbol{w} = (w_1, w_2, \dots, w_m)^T$ is an m-dimensional vector of unknown parameters. Then a regression model with Gaussian noise is expressed as a probability density function

$$f(y_\alpha | \boldsymbol{x}_\alpha; \boldsymbol{\theta}) = \frac{1}{\sqrt{2\pi\sigma^2}} \exp\left[-\frac{\{y_\alpha - \boldsymbol{w}^T \boldsymbol{b}(\boldsymbol{x}_\alpha)\}^2}{2\sigma^2} \right], \tag{6.6}$$

where $\boldsymbol{\theta} = (\boldsymbol{w}^T, \sigma^2)^T$.

The unknown parameter vector $\boldsymbol{\theta}$ is estimated by maximizing the penalized log-likelihood function:

$$\begin{aligned} \ell_\lambda(\boldsymbol{\theta}) &= \sum_{\alpha=1}^{n} \log f(y_\alpha|\boldsymbol{x}_\alpha;\boldsymbol{\theta}) - \frac{n\lambda}{2}\boldsymbol{w}^T K \boldsymbol{w} \qquad (6.7) \\ &= -\frac{n}{2}\log(2\pi\sigma^2) - \frac{1}{2\sigma^2}\sum_{\alpha=1}^{n}\left\{y_\alpha - \boldsymbol{w}^T\boldsymbol{b}(\boldsymbol{x}_\alpha)\right\}^2 - \frac{n\lambda}{2}\boldsymbol{w}^T K\boldsymbol{w} \\ &= -\frac{n}{2}\log(2\pi\sigma^2) - \frac{1}{2\sigma^2}(\boldsymbol{y} - B\boldsymbol{w})^T(\boldsymbol{y} - B\boldsymbol{w}) - \frac{n\lambda}{2}\boldsymbol{w}^T K\boldsymbol{w}, \end{aligned}$$

where $\boldsymbol{y} = (y_1, y_2, \dots, y_n)^T$ and B is an $n \times m$ matrix composed of the following basis functions:

$$B = \begin{bmatrix} \boldsymbol{b}(\boldsymbol{x}_1)^T \\ \boldsymbol{b}(\boldsymbol{x}_2)^T \\ \vdots \\ \boldsymbol{b}(\boldsymbol{x}_n)^T \end{bmatrix} = \begin{bmatrix} b_1(\boldsymbol{x}_1) & b_2(\boldsymbol{x}_1) & \cdots & b_m(\boldsymbol{x}_1) \\ b_1(\boldsymbol{x}_2) & b_2(\boldsymbol{x}_2) & \cdots & b_m(\boldsymbol{x}_2) \\ \vdots & \vdots & \ddots & \vdots \\ b_1(\boldsymbol{x}_n) & b_2(\boldsymbol{x}_n) & \cdots & b_m(\boldsymbol{x}_n) \end{bmatrix}. \qquad (6.8)$$

By differentiating $\ell_\lambda(\boldsymbol{\theta})$ with respect to $\boldsymbol{\theta} = (\boldsymbol{\beta}^T, \sigma^2)^T$ and setting the result equal to $\boldsymbol{0}$, we have the maximum penalized likelihood estimators for $\boldsymbol{w}$ and σ^2 respectively given by

$$\hat{\boldsymbol{w}} = (B^T B + n\lambda\hat{\sigma}^2 K)^{-1} B^T \boldsymbol{y} \quad \text{and} \quad \hat{\sigma}^2 = \frac{1}{n}(\boldsymbol{y} - B\hat{\boldsymbol{w}})^T(\boldsymbol{y} - B\hat{\boldsymbol{w}}). \quad (6.9)$$

Since the estimator $\hat{\boldsymbol{w}}$ in (6.9) depends on the variance estimator $\hat{\sigma}^2$, in practice it is calculated using the following method. First, put $\beta = \lambda\hat{\sigma}^2$ and determine $\hat{\boldsymbol{w}} = (B^T B + n\beta_0 K)^{-1} B^T \boldsymbol{y}$ for a given $\beta = \beta_0$. Then, after determining the variance estimator $\hat{\sigma}^2$, obtain the value of the smoothing parameter as $\lambda = \beta/\hat{\sigma}^2$.

The statistical model is obtained by replacing the unknown parameters $\boldsymbol{w}$ and σ^2 in (6.6) with their estimators $\hat{\boldsymbol{w}}$ and $\hat{\sigma}^2$ and is of the form

$$f(y_\alpha|\boldsymbol{x}_\alpha;\hat{\boldsymbol{\theta}}_P) = \frac{1}{\sqrt{2\pi\hat{\sigma}^2}}\exp\left[-\frac{\left\{y_\alpha - \hat{\boldsymbol{w}}^T\boldsymbol{b}(\boldsymbol{x}_\alpha)\right\}^2}{2\hat{\sigma}^2}\right]. \qquad (6.10)$$

The estimators $\hat{\boldsymbol{w}}$ and $\hat{\sigma}^2$ depend on the smoothing parameter λ (or β) and also the number m of basis functions. The optimal values of these adjusted parameters have to be chosen by a suitable criterion, for which we use an information criterion for evaluating the statistical model $f(y_\alpha|\boldsymbol{x}_\alpha;\hat{\boldsymbol{\theta}}_P)$.

Writing $\log f(y_\alpha|\boldsymbol{x}_\alpha;\boldsymbol{\theta}) = \ell_\alpha(\boldsymbol{\theta})$, the first and second partial derivatives with respect to $\boldsymbol{\theta} = (\boldsymbol{w}^T, \sigma^2)^T$ are given by

$$\frac{\partial \ell_\alpha(\boldsymbol{\theta})}{\partial \sigma^2} = -\frac{1}{2\sigma^2} + \frac{1}{2\sigma^4}\{y_\alpha - \boldsymbol{w}^T \boldsymbol{b}(\boldsymbol{x}_\alpha)\}^2,$$
$$\frac{\partial \ell_\alpha(\boldsymbol{\theta})}{\partial \boldsymbol{w}} = \frac{1}{\sigma^2}\{y_\alpha - \boldsymbol{w}^T \boldsymbol{b}(\boldsymbol{x}_\alpha)\}\boldsymbol{b}(\boldsymbol{x}_\alpha), \quad (6.11)$$

and

$$\frac{\partial^2 \ell_\alpha(\boldsymbol{\theta})}{\partial \sigma^2 \partial \sigma^2} = \frac{1}{2\sigma^4} - \frac{1}{\sigma^6}\{y_\alpha - \boldsymbol{w}^T \boldsymbol{b}(\boldsymbol{x}_\alpha)\}^2,$$
$$\frac{\partial^2 \ell_\alpha(\boldsymbol{\theta})}{\partial \boldsymbol{w} \partial \boldsymbol{w}^T} = -\frac{1}{\sigma^2}\boldsymbol{b}(\boldsymbol{x}_\alpha)\boldsymbol{b}(\boldsymbol{x}_\alpha)^T,$$
$$\frac{\partial^2 \ell_\alpha(\boldsymbol{\theta})}{\partial \sigma^2 \partial \boldsymbol{w}} = -\frac{1}{\sigma^4}\{y_\alpha - \boldsymbol{w}^T \boldsymbol{b}(\boldsymbol{x}_\alpha)\}\boldsymbol{b}(\boldsymbol{x}_\alpha). \quad (6.12)$$

From the results (5.142), (5.144), and (5.145), we have the following:

Information criterion for a statistical model constructed by regularized basis expansions. Suppose that $f(y_\alpha|\boldsymbol{x}_\alpha;\boldsymbol{\theta})$ in (6.10) is the Gaussian nonlinear regression model based on basis functions. Then an information criterion for the model $f(y_\alpha|\boldsymbol{x}_\alpha;\hat{\boldsymbol{\theta}}_P)$ estimated by regularization is given by

$$\mathrm{GIC_{PB}} = n(\log 2\pi + 1) + n\log(\hat{\sigma}^2) + 2\mathrm{tr}\left\{R(\boldsymbol{\psi}_P, \hat{G})^{-1}Q(\boldsymbol{\psi}_P, \hat{G})\right\}, \quad (6.13)$$

where $\hat{\sigma}^2$ is given in (6.9), and the $(m+1)\times(m+1)$ matrices $R(\boldsymbol{\psi}_P, \hat{G})$ and $Q(\boldsymbol{\psi}_P, \hat{G})$ are, respectively,

$$R(\boldsymbol{\psi}_P, \hat{G}) = \frac{1}{n\hat{\sigma}^2}\begin{bmatrix} B^T B + n\lambda\hat{\sigma}^2 K & \frac{1}{\hat{\sigma}^2}B^T \Lambda \mathbf{1}_n \\ \frac{1}{\hat{\sigma}^2}\mathbf{1}_n^T \Lambda B & \frac{n}{2\hat{\sigma}^2} \end{bmatrix}, \quad (6.14)$$
$$Q(\boldsymbol{\psi}_P, \hat{G}) = \frac{1}{n\hat{\sigma}^2}\begin{bmatrix} \frac{1}{\hat{\sigma}^2}B^T \Lambda^2 B - \lambda K \boldsymbol{w}\mathbf{1}_n^T \Lambda B & \frac{1}{2\hat{\sigma}^4}B^T \Lambda^3 \mathbf{1}_n - \frac{1}{2\hat{\sigma}^2}B^T \Lambda \mathbf{1}_n \\ \frac{1}{2\hat{\sigma}^4}\mathbf{1}_n^T \Lambda^3 B - \frac{1}{2\hat{\sigma}^2}\mathbf{1}_n^T \Lambda B & \frac{1}{4\hat{\sigma}^6}\mathbf{1}_n^T \Lambda^4 \mathbf{1}_n - \frac{n}{4\hat{\sigma}^2} \end{bmatrix},$$

where $\mathbf{1}_n = (1, 1, \ldots, 1)^T$ is an n-dimensional vector, the elements of which are all 1, and Λ is an $n \times n$ diagonal matrix defined by

$$\Lambda = \mathrm{diag}\left[y_1 - \hat{\boldsymbol{w}}^T \boldsymbol{b}(\boldsymbol{x}_1), y_2 - \hat{\boldsymbol{w}}^T \boldsymbol{b}(\boldsymbol{x}_2), \ldots, y_n - \hat{\boldsymbol{w}}^T \boldsymbol{b}(\boldsymbol{x}_n)\right]. \quad (6.15)$$

With respect to the number m of basis functions and the values of the smoothing parameter λ (or β), we select the values of $(\hat{m}, \hat{\lambda})$ that minimize the information criterion $\mathrm{GIC_{PB}}$ as the optimal values. In applying this technique to practical problems, the smoothness can also be controlled using λ, by fixing the number of basis functions.

6.2 Basis Functions

6.2.1 *B*-Splines

Suppose that we have n sets of observations $\{(y_\alpha, x_\alpha);\ \alpha = 1, 2, \ldots, n\}$ and that the responses y_α are generated from an unknown true distribution $G(y|x)$ having probability density $g(y|x)$. It is assumed that the observations on the explanatory variable are sorted by magnitude as $x_1 < x_2 < \cdots < x_n$.@

Consider the regression model based on B-spline basis functions

$$\begin{aligned} y_\alpha &= \sum_{i=1}^{m} w_i b_i(x_\alpha) + \varepsilon_\alpha \\ &= \boldsymbol{w}^T \boldsymbol{b}(x_\alpha) + \varepsilon_\alpha, \qquad\qquad \alpha = 1, 2, \ldots, n, \end{aligned} \tag{6.16}$$

where $\boldsymbol{b}(x) = (b_1(x),\ b_2(x),\ \ldots,\ b_m(x))^T$ is an m-dimensional vector of B-spline basis functions and $\boldsymbol{w} = (w_1, w_2, \ldots, w_m)^T$ is an m-dimensional vector of unknown parameters. We consider B-splines of degree 3, constructed from polynomial functions. The B-spline basis function $b_j(x)$ is composed of known piecewise polynomials that are smoothly connected at points t_i, called *knots* [see de Boor (1978), Eilers and Marx (1996), Imoto (2001), and Imoto and Koishi (2003)].

Let us set up the knots required to construct m basis functions $\{b_1(x), b_2(x), \ldots, b_m(x)\}$ as follows:

$$t_1 < t_2 < t_3 < t_4 = x_1 < \cdots < t_{m+1} = x_n < \cdots t_{m+4}. \tag{6.17}$$

By setting the knots in this way, the n observations are partitioned into $m-3$ intervals $[t_4, t_5]$, $[t_5, t_6]$, $\ldots$, $[t_m, t_{m+1}]$. Furthermore, each interval $[t_i, t_{i+1}]$ $(i = 4, \ldots, m)$ is covered by four B-spline basis functions. The algorithm developed by de Boor (1978) can be conveniently used in constructing the B-spline basis functions.

Generally, we write a B-spline function of degree r as $b_j(x; r)$. First, let us define a B-spline function of degree 0 as follows:

$$b_j(x; 0) = \begin{cases} 1, & \text{for} \quad t_j \le x < t_{j+1}, \\ 0, & \text{otherwise}. \end{cases} \tag{6.18}$$

Starting from the B-spline function of degree 0, a B-spline function of degree r can be obtained using the recursive formula:

$$b_j(x; r) = \frac{x - t_j}{t_{j+r} - t_j} b_j(x; r-1) + \frac{t_{j+r+1} - x}{t_{j+r+1} - t_{j+1}} b_{j+1}(x; r-1). \tag{6.19}$$

Let $b_j(x) = b_j(x; 3)$ be the B-spline basis function of degree 3. Then the Gaussian nonlinear regression model based on a *cubic B-splines* can be expressed as

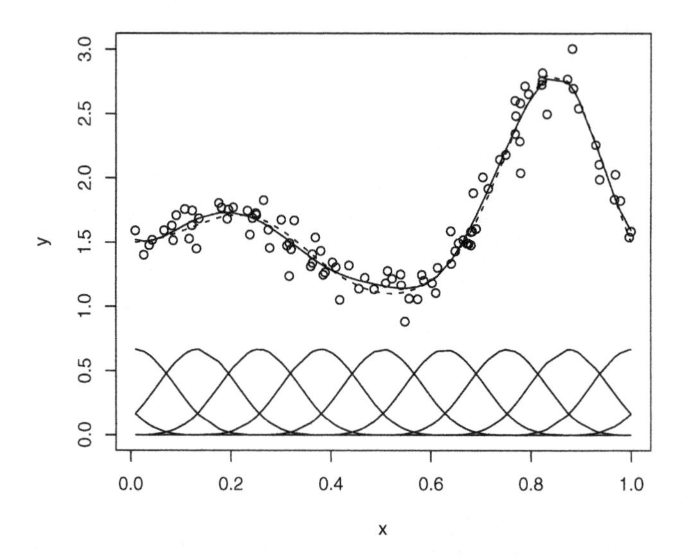

Fig. 6.1. B-spline bases and the true (dashed line) and smoothed (solid line) curves.

$$f(y_\alpha|x_\alpha;\boldsymbol{\theta}) = \frac{1}{\sqrt{2\pi\sigma^2}} \exp\left\{ -\frac{\left(y_\alpha - \boldsymbol{w}^T\boldsymbol{b}(x_\alpha)\right)^2}{2\sigma^2} \right\}, \tag{6.20}$$

where $\boldsymbol{b}(x_\alpha) = (b_1(x_\alpha;3), b_2(x_\alpha;3), \ldots, b_m(x_\alpha;3))^T$ and $\boldsymbol{\theta} = (\boldsymbol{w}^T, \sigma^2)^T$. Estimating the unknown parameters $\boldsymbol{\theta}$ by the regularization method, we obtain the nonlinear regression model and the predicted values as follows:

$$y = \hat{\boldsymbol{w}}^T\boldsymbol{b}(x) \quad \text{and} \quad \hat{\boldsymbol{y}} = B(B^TB + n\lambda\hat{\sigma}^2K)^{-1}B^T\boldsymbol{y}. \tag{6.21}$$

Example 1 (Numerical result) For illustration, data $\{(y_\alpha, x_\alpha),\ \alpha = 1, \ldots, 100\}$ were generated from the true model

$$y_\alpha = \exp\left\{-x_\alpha \sin(2\pi x_\alpha)\right\} + 1 + \varepsilon_\alpha, \tag{6.22}$$

with Gaussian noise $N(0, 0.3^2)$, where the design points are uniformly distributed in $[0, 1]$. Figure 6.1 gives B-spline basis functions of degree 3 with knots 0.0, 0.1, ..., 1.0 and the true and fitted curves. We see that B-splines give a good representation of the underlying function over the region $[0, 1]$ by taking the number of basis functions and the value of the smoothing parameter.

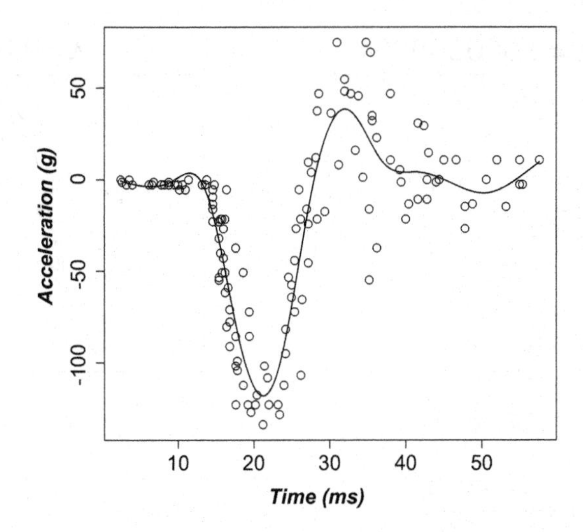

Fig. 6.2. Data and B-spline function

Example 2 (Motorcycle impact data) The motorcycle impact data [Härdle (1990)] were simulated to investigate the efficiency of crash helmets and comprise a series of measurements of the head acceleration in units of gravity (g) as a function of the time in milliseconds (ms) after impact. Figure 6.2 shows a plot of 133 observations. When dealing with data containing such a complex nonlinear structure, polynomial models or models that use specific nonlinear functions are not flexible enough to effectively capture the structure of the phenomena at hand. When addressing data containing a complex, nonlinear structure, we need to set up a model that provides flexibility in describing the true structure. The solid curve in Figure 6.2 shows the fitted model based on cubic B-splines. Selecting the number of basis functions and the value of the smoothing parameter using the $\mathrm{GIC_{PB}}$ in (6.13) yields $m = 16$ and $\lambda = 7.74 \times 10^{-7}$.

Example 3 (The role of the smoothing parameter) Figure 6.3 shows the role of the smoothing parameter in the regularization method for curve fitting. The figure shows that as λ becomes large, the penalty term in the second term also increases considerably. In order to increase the regularized log-likelihood function $\ell_\lambda(\boldsymbol{\theta})$, the B-spline function approaches a linear function. When the value of λ is small, the term containing the log-likelihood function dominates, and the function passes through the vicinity of the data even at the expense of increase of variation in the curve. See Eilers and Marx (1996) and Imoto and Konishi (2003) for regression models based on B-splines.

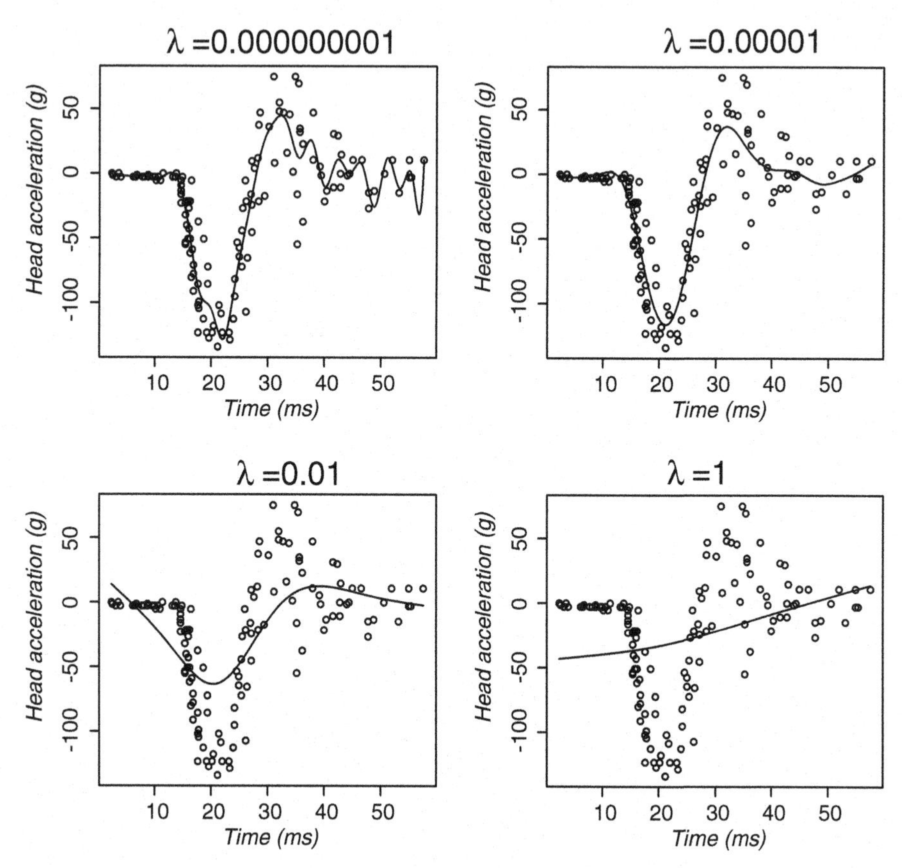

Fig. 6.3. The effect of the smoothing parameter in the regularization method. $\lambda = 0.00001$ yields the best estimate.

6.2.2 Radial Basis Functions

Given n sets of data $\{(y_\alpha, \boldsymbol{x}_\alpha);\ \alpha = 1, 2, \ldots, n\}$ observed on a response variable y and a p-dimensional vector of explanatory variables $\boldsymbol{x}$, a regression model based on radial basis functions is generally given by

$$y_\alpha = w_0 + \sum_{i=1}^{m} w_i \phi\left(||\boldsymbol{x}_\alpha - \boldsymbol{\mu}_i||\right) + \varepsilon_\alpha, \quad \alpha = 1, 2, \ldots, n \tag{6.23}$$

[Bishop (1995, Chapter 5), Ripley (1996, Section 4.2), and Webb (1999, Chapter 5)], where $\boldsymbol{\mu}_i$ is a p-dimensional vector of centers that determines the position of the basis function, and $||\cdot||$ is the Euclidean norm. The following Gaussian basis function is frequently employed in practice:

$$\phi_i(\boldsymbol{x}) = \exp\left(-\frac{||\boldsymbol{x} - \boldsymbol{\mu}_i||^2}{2h_i^2}\right), \qquad i = 1, 2, \ldots, m, \tag{6.24}$$

where h_i^2 is a quantity that represents the spread of the function.

The unknown parameters included in the nonlinear regression model with Gaussian basis functions are $\{\boldsymbol{\mu}_1, \ldots, \boldsymbol{\mu}_m, h_1^2, \ldots, h_m^2\}$ in addition to the coefficients $\{w_0, w_1, \ldots, w_m\}$. Although a method of simultaneously estimating these parameters is conceivable, the multiplicity of local maxima causes problems when performing numerical optimization. Furthermore, when the number of basis functions involved and the problem of selecting regularization parameters are taken into consideration, the number of computations required becomes enormous.

A useful technique from a practical point of view for overcoming these problems is the method of determining basis functions on an a priori basis by first applying a clustering technique to the data related to explanatory variables [Moody and Darken (1989)]. In the first stage, the centers $\boldsymbol{\mu}_i$ and width parameters h_i^2 are determined by using only the input data set $\{\boldsymbol{x}_\alpha; \alpha = 1, \ldots, n\}$ for explanatory variables. In the second stage, the weights w_i are estimated using appropriate estimation procedures like the method of regularization.

Among the various possible strategies for determining the centers and widths of the basis functions, we use a k-means clustering algorithm. This algorithm divides the input data set $\{\boldsymbol{x}_\alpha; \alpha = 1, \ldots, n\}$ into m clusters $C_1, \ldots, C_m$ that correspond to the number of the basis functions. The centers and width parameters are then determined using

$$\hat{\boldsymbol{\mu}}_i = \frac{1}{n_i} \sum_{\boldsymbol{x}_\alpha \in C_i} \boldsymbol{x}_\alpha \quad \text{and} \quad \hat{h}_i^2 = \frac{1}{n_i} \sum_{\boldsymbol{x}_\alpha \in C_i} ||\boldsymbol{x}_\alpha - \hat{\boldsymbol{\mu}}_i||^2, \tag{6.25}$$

where n_i is the number of the observations that belong to the ith cluster C_i. Substituting these estimates into the Gaussian basis function (6.24) gives us a set of m basis functions

$$\phi_i(\boldsymbol{x}) \equiv \exp\left(-\frac{||\boldsymbol{x} - \hat{\boldsymbol{\mu}}_i||^2}{2\hat{h}_i^2}\right), \qquad i = 1, 2, \ldots, m. \tag{6.26}$$

We use the nonlinear regression model with the Gaussian basis functions given by

$$\begin{aligned} y_\alpha &= w_0 + \sum_{i=1}^{m} w_i \phi_i(\boldsymbol{x}_\alpha) + \varepsilon_\alpha \\ &= \boldsymbol{w}^T \boldsymbol{\phi}(\boldsymbol{x}_\alpha) + \varepsilon_\alpha, \qquad \alpha = 1, 2, \ldots, n, \end{aligned} \tag{6.27}$$

where $\boldsymbol{\phi}(\boldsymbol{x}) = (1, \phi_1(\boldsymbol{x}), \phi_2(\boldsymbol{x}), \ldots, \phi_m(\boldsymbol{x}))^T$ is an $(m+1)$-dimensional vector of the Gaussian bases and $\boldsymbol{w} = (w_0, w_1, \ldots, w_m)^T$ is an $(m+1)$-dimensional vector of unknown parameters. Then the nonlinear regression model with Gaussian noise can be expressed as a probability density function

$$f(y_\alpha|\boldsymbol{x}_\alpha;\boldsymbol{\theta}) = \frac{1}{\sqrt{2\pi\sigma^2}} \exp\left\{ -\frac{\left(y_\alpha - \boldsymbol{w}^T\boldsymbol{\phi}(\boldsymbol{x}_\alpha)\right)^2}{2\sigma^2} \right\}, \tag{6.28}$$

where $\boldsymbol{\theta} = (\boldsymbol{w}^T, \sigma^2)^T$.

By estimating the unknown parameter vector $\boldsymbol{\theta}$ using the regularization method, we obtain the special case of (6.9)

$$\hat{\boldsymbol{w}} = (B^TB + n\lambda\hat{\sigma}^2K)^{-1}B^T\boldsymbol{y}, \quad \hat{\sigma}^2 = \frac{1}{n}(\boldsymbol{y} - B\hat{\boldsymbol{w}})^T(\boldsymbol{y} - B\hat{\boldsymbol{w}}), \tag{6.29}$$

in which B is an $n \times (m+1)$ matrix consisting of values of the Gaussian basis functions in (6.26):

$$B = \begin{bmatrix} \boldsymbol{\phi}(\boldsymbol{x}_1)^T \\ \boldsymbol{\phi}(\boldsymbol{x}_2)^T \\ \vdots \\ \boldsymbol{\phi}(\boldsymbol{x}_n)^T \end{bmatrix} = \begin{bmatrix} 1 & \phi_1(\boldsymbol{x}_1) & \phi_2(\boldsymbol{x}_1) & \cdots & \phi_m(\boldsymbol{x}_1) \\ 1 & \phi_1(\boldsymbol{x}_2) & \phi_2(\boldsymbol{x}_2) & \cdots & \phi_m(\boldsymbol{x}_2) \\ \vdots & \vdots & \vdots & \ddots & \vdots \\ 1 & \phi_1(\boldsymbol{x}_n) & \phi_2(\boldsymbol{x}_n) & \cdots & \phi_m(\boldsymbol{x}_n) \end{bmatrix}. \tag{6.30}$$

In addition, the information criterion for evaluating the statistical model constructed by the regularized Gaussian basis expansion is given by a formula in which matrix B in (6.14) is replaced with the Gaussian basis function matrix (6.30).

The radial basis functions overlap each other to capture the information from the input data, and the width parameters control the amount of overlapping between basis functions. Hence, the values of width parameters play an essential role in determining the smoothness of the estimated regression function.

Moody and Darken (1989) used k-means clustering algorithm and adopted the P nearest neighbor heuristically, determining the width as the average Euclidean distance of the P nearest neighbor of each basis function. The maximum Euclidean distance among the selected centers of the basis functions was also employed by Broomhead and Lowe (1988), where they randomly selected the centers from the input data set. Such a heuristic approach does not always yield sufficiently accurate results [Ando et al. (2005)]. To overcome this problem, Ando et al. (2005) introduced the Gaussian basis functions with hyperparameter ν given by the following:

$$\phi_i(\boldsymbol{x};\boldsymbol{\mu}_i,\sigma_i,\nu) = \exp\left(-\frac{||\boldsymbol{x}-\boldsymbol{\mu}_i||^2}{2\nu\sigma_i^2}\right), \quad i = 1,\ldots,m. \tag{6.31}$$

The hyperparameter ν adjusts the amount of overlapping between basis functions so that the estimated regression function captures the structure in the data over the region of the input space and incorporates this information in the response variables.

Fujii and Konishi (2006) proposed a regularized wavelet-based method for nonlinear regression modeling when design points are not equally spaced and derived an information criterion to choose smoothing parameters, using GIC_P in (5.142). Regularized local likelihood method for nonlinear regression modeling was investigated by Nonaka and Konishi (2005), in which they used GIC_P in (5.142) for selecting the degree of polynomial and a smoothing parameter. For local likelihood estimation, we refer to Fan and Gijbels (1996) and Loader (1999).

6.3 Logistic Regression Models for Discrete Data

The logistic model is used to predict a discrete outcome from a set of explanatory variables that may be continuous and/or categorical. The response variable is generally dichotomous such as success or failure and takes the value 1 with probability of success π or the value 0 with probability of failure $1 - \pi$. Logistic modeling enables us to model the relationship between the explanatory and response variables.

6.3.1 Linear Logistic Regression Model

Suppose that we have n sets of observations $\{(y_\alpha, \boldsymbol{x}_\alpha);\ \alpha = 1, \ldots, n\}$, where y_α are independent random variables coded as either 0 or 1 and $\boldsymbol{x}_\alpha = (1, x_{\alpha 1}, x_{\alpha 2}, \ldots, x_{\alpha p})^T$ is a vector of p covariates. The logistic model assumes that

$$\Pr(Y_\alpha = 1|\boldsymbol{x}_\alpha) = \pi(\boldsymbol{x}_\alpha) \quad \text{and} \quad \Pr(Y_\alpha = 0|\boldsymbol{x}_\alpha) = 1 - \pi(\boldsymbol{x}_\alpha), \tag{6.32}$$

where Y_α is a random variable distributed according to the Bernoulli distribution

$$f(y_\alpha|\boldsymbol{x}_\alpha; \boldsymbol{\beta}) = \pi(\boldsymbol{x}_\alpha)^{y_\alpha} \{1 - \pi(\boldsymbol{x}_\alpha)\}^{1-y_\alpha}, \qquad y_\alpha = 0, 1. \tag{6.33}$$

The linear logistic model further assumes that

$$\pi(\boldsymbol{x}_\alpha) = \frac{\exp(\boldsymbol{x}_\alpha^T \boldsymbol{\beta})}{1 + \exp(\boldsymbol{x}_\alpha^T \boldsymbol{\beta})} \quad \text{or} \quad \log \frac{\pi(\boldsymbol{x}_\alpha)}{1 - \pi(\boldsymbol{x}_\alpha)} = \boldsymbol{x}_\alpha^T \boldsymbol{\beta}, \tag{6.34}$$

which links level $\boldsymbol{x}_\alpha$ stimuli to the conditional probability $\pi(\boldsymbol{x}_\alpha)$, where $\boldsymbol{x}_\alpha^T \boldsymbol{\beta} = \beta_0 + \beta_1 x_{\alpha 1} + \beta_2 x_{\alpha 2} + \cdots + \beta_p x_{\alpha p}$. Under this model, the log-likelihood function for y_α in terms of $\boldsymbol{\beta}$ is

$$\begin{aligned}
\ell(\boldsymbol{\beta}) &= \sum_{\alpha=1}^{n} \left[y_\alpha \log \pi(\boldsymbol{x}_\alpha) + (1 - y_\alpha) \log \{1 - \pi(\boldsymbol{x}_\alpha)\} \right] \\
&= \sum_{\alpha=1}^{n} \left[y_\alpha \log \frac{\pi(\boldsymbol{x}_\alpha)}{1 - \pi(\boldsymbol{x}_\alpha)} + \log \{1 - \pi(\boldsymbol{x}_\alpha)\} \right] \\
&= \sum_{\alpha=1}^{n} \left[y_\alpha \boldsymbol{x}_\alpha^T \boldsymbol{\beta} - \log\{1 + \exp(\boldsymbol{x}_\alpha^T \boldsymbol{\beta})\} \right].
\end{aligned} \tag{6.35}$$

The maximum likelihood method frequently yields unstable parameter estimates with significant variation when the explanatory variables are highly correlated or when there are an insufficient number of observations relative to the number of explanatory variables. In such a case, the $(p+1)$-dimensional parameter vector $\boldsymbol{\beta}$ may be estimated by maximizing the penalized log-likelihood function:

$$\ell_\lambda(\boldsymbol{\beta}) = \sum_{\alpha=1}^{n} \left[y_\alpha \boldsymbol{x}_\alpha^T \boldsymbol{\beta} - \log\left\{1 + \exp(\boldsymbol{x}_\alpha^T \boldsymbol{\beta})\right\}\right] - \frac{n\lambda}{2} \boldsymbol{\beta}^T K \boldsymbol{\beta}, \tag{6.36}$$

where K is a $(p+1) \times (p+1)$ nonnegative definite matrix (see Subsection 5.2.4). The shrinkage estimator can be obtained by setting $K = I_{p+1}$.

The optimization process with respect to unknown parameter vector $\boldsymbol{\beta}$ is nonlinear, and the equation does not have an explicit solution. The solution, $\hat{\boldsymbol{\beta}}$, in this case may be obtained using an iterative algorithm.

Fisher's scoring method. The first and second derivatives of the penalized log-likelihood function with respect to $\boldsymbol{\beta}$ are given by

$$\begin{aligned} \frac{\partial \ell_\lambda(\boldsymbol{\beta})}{\partial \boldsymbol{\beta}} &= \sum_{\alpha=1}^{n} \{y_\alpha - \pi(\boldsymbol{x}_\alpha)\} \boldsymbol{x}_\alpha - n\lambda K \boldsymbol{\beta} \\ &= X^T \Lambda \mathbf{1}_n - n\lambda K \boldsymbol{\beta}, \end{aligned} \tag{6.37}$$

$$\begin{aligned} \frac{\partial^2 \ell_\lambda(\boldsymbol{\beta})}{\partial \boldsymbol{\beta} \partial \boldsymbol{\beta}^T} &= -\sum_{\alpha=1}^{n} \pi(\boldsymbol{x}_\alpha)\{1 - \pi(\boldsymbol{x}_\alpha)\} \boldsymbol{x}_\alpha \boldsymbol{x}_\alpha^T - n\lambda K \\ &= -X^T \Pi (I_n - \Pi) X - n\lambda K, \end{aligned} \tag{6.38}$$

where $X = (\boldsymbol{x}_1, \boldsymbol{x}_2, \ldots, \boldsymbol{x}_n)^T$ is an $n \times (p+1)$ matrix, I_n is an $n \times n$ identity matrix, $\mathbf{1}_n = (1, 1, \ldots, 1)^T$ is an n-dimensional vector, the elements of which are all 1, and Λ and Π are $n \times n$ diagonal matrices defined as

$$\begin{aligned} \Lambda &= \operatorname{diag}\left[y_1 - \pi(\boldsymbol{x}_1), y_2 - \pi(\boldsymbol{x}_2), \ldots, y_n - \pi(\boldsymbol{x}_n)\right], \\ \Pi &= \operatorname{diag}\left[\pi(\boldsymbol{x}_1), \pi(\boldsymbol{x}_2), \ldots, \pi(\boldsymbol{x}_n)\right]. \end{aligned} \tag{6.39}$$

Starting from an initial value, we numerically obtain a solution using the following update formula:

$$\boldsymbol{\beta}^{\text{new}} = \boldsymbol{\beta}^{\text{old}} - \left[E\left\{ \frac{\partial^2 \ell_\lambda(\boldsymbol{\beta})}{\partial \boldsymbol{\beta} \partial \boldsymbol{\beta}^T} \right\} \right]^{-1} \frac{\partial \ell_\lambda(\boldsymbol{\beta}^{\text{old}})}{\partial \boldsymbol{\beta}}. \tag{6.40}$$

This update formula is referred to as *Fisher's scoring algorithm* [Nelder and Wedderburn (1972), Green and Silverman (1994)], and the $(r+1)^{st}$ estimator, $\hat{\boldsymbol{\beta}}^{(r+1)}$, is updated by

$$\hat{\boldsymbol{\beta}}^{(r+1)} = \left\{ X^T \Pi^{(r)} (I_n - \Pi^{(r)}) X + n\lambda K \right\}^{-1} X^T \Pi^{(r)} (I_n - \Pi^{(r)}) \boldsymbol{\xi}^{(r)}, \tag{6.41}$$

where $\boldsymbol{\xi}^{(r)} = X\boldsymbol{\beta}^{(r)} + \{\Pi^{(r)}(I_n - \Pi^{(r)})\}^{-1}(\boldsymbol{y} - \Pi^{(r)}\mathbf{1}_n)$ and $\Pi^{(r)}$ is an $n \times n$ diagonal matrix having $\pi(\boldsymbol{x}_\alpha)$ for the rth estimator $\hat{\boldsymbol{\beta}}^{(r)}$ in the αth diagonal element.

Thus, the statistical model is obtained by substituting the estimator $\hat{\boldsymbol{\beta}}$ determined by the numerical optimization procedure into the probability model of (6.33)

$$f(y_\alpha|\boldsymbol{x}_\alpha;\hat{\boldsymbol{\beta}}) = \hat{\pi}(\boldsymbol{x}_\alpha)^{y_\alpha}\{1 - \hat{\pi}(\boldsymbol{x}_\alpha)\}^{1-y_\alpha}, \tag{6.42}$$

where

$$\hat{\pi}(\boldsymbol{x}_\alpha) = \frac{\exp(\boldsymbol{x}_\alpha^T\hat{\boldsymbol{\beta}})}{1 + \exp(\boldsymbol{x}_\alpha^T\hat{\boldsymbol{\beta}})}. \tag{6.43}$$

The statistical model (6.42) estimated by maximizing the penalized log-likelihood function depends on the regularization parameter λ. The problem is how to select the optimal value of λ by using a suitable criterion. We overcome this problem by obtaining a criterion within the framework of an M-estimator.

Noting that the derivative of the penalized log-likelihood function with respect to $\boldsymbol{\beta}$ is

$$\frac{\partial \ell_\lambda(\boldsymbol{\beta})}{\partial \boldsymbol{\beta}} = \sum_{\alpha=1}^{n} \{y_\alpha - \pi(\boldsymbol{x}_\alpha)\}\,\boldsymbol{x}_\alpha - n\lambda K\boldsymbol{\beta}, \tag{6.44}$$

we see that the regularized estimator $\hat{\boldsymbol{\beta}}$ is given as the solution of the implicit equation

$$\frac{\partial \ell_\lambda(\boldsymbol{\beta})}{\partial \boldsymbol{\beta}} = \sum_{\alpha=1}^{n} \boldsymbol{\psi}_L(y_\alpha, \boldsymbol{\beta}) = \mathbf{0}, \tag{6.45}$$

where

$$\boldsymbol{\psi}_L(y_\alpha, \boldsymbol{\beta}) = \{y_\alpha - \pi(\boldsymbol{x}_\alpha)\}\,\boldsymbol{x}_\alpha - \lambda K\boldsymbol{\beta}. \tag{6.46}$$

By taking $\boldsymbol{\psi}_L(y_\alpha, \boldsymbol{\beta})$ as the $\boldsymbol{\psi}$-function in (5.143), the two matrices required in the calculation of the bias correction term can be obtained as

$$\begin{aligned} R(\boldsymbol{\psi}_L, \hat{G}) &= -\frac{1}{n}\sum_{\alpha=1}^{n} \left.\frac{\partial \boldsymbol{\psi}_L(y_\alpha, \boldsymbol{\beta})^T}{\partial \boldsymbol{\beta}}\right|_{\hat{\boldsymbol{\beta}}} \\ &= \frac{1}{n}\sum_{\alpha=1}^{n} \hat{\pi}(\boldsymbol{x}_\alpha)\{1 - \hat{\pi}(\boldsymbol{x}_\alpha)\}\boldsymbol{x}_\alpha\boldsymbol{x}_\alpha^T + \lambda K \\ &= \frac{1}{n}X^T\hat{\Pi}(I_n - \hat{\Pi})X + \lambda K, \end{aligned} \tag{6.47}$$

$$
\begin{aligned}
Q(\boldsymbol{\psi}_L, \hat{G}) &= \frac{1}{n} \sum_{\alpha=1}^{n} \boldsymbol{\psi}_L(y_\alpha, \boldsymbol{\beta}) \left. \frac{\partial \log f(y_\alpha | \boldsymbol{x}_\alpha; \boldsymbol{\beta})}{\partial \boldsymbol{\beta}^T} \right|_{\hat{\boldsymbol{\beta}}} \\
&= \frac{1}{n} \sum_{\alpha=1}^{n} \left[\{y_\alpha - \hat{\pi}(\boldsymbol{x}_\alpha)\} \boldsymbol{x}_\alpha - \lambda K \hat{\boldsymbol{\beta}} \right] \{y_\alpha - \hat{\pi}(\boldsymbol{x}_\alpha)\} \boldsymbol{x}_\alpha^T \\
&= \frac{1}{n} \left\{ X^T \hat{\Lambda}^2 X - \lambda K \hat{\boldsymbol{\beta}} \mathbf{1}_n^T \hat{\Lambda} X \right\},
\end{aligned}
\tag{6.48}
$$

where $\hat{\Lambda}$ and $\hat{\Pi}$ are $n \times n$ diagonal matrices defined by

$$
\begin{aligned}
\hat{\Lambda} &= \mathrm{diag}\left[y_1 - \hat{\pi}(\boldsymbol{x}_1), y_2 - \hat{\pi}(\boldsymbol{x}_2), \ldots, y_n - \hat{\pi}(\boldsymbol{x}_n) \right], \\
\hat{\Pi} &= \mathrm{diag}\left[\hat{\pi}(\boldsymbol{x}_1), \hat{\pi}(\boldsymbol{x}_2), \ldots, \hat{\pi}(\boldsymbol{x}_n) \right].
\end{aligned}
\tag{6.49}
$$

We then have the following result:

Information criterion for a linear logistic model estimated by regularization. Let $f(y|\boldsymbol{x}; \boldsymbol{\beta})$ be a linear logistic model given in (6.33) and (6.34). Then an information criterion for evaluating the model $f(y|\boldsymbol{x}; \hat{\boldsymbol{\beta}})$ in (6.42) estimated by regularization is given by the following:

$$
\begin{aligned}
\mathrm{GIC}_L = &-2 \sum_{\alpha=1}^{n} \left[y_\alpha \log \hat{\pi}(\boldsymbol{x}_\alpha) + (1 - y_\alpha) \log \{1 - \hat{\pi}(\boldsymbol{x}_\alpha)\} \right] \\
&+ 2\mathrm{tr} \left\{ R(\boldsymbol{\psi}_L, \hat{G})^{-1} Q(\boldsymbol{\psi}_L, \hat{G}) \right\},
\end{aligned}
\tag{6.50}
$$

where $R(\boldsymbol{\psi}_L, \hat{G})$ and $Q(\boldsymbol{\psi}_L, \hat{G})$ are $(p+1) \times (p+1)$ matrices given, respectively, by

$$
\begin{aligned}
R(\boldsymbol{\psi}_L, \hat{G}) &= \frac{1}{n} X^T \hat{\Pi} (I_n - \hat{\Pi}) X + \lambda K, \\
Q(\boldsymbol{\psi}_L, \hat{G}) &= \frac{1}{n} \left\{ X^T \hat{\Lambda}^2 X - \lambda K \hat{\boldsymbol{\beta}} \mathbf{1}_n^T \hat{\Lambda} X \right\},
\end{aligned}
\tag{6.51}
$$

with $\hat{\Lambda}$ and $\hat{\Pi}$ given by (6.49).

We choose the value of the regularization parameter λ that minimizes the information criterion GIC_L for various statistical models determined in correspondence to the values of λ.

6.3.2 Nonlinear Logistic Regression Models

We now extend the linear logistic model developed in the previous section into a model having a more complex nonlinear structure using a basis expansion method.

Let $y_1, \ldots, y_n$ be an independent sequence of binary random variables taking values of 0 and 1 with conditional probabilities

$$\Pr(Y = 1|\boldsymbol{x}_\alpha) = \pi(\boldsymbol{x}_\alpha) \quad \text{and} \quad \Pr(Y = 0|\boldsymbol{x}_\alpha) = 1 - \pi(\boldsymbol{x}_\alpha), \tag{6.52}$$

where $\boldsymbol{x}_\alpha$ are vectors of p explanatory variables. Using the basis expansions as a device to approximate the mean structure, we consider the nonlinear logistic model

$$\log \frac{\pi(\boldsymbol{x}_\alpha)}{1 - \pi(\boldsymbol{x}_\alpha)} = w_0 + \sum_{i=1}^{m} w_i b_i(\boldsymbol{x}_\alpha), \tag{6.53}$$

where $b_i(\boldsymbol{x}_\alpha)$ is a basis function. The conditional probability $\pi(\boldsymbol{x}_\alpha)$ can be rewritten as

$$\pi(\boldsymbol{x}_\alpha) = \frac{\exp\left\{\boldsymbol{w}^T \boldsymbol{b}(\boldsymbol{x}_\alpha)\right\}}{1 + \exp\left\{\boldsymbol{w}^T \boldsymbol{b}(\boldsymbol{x}_\alpha)\right\}}, \tag{6.54}$$

where $\boldsymbol{w} = (w_0, \ldots, w_m)^T$ and $\boldsymbol{b}(\boldsymbol{x}_\alpha) = (1, b_1(\boldsymbol{x}_\alpha), \ldots, b_m(\boldsymbol{x}_\alpha))^T$. The nonlinear logistic model can be expressed as the probability model

$$f(y_\alpha|\boldsymbol{x}_\alpha; \boldsymbol{w}) = \pi(\boldsymbol{x}_\alpha)^{y_\alpha} \left\{1 - \pi(\boldsymbol{x}_\alpha)\right\}^{1-y_\alpha}, \quad y_\alpha = 0, 1. \tag{6.55}$$

Hence, the log-likelihood function for y_α in terms of $\boldsymbol{w} = (w_0, \ldots, w_m)^T$ is

$$\begin{aligned} \ell(\boldsymbol{w}) &= \sum_{\alpha=1}^{n} \left\{ y_\alpha \log \pi(\boldsymbol{x}_\alpha) + (1 - y_\alpha) \log(1 - \pi(\boldsymbol{x}_\alpha)) \right\} \\ &= -\sum_{\alpha=1}^{n} \left[\log\left\{1 + \exp(\boldsymbol{w}^T \boldsymbol{b}(\boldsymbol{x}_\alpha))\right\} - y_\alpha \boldsymbol{w}^T \boldsymbol{b}(\boldsymbol{x}_\alpha) \right]. \end{aligned} \tag{6.56}$$

The unknown parameter vector $\boldsymbol{w}$ is estimated by maximizing the penalized log-likelihood

$$\ell_\lambda(\boldsymbol{w}) = \ell(\boldsymbol{w}) - \frac{n\lambda}{2} \boldsymbol{w}^T K \boldsymbol{w}, \tag{6.57}$$

where the penalty term is given by (5.135). The optimization process with respect to the unknown parameter vector $\boldsymbol{w}$ is nonlinear and the equation does not have an explicit solution. The solution $\boldsymbol{w} = \hat{\boldsymbol{w}}_\lambda$, which maximizes $\ell_\lambda(\boldsymbol{w})$ with respect to a given λ, is estimated using the numerical optimization method described in Subsection 6.3.1. In the estimation process, the following substitutions are made in (6.37) and (6.38):

$$\boldsymbol{\beta} \Rightarrow \boldsymbol{w}, \qquad X \Rightarrow B,$$

$$\pi(\boldsymbol{x}_\alpha) = \frac{\exp(\boldsymbol{x}_\alpha^T \boldsymbol{\beta})}{1 + \exp(\boldsymbol{x}_\alpha^T \boldsymbol{\beta})} \Rightarrow \pi(\boldsymbol{x}_\alpha) = \frac{\exp\left\{\boldsymbol{w}^T \boldsymbol{b}(\boldsymbol{x}_\alpha)\right\}}{1 + \exp\left\{\boldsymbol{w}^T \boldsymbol{b}(\boldsymbol{x}_\alpha)\right\}}, \tag{6.58}$$

where B is an $n \times (m+1)$ basis function matrix $B = (\boldsymbol{b}(\boldsymbol{x}_1), \boldsymbol{b}(\boldsymbol{x}_2), \ldots, \boldsymbol{b}(\boldsymbol{x}_n))^T$.

Substituting the estimator $\hat{\boldsymbol{w}}_\lambda$ obtained by the numerical optimization method into the probability model (6.55), we obtain the following statistical model:

$$f(y_\alpha|\boldsymbol{x}_\alpha;\hat{\boldsymbol{w}}_\lambda) = \hat{\pi}(\boldsymbol{x}_\alpha)^{y_\alpha}\{1-\hat{\pi}(\boldsymbol{x}_\alpha)\}^{1-y_\alpha}, \tag{6.59}$$

where

$$\hat{\pi}(\boldsymbol{x}_\alpha) = \frac{\exp\left\{\hat{\boldsymbol{w}}_\lambda^T\boldsymbol{b}(\boldsymbol{x}_\alpha)\right\}}{1+\exp\left\{\hat{\boldsymbol{w}}_\lambda^T\boldsymbol{b}(\boldsymbol{x}_\alpha)\right\}}. \tag{6.60}$$

An information criterion for the statistical model estimated by the regularization method can easily be determined within the framework of M-estimation. Specifically, it can be seen from (6.56) and (6.57) that the estimator $\hat{\boldsymbol{w}}_\lambda$ can be given as the solution of the implicit equation

$$\frac{\partial \ell_\lambda(\boldsymbol{w})}{\partial \boldsymbol{w}} = \sum_{\alpha=1}^n \psi_{\mathrm{LB}}(y_\alpha,\boldsymbol{w}) = \boldsymbol{0}, \tag{6.61}$$

where

$$\psi_{\mathrm{LB}}(y_\alpha,\boldsymbol{w}) = \{y_\alpha - \pi(\boldsymbol{x}_\alpha)\}\,\boldsymbol{b}(\boldsymbol{x}_\alpha) - \lambda K\boldsymbol{w}. \tag{6.62}$$

Taking $\psi_{LB}(y_\alpha,\boldsymbol{w})$ as the ψ-function in (5.143) gives the matrices required for calculating the bias correction term in the form

$$\begin{aligned}
R(\psi_{LB},\hat{G}) &= -\frac{1}{n}\sum_{\alpha=1}^n \left.\frac{\partial\psi_{LB}(y_\alpha,\boldsymbol{w})^T}{\partial\boldsymbol{w}}\right|_{\hat{\boldsymbol{w}}_\lambda} \\
&= \frac{1}{n}\sum_{\alpha=1}^n \hat{\pi}(\boldsymbol{x}_\alpha)\{1-\hat{\pi}(\boldsymbol{x}_\alpha)\}\boldsymbol{b}(\boldsymbol{x}_\alpha)\boldsymbol{b}(\boldsymbol{x}_\alpha)^T + \lambda K \\
&= \frac{1}{n}B^T\hat{\Pi}(I_n-\hat{\Pi})B + \lambda K,
\end{aligned} \tag{6.63}$$

$$\begin{aligned}
Q(\psi_{LB},\hat{G}) &= \frac{1}{n}\sum_{\alpha=1}^n \psi_{LB}(y_\alpha,\boldsymbol{w})\left.\frac{\partial\log f(y_\alpha|\boldsymbol{x}_\alpha;\boldsymbol{w})}{\partial\boldsymbol{w}^T}\right|_{\hat{\boldsymbol{w}}_\lambda} \\
&= \frac{1}{n}\sum_{\alpha=1}^n \left[\{y_\alpha-\hat{\pi}(\boldsymbol{x}_\alpha)\}\boldsymbol{b}(\boldsymbol{x}_\alpha) - \lambda K\hat{\boldsymbol{w}}_\lambda\right]\{y_\alpha-\hat{\pi}(\boldsymbol{x}_\alpha)\}\boldsymbol{b}(\boldsymbol{x}_\alpha)^T \\
&= \frac{1}{n}\left\{B^T\hat{\Lambda}^2B - \lambda K\hat{\boldsymbol{w}}_\lambda\boldsymbol{1}_n^T\hat{\Lambda}B\right\},
\end{aligned} \tag{6.64}$$

where $\hat{\Lambda}$ and $\hat{\Pi}$ are $n\times n$ diagonal matrices defined by

$$\begin{aligned}
\hat{\Lambda} &= \mathrm{diag}\left[y_1-\hat{\pi}(\boldsymbol{x}_1), y_2-\hat{\pi}(\boldsymbol{x}_2),\ldots,y_n-\hat{\pi}(\boldsymbol{x}_n)\right], \\
\hat{\Pi} &= \mathrm{diag}\left[\hat{\pi}(\boldsymbol{x}_1),\hat{\pi}(\boldsymbol{x}_2),\ldots,\hat{\pi}(\boldsymbol{x}_n)\right],
\end{aligned} \tag{6.65}$$

with $\hat{\pi}(\boldsymbol{x}_\alpha)$ given by (6.60). Then we have the following result:

Information criterion for a nonlinear logistic model by regularized basis expansions. Let $f(y_\alpha|\boldsymbol{x}_\alpha;\boldsymbol{w})$ be the nonlinear logistic model in (6.55). Then an information criterion for the statistical model $f(y_\alpha|\boldsymbol{x}_\alpha;\hat{\boldsymbol{w}}_\lambda)$ in (6.59) constructed by the regularized basis expansion is given by

$$\begin{aligned}\mathrm{GIC_{LB}} &= -2\sum_{\alpha=1}^{n}\left[y_\alpha \log\hat{\pi}(\boldsymbol{x}_\alpha) + (1-y_\alpha)\log\{1-\hat{\pi}(\boldsymbol{x}_\alpha)\}\right] \\ &\quad + 2\mathrm{tr}\left\{R(\boldsymbol{\psi}_{\mathrm{LB}},\hat{G})^{-1}Q(\boldsymbol{\psi}_{\mathrm{LB}},\hat{G})\right\},\end{aligned} \tag{6.66}$$

where

$$\begin{aligned}R(\boldsymbol{\psi}_{\mathrm{LB}},\hat{G}) &= \frac{1}{n}B^T\hat{\Pi}(I_n-\hat{\Pi})B+\lambda K, \\ Q(\boldsymbol{\psi}_{\mathrm{LB}},\hat{G}) &= \frac{1}{n}\left\{B^T\hat{\Lambda}^2B-\lambda K\hat{\boldsymbol{w}}_\lambda\boldsymbol{1}_n^T\hat{\Lambda}B\right\},\end{aligned} \tag{6.67}$$

with $n\times n$ diagonal matrices $\hat{\Lambda}$ and $\hat{\Pi}$ defined by (6.65).

Out of the statistical models generated by the various values of the smoothing parameter λ, the optimal value is selected by minimizing the information criterion $\mathrm{GIC_{LB}}$.

Example 4 (Probability of occurrence of kyphosis) Figure 6.4 shows a plot of data for 83 patients who received laminectomy, in terms of their age (x, in months) at the time of operation, and $Y=1$ if the patient developed kyphosis and $Y=0$ otherwise [Hastie and Tibshirani (1990, p. 301)]. The objective here is to predict a decrease in the probability of the onset of kyphosis, $\Pr(Y=1|x)=\pi(x)$, as a function of the time of laminectomy.

If the probability of onset of kyphosis was monotonic with respect to the age of the patients in months, it would suffice to assume the logistic model:

$$\log\frac{\pi(x_\alpha)}{1-\pi(x_\alpha)} = \beta_0+\beta_1 x_\alpha, \qquad \alpha=1,2,\ldots,83.$$

However, as the figure indicates, the probability of onset is not necessarily monotone with respect to age expressed in months.

Therefore, let us consider fitting the following logistic model based on a B-spline:

$$\log\left\{\frac{\pi(x_\alpha)}{1-\pi(x_\alpha)}\right\} = \sum_{i=1}^{m} w_i b_i(x_\alpha), \qquad \alpha=1,2,\ldots,83. \tag{6.68}$$

We estimated the parameters $\boldsymbol{w}=(w_1,w_2,\ldots,w_m)^T$ using the regularization method with a difference matrix of degree 2 given by (5.138) as a

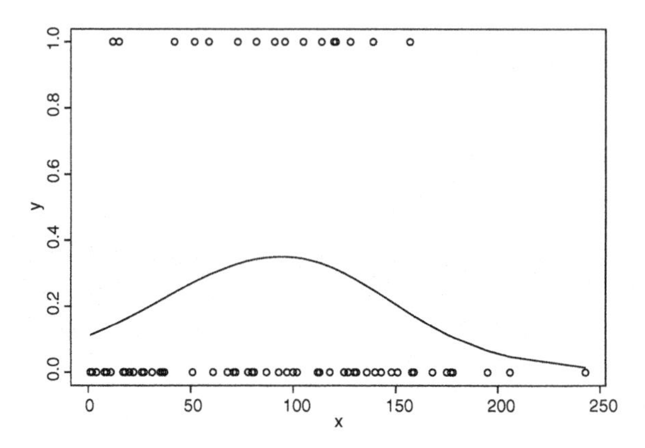

Fig. 6.4. Probability of onset of kyphosis.

regularization term. By applying the information criterion $\mathrm{GIC_{LB}}$ given by (6.66), we determined the optimum number of basis functions m to be 10, and the value of the smoothing parameter is $\lambda = 0.0159$. The corresponding logistic curve is given by

$$y = \frac{\exp\left(\sum_{i=1}^{m} \hat{w}_i b_i(x)\right)}{1 + \exp\left(\sum_{i=1}^{m} \hat{w}_i b_i(x)\right)}. \tag{6.69}$$

The estimates of the coefficients of the basis functions were $\hat{\boldsymbol{w}} = (-2.48, -1.59, -0.92, -0.63, -0.84, -1.60, -2.65, -3.76, -4.88, -6.00)^T$. The curve in the figure represents the estimated curve. It can be seen from the estimated logistic curve that while the rate of onset increases with the patient's age in months at the time of surgery, a peak occurs at approximately 100 months, and the rate of onset begins to decrease thereafter.

6.4 Logistic Discriminant Analysis

Classification or discrimination techniques are some of the most widely used statistical tools in various fields of natural and social sciences. The primary aim in discriminant analysis is to assign an individual to one of two or more groups on the basis of measurements on feature variables. In recent years, several techniques have been proposed for analyzing multivariate observations with complex structure [see, for example, Hastie et al. (2001) and McLachlan (2004)].

This section introduces linear and nonlinear discriminant analyses using basis expansions with the help of regularization. We consider the two-group discrimination. It is designed to construct a decision rule based on a set of training data, each of which is assigned to one of two groups.

6.4.1 Linear Logistic Discrimination

Suppose we have n independent observations $\{(\boldsymbol{x}_\alpha, g_\alpha);\ \alpha = 1, 2, \ldots, n\}$, where $\boldsymbol{x}_\alpha = (x_{\alpha 1}, x_{\alpha 2}, \ldots, x_{\alpha p})^T$ are the p-dimensional observed feature vectors and g_α are indicators of the group membership. A Bayes rule of allocation is to assign $\boldsymbol{x}_\alpha$ to group G_k $(k = 1, 2)$ with maximum posterior probability $\Pr(g = k|\boldsymbol{x}_\alpha)$. We first consider the log-odds of the posterior probabilities given by the linear combination of p feature variables

$$\log \frac{\Pr(g = 1|\boldsymbol{x}_\alpha)}{\Pr(g = 2|\boldsymbol{x}_\alpha)} = w_0 + \sum_{i=1}^{p} w_i x_{\alpha i}. \tag{6.70}$$

Denote the posterior probability $\Pr(g = 1|\boldsymbol{x}_\alpha) = \pi(\boldsymbol{x}_\alpha)$, so that $\Pr(g = 2|\boldsymbol{x}_\alpha) = 1 - \pi(\boldsymbol{x}_\alpha)$. The log-odds model (6.70) can then be written as

$$\log \frac{\pi(\boldsymbol{x}_\alpha)}{1 - \pi(\boldsymbol{x}_\alpha)} = w_0 + \sum_{i=1}^{p} w_i x_{\alpha i}. \tag{6.71}$$

We define the binary variable y_α coded as either 0 or 1 to indicate the group membership of the αth observed feature vector $\boldsymbol{x}_\alpha$, that is,

$$y_\alpha = 1 \quad \text{if} \quad g_\alpha = 1 \quad \text{and} \quad y_\alpha = 0 \quad \text{if} \quad g_\alpha = 2. \tag{6.72}$$

The group-indicator variables $y_1, y_2, \ldots, y_n$ are distributed independently according to the Bernoulli distribution

$$f(y_\alpha|\boldsymbol{x}_\alpha; \boldsymbol{w}) = \pi(\boldsymbol{x}_\alpha)^{y_\alpha} \{1 - \pi(\boldsymbol{x}_\alpha)\}^{1-y_\alpha}, \quad y_\alpha = 0, 1, \tag{6.73}$$

conditional on $\boldsymbol{x}_\alpha$, where

$$\pi(\boldsymbol{x}_\alpha) = \frac{\exp\left(w_0 + \sum_{i=1}^{p} w_i x_{\alpha i}\right)}{1 + \exp\left(w_0 + \sum_{i=1}^{p} w_i x_{\alpha i}\right)}. \tag{6.74}$$

By maximizing the log-likelihood function

$$l(\boldsymbol{w}) = \sum_{\alpha=1}^{n} \left[y_\alpha \log \pi(\boldsymbol{x}_\alpha) + (1 - y_\alpha) \log\{1 - \pi(\boldsymbol{x}_\alpha)\} \right], \tag{6.75}$$

we obtain the maximum likelihood estimates of the unknown parameters $\{w_0, w_1, w_2, \ldots, w_p\}$.

Often the maximum likelihood method yields unstable estimates of weight parameters and so leads to large errors in predicting future observations. In such cases, the regularization method is used for parameter estimation in logistic modeling. We obtain the solution by employing a nonlinear optimization scheme discussed in Subsection 6.3.1, and the value of a smoothing parameter is chosen as the minimizer of GIC_L in (6.50).

The estimated posterior probabilities of group membership for the future observation $\boldsymbol{z} = (z_1, z_2, \ldots, z_p)^T$ are given by

$$\begin{aligned} \Pr(g=1|\boldsymbol{z}) = \hat{\pi}(\boldsymbol{z}) &= \frac{\exp\left(\hat{w}_0 + \sum_{i=1}^{p} \hat{w}_i z_i\right)}{1 + \exp\left(\hat{w}_0 + \sum_{i=1}^{p} \hat{w}_i z_i\right)}, \\ \Pr(g=2|\boldsymbol{z}) = 1 - \hat{\pi}(\boldsymbol{z}) &= \frac{1}{1 + \exp\left(\hat{w}_0 + \sum_{i=1}^{p} \hat{w}_i z_i\right)}, \end{aligned} \tag{6.76}$$

where $\hat{\pi}(\boldsymbol{z})$ is the estimated conditional probability. Allocation is then carried out by evaluating the posterior probabilities, and the future observation $\boldsymbol{z}$ is assigned according to the following decision rule:

$$\begin{aligned} &\text{assign } \boldsymbol{z} \text{ to } G_1 \quad \text{if} \quad \Pr(g=1|\boldsymbol{z}) \geq \Pr(g=2|\boldsymbol{z}), \\ &\text{assign } \boldsymbol{z} \text{ to } G_2 \quad \text{if} \quad \Pr(g=1|\boldsymbol{z}) < \Pr(g=2|\boldsymbol{z}). \end{aligned} \tag{6.77}$$

By taking the logit transformation

$$\log \frac{\hat{\pi}(\boldsymbol{z})}{1 - \hat{\pi}(\boldsymbol{z})} = \hat{w}_0 + \sum_{i=1}^{p} \hat{w}_i z_i, \tag{6.78}$$

we see that the decision rule is equivalent to the rule

$$\begin{aligned} &\text{assign } \boldsymbol{z} \text{ to } G_1 \quad \text{if} \quad \hat{w}_0 + \sum_{i=1}^{p} \hat{w}_i z_i \geq 0, \\ &\text{assign } \boldsymbol{z} \text{ to } G_2 \quad \text{if} \quad \hat{w}_0 + \sum_{i=1}^{p} \hat{w}_i z_i < 0. \end{aligned} \tag{6.79}$$

In general, the function defined by a linear combination of the feature variables is called a *linear discriminant function*. In practice, Fisher's linear discriminant analysis is a commonly used technique for data classification. This approach involves maximizing the ratio of the between-groups sum of square to the within-groups sum of square. In cases where a linear discriminant rule is not suitable for allocating a randomly selected future observation, we may use a nonlinear discriminant procedure. The linear logistic discriminant analysis can be extended naturally for use in nonlinear discrimination via basis expansions, which will be described in the next subsection.

6.4.2 Nonlinear Logistic Discrimination

We assume that the log-odds of the posterior probabilities are given by the linear combination of basis functions as follows:

$$\log \frac{\Pr(g=1|\boldsymbol{x}_\alpha)}{\Pr(g=2|\boldsymbol{x}_\alpha)} = \sum_{i=1}^{m} w_i b_i(\boldsymbol{x}_\alpha) \tag{6.80}$$

or, writing $\Pr(g=1|\boldsymbol{x}_\alpha) = \pi(\boldsymbol{x}_\alpha)$,

$$\log \frac{\pi(\boldsymbol{x}_\alpha)}{1-\pi(\boldsymbol{x}_\alpha)} = \sum_{i=1}^{m} w_i b_i(\boldsymbol{x}_\alpha). \tag{6.81}$$

Since the group indicator variables $y_1, y_2, \ldots, y_n$ defined in (6.72) are distributed independently according to the Bernoulli distribution, the nonlinear logistic discrimination model can be expressed as

$$\begin{aligned} f(y_\alpha|\boldsymbol{x}_\alpha;\boldsymbol{w}) &= \pi(\boldsymbol{x}_\alpha)^{y_\alpha}\{1-\pi(\boldsymbol{x}_\alpha)\}^{1-y_\alpha} \\ &= \left\{\exp\left(\sum_{i=1}^{m} w_i b_i(\boldsymbol{x}_\alpha)\right)\right\}^{y_\alpha}\left\{1+\exp\left(\sum_{i=1}^{m} w_i b_i(\boldsymbol{x}_\alpha)\right)\right\}^{-1}, \\ &\qquad\qquad y_\alpha = 0, 1, \end{aligned} \tag{6.82}$$

conditional on $\boldsymbol{x}_\alpha$. The statistical model is obtained by replacing the unknown parameters with their estimates, and we have

$$f(y_\alpha|\boldsymbol{x}_\alpha;\hat{\boldsymbol{w}}) = \hat{\pi}(\boldsymbol{x}_\alpha)^{y_\alpha}\{1-\hat{\pi}(\boldsymbol{x}_\alpha)\}^{1-y_\alpha}, \tag{6.83}$$

where $\hat{\pi}(\boldsymbol{x}_\alpha)$ is the estimated conditional probability given by

$$\hat{\pi}(\boldsymbol{x}_\alpha) = \frac{\exp\left\{\sum_{i=1}^{m} \hat{w}_i b_i(\boldsymbol{x}_\alpha)\right\}}{1+\exp\left\{\sum_{i=1}^{m} \hat{w}_i b_i(\boldsymbol{x}_\alpha)\right\}}. \tag{6.84}$$

The model estimated by the maximum likelihood method can be evaluated by the AIC, and the number of basis functions is determined by minimizing the value of the AIC. If the model is constructed by regularization, then the number of basis functions and the value of a smoothing parameter are chosen by evaluating the estimated model by GIC_{LB} in (6.66). The future observation $\boldsymbol{z}$ is assigned by the nonlinear discriminant function as follows:

$$\begin{aligned}
&\text{assign } \boldsymbol{z} \text{ to } G_1 \quad \text{if} \quad \sum_{i=1}^{m} \hat{w}_i b_i(\boldsymbol{x}) \geq 0, \\
&\text{assign } \boldsymbol{z} \text{ to } G_2 \quad \text{if} \quad \sum_{i=1}^{m} \hat{w}_i b_i(\boldsymbol{x}) < 0.
\end{aligned} \tag{6.85}$$

Example 5 (Synthetic data) We illustrate the nonlinear logistic discriminant analysis using synthetic data taken from Ripley (1994). The data are generated from a mixture of two bivariate normal distributions $N_2(\boldsymbol{\mu}, \Sigma)$:

$$\begin{aligned}
G_1: \quad & g_1(\boldsymbol{x}) = (1-\varepsilon)N_2(\boldsymbol{\mu}_1^{(1)}, \sigma^2 I_2) + \varepsilon N_2(\boldsymbol{\mu}_2^{(1)}, \sigma^2 I_2), \\
G_2: \quad & g_2(\boldsymbol{x}) = (1-\varepsilon)N_2(\boldsymbol{\mu}_1^{(2)}, \sigma^2 I_2) + \varepsilon N_2(\boldsymbol{\mu}_2^{(2)}, \sigma^2 I_2),
\end{aligned} \tag{6.86}$$

where $\boldsymbol{\mu}_1^{(1)} = (-0.3, 0.7)^T$, $\boldsymbol{\mu}_2^{(1)} = (0.4, 0.7)^T$ and $\boldsymbol{\mu}_1^{(2)} = (-0.7, 0.3)^T$, $\boldsymbol{\mu}_2^{(2)} = (0.3, 0.3)^T$, with common variance $\sigma^2 = 0.03$.

The decision boundaries in Figure 6.5 were constructed using the model based on Gaussian basis functions:

$$\log \frac{\pi(\boldsymbol{x}_\alpha)}{1-\pi(\boldsymbol{x}_\alpha)} = w_0 + \sum_{i=1}^{15} w_i \phi_i(\boldsymbol{x}_\alpha), \tag{6.87}$$

with

$$\phi_i(\boldsymbol{x}) = \exp\left(-\frac{||\boldsymbol{x} - \hat{\boldsymbol{\mu}}_i||^2}{2 \times 15\hat{h}_i^2}\right), \quad i = 1, 2, \ldots, 15, \tag{6.88}$$

where $\hat{\boldsymbol{\mu}}_i$ is the two-dimensional vector that determines the location of the basis function and $15\hat{h}_i^2$ is the adjusted scale parameter (see Subsection 6.2.2 for the Gaussian basis functions). The model was estimated using the regularization method. Figure 6.5 shows the decision boundaries for various values of the smoothing parameter λ. The optimum value of λ was chosen by evaluating the estimated model by GIC_{LB} in (6.66), and the corresponding decision boundary is given in Figure 6.5. We see that the nonlinearity of the decision boundary can be controlled by the smoothing parameter; the decision boundary approaches a linear function for larger values of λ.

6.5 Penalized Least Squares Methods

Consider the regression model expressed as a linear combination of a prescribed set of m basis functions as follows:

$$y_\alpha = \sum_{i=1}^{m} w_i b_i(\boldsymbol{x}_\alpha) + \varepsilon_\alpha, \quad \alpha = 1, 2, \ldots, n, \tag{6.89}$$

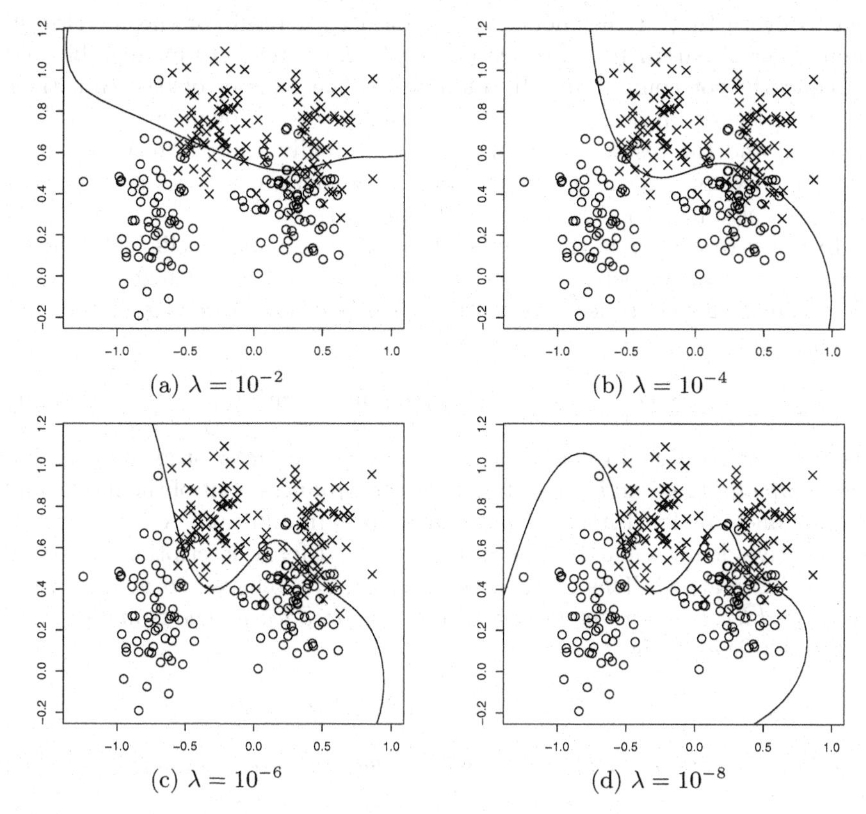

Fig. 6.5. The role of a smoothing parameter in nonlinear logistic discriminant analysis.

where y_α are random response variables and $\boldsymbol{x}_\alpha$ are p-dimensional vectors of explanatory variables. It is assumed that the noise ε_α are uncorrelated, $E[\varepsilon_\alpha] = 0$, and $E[\varepsilon_\alpha^2] = \sigma^2$. The least squares estimates are those that minimize the sum of squares of noise ε_α

$$
\begin{aligned}
S(\boldsymbol{w}) &= \sum_{\alpha=1}^{n} \left\{ y_\alpha - \sum_{i=1}^{m} w_i b_i(\boldsymbol{x}_\alpha) \right\}^2 \\
&= \sum_{\alpha=1}^{n} \left\{ y_\alpha - \boldsymbol{w}^T \boldsymbol{b}(\boldsymbol{x}_\alpha) \right\}^2 \qquad (6.90) \\
&= (\boldsymbol{y} - B\boldsymbol{w})^T (\boldsymbol{y} - B\boldsymbol{w}),
\end{aligned}
$$

where $B = (\boldsymbol{b}(\boldsymbol{x}_1), \boldsymbol{b}(\boldsymbol{x}_2), \ldots, \boldsymbol{b}(\boldsymbol{x}_n))^T$, and are given by $\hat{\boldsymbol{w}} = (B^T B)^{-1} B^T \boldsymbol{y}$.

One conceivable approach for analyzing phenomena having a complex nonlinear structure is to capture the structure by increasing the number of basis functions employed. However, increasing the number of basis functions can lead to overfitting of the model to the data as a result of the increase in the number of parameters. In such cases, $(B^TB)^{-1}$ tends to be unstable and is frequently not computable. In addition, as the number of basis functions increases, the estimated curve or the surface passes through space closer to the data, and the residual sum of squares gradually approaches 0. The fact that the curve passes through space close to the data indicates that the curve undergoes significant local variation (fluctuation).

In order to overcome these difficulties, the regression coefficients are estimated by adding a penalty term (regularization term) designed to increase with decreasing smoothness when fitting. The solution for $\boldsymbol{w}$ is given by minimizing the penalized sum of squares

$$S_\gamma(\boldsymbol{w}) = (\boldsymbol{y} - B\boldsymbol{w})^T(\boldsymbol{y} - B\boldsymbol{w}) + \gamma \boldsymbol{w}^T K \boldsymbol{w}, \tag{6.91}$$

where $\gamma > 0$ is referred to either as a *smoothing parameter* or as a *regularization parameter* that can be used to adjust the goodness of fit of the model and the roughness or local fluctuation of the curve. In addition, K is an $m \times m$ nonnegative definite matrix (see Subsection 5.2.4 for a description of how to set up this matrix). This method of estimation is referred to as either the *regularized least squares method* or as the *penalized least squares method*, and its solution is given by

$$\hat{\boldsymbol{w}} = (B^TB + \gamma K)^{-1}B^T\boldsymbol{y}. \tag{6.92}$$

By taking $K = I_m$ in (6.92), we have the *ridge regression estimate* of $\boldsymbol{w}$ given by

$$\hat{\boldsymbol{w}} = (B^TB + \gamma I_m)^{-1}B^T\boldsymbol{y}. \tag{6.93}$$

We also notice that the penalized log-likelihood function for the regression model with Gaussian noise in (6.7) can be rewritten as

$$\begin{aligned} \ell_\lambda(\boldsymbol{\theta}) &= -\frac{n}{2}\log(2\pi\sigma^2) - \frac{1}{2\sigma^2}(\boldsymbol{y} - B\boldsymbol{w})^T(\boldsymbol{y} - B\boldsymbol{w}) - \frac{n\lambda}{2}\boldsymbol{w}^T K \boldsymbol{w} \\ &= -\frac{n}{2}\log(2\pi\sigma^2) - \frac{1}{2\sigma^2}\{(\boldsymbol{y} - B\boldsymbol{w})^T(\boldsymbol{y} - B\boldsymbol{w}) + n\lambda\sigma^2\boldsymbol{w}^T K\boldsymbol{w}\}. \end{aligned} \tag{6.94}$$

Therefore, by setting $n\lambda\sigma^2 = \gamma$, we see that maximizing the regularized log-likelihood function is equivalent to minimizing the penalized sum of squares $S_\gamma(\boldsymbol{w})$ in (6.91).

6.6 Effective Number of Parameters

In Section 6.1, we discussed Gaussian regression modeling based on the basis expansion

$$y_\alpha = \sum_{i=1}^{m} w_i b_i(\boldsymbol{x}_\alpha) + \varepsilon_\alpha = \boldsymbol{w}^T \boldsymbol{b}(\boldsymbol{x}_\alpha) + \varepsilon_\alpha, \quad \alpha = 1, \ldots, n, \tag{6.95}$$

where it is assumed that ε_α, $\alpha = 1, \ldots, n$, are independently distributed according to the normal distribution $N(0, \sigma^2)$. The maximum likelihood estimates of the unknown parameters $\boldsymbol{w} = (w_1, w_2, \ldots, w_m)^T$ and σ^2 are, respectively,

$$\hat{\boldsymbol{w}} = (B^T B)^{-1} B^T \boldsymbol{y} \quad \text{and} \quad \hat{\sigma}^2 = \frac{1}{n}(\boldsymbol{y} - \hat{\boldsymbol{y}})^T (\boldsymbol{y} - \hat{\boldsymbol{y}}), \tag{6.96}$$

where $\boldsymbol{y} = (y_1, y_2, \ldots, y_n)^T$, $B = (\boldsymbol{b}(\boldsymbol{x}_1), \boldsymbol{b}(\boldsymbol{x}_2), \ldots, \boldsymbol{b}(\boldsymbol{x}_n))^T$ ($n \times m$ matrix), and $\hat{\boldsymbol{y}}$ is an n-dimensional vector of predicted values given by

$$\hat{\boldsymbol{y}} = B\hat{\boldsymbol{w}} = B(B^T B)^{-1} B^T \boldsymbol{y}. \tag{6.97}$$

In this case, the AIC is given by

$$\text{AIC} = n(\log 2\pi + 1) + n \log \hat{\sigma}^2 + 2(m + 1). \tag{6.98}$$

The number of parameters or the degrees of freedom for the model is $m + 1$, which is equal to the number m of basis functions plus 1, corresponding to the error variance σ^2. In particular, the model in (6.95) becomes complex as the number of basis functions increases. The number of parameters related to the basis functions gives an indication of the model complexity. For example, the number of explanatory variables measures complexity for linear regression models, and the order of the polynomial measures the complexity for polynomial models. By contrast, if a model is estimated using the regularization method, then the model's complexity is also controlled by a smoothing parameter in addition to the number of basis functions involved. Hence, the number of parameters is no longer adequate for characterizing the complexity of the model. In view of this problem, Hastie and Tibshirani (1990) defined the complexity of models controlled with smoothing parameters as follows [see also Wahba (1990) and Moody (1992)]:

First, note that $\hat{\boldsymbol{y}}$ of (6.97) is the projection of $\boldsymbol{y}$ onto the m-dimensional space that is spanned by the m column vectors of the $n \times m$ matrix B,

$$\hat{\boldsymbol{y}} = H\boldsymbol{y}, \qquad H = B(B^T B)^{-1} B^T, \tag{6.99}$$

where H is the projection matrix. Next note that

$$\text{number of free parameters} = \text{tr}(H) = \text{tr}\left\{B(B^T B)^{-1} B^T\right\} = m. \tag{6.100}$$

On the other hand, the predicted values estimated by regularization are, from (6.9),

$$\hat{\boldsymbol{y}} = H(\lambda, m)\boldsymbol{y}; \qquad H(\lambda, m) = B(B^T B + n\lambda\hat{\sigma}^2 K)^{-1} B^T. \tag{6.101}$$

Hastie and Tibshirani (1990) defined the complexity of models controlled with smoothing parameters as

$$\mathrm{enp} = \mathrm{tr}\left\{H(\lambda, m)\right\} = \mathrm{tr}\left\{B(B^T B + n\lambda\hat{\sigma}^2 K)^{-1} B^T\right\} \tag{6.102}$$

and called it the *effective number of parameters*. Consequently, the information criterion for the Gaussian nonlinear regression model (6.95) estimated by regularization is given as

$$\begin{aligned} \mathrm{AIC}_M &= n(\log 2\pi + 1) + n \log(\hat{\sigma}^2) \\ &\quad + 2\left[\mathrm{tr}\left\{B(B^T B + n\lambda\hat{\sigma}^2 K)^{-1} B^T\right\} + 1\right], \end{aligned} \tag{6.103}$$

in which the number m of basis functions of the AIC in (6.98) is formally replaced with the effective number of parameters. An optimal model can be obtained by selecting λ and m that minimize the information criterion AIC_M.

Generally, since H and $H(\lambda, m)$ are matrices that transform the observation vector $\boldsymbol{y}$ into a predicted value vector $\hat{\boldsymbol{y}}$, it is referred to as a *hat matrix*, or for the estimation of a curve (or surface), this matrix is called a *smoother matrix*. The use of the trace of the hat matrix as the effective number of parameters has been investigated in smoothing methods [Wahba (1990)] and generalized additive models [Hastie and Tibshirani (1990)]. Ye (1998) developed a concept of the effective number of parameters that is applicable to complex modeling procedures.

Example 6 (Numerical result) Figure 6.6 shows a plot of 100 observations that are generated according to the model

$$y_\alpha = \sin(2\pi x_\alpha^3) + \varepsilon_\alpha, \qquad \varepsilon_\alpha \sim N(0, 10^{-1.3}),$$

where x_α is generated by uniform random numbers over $[0, 1)$. We fitted a B-spline regression model with 10 basis functions to the simulated data. The parameters of the model were estimated by using the regularization method with a difference matrix of degree 2 given in Subsection 5.2.4 as the penalty term. Figure 6.7 shows the relationship between the value of the smoothing parameter λ and the effective number of parameters:

$$\mathrm{enp} = \mathrm{tr}\left\{B(B^T B + n\lambda\hat{\sigma}^2 D_2^T D_2)^{-1} B^T\right\}.$$

It can be seen from Figure 6.7 that the effective number of parameters becomes $\mathrm{tr}\{B(B^T B)^{-1} B^T\} = 10$ (number of basis functions) when the value of the smoothing parameter is 0 and that the effective number of parameters (enp) approaches 2 as the value of the smoothing parameter increases. In Figure 6.6, the solid and dashed curves represent the estimated regression curves corresponding to $\lambda = 0$ and $\lambda = 80$, respectively, and when λ is sufficiently large, the model approximates a straight line (number of parameters: 2). Therefore, we see that the effective number of parameters is a real number between 2 and the number of basis functions.

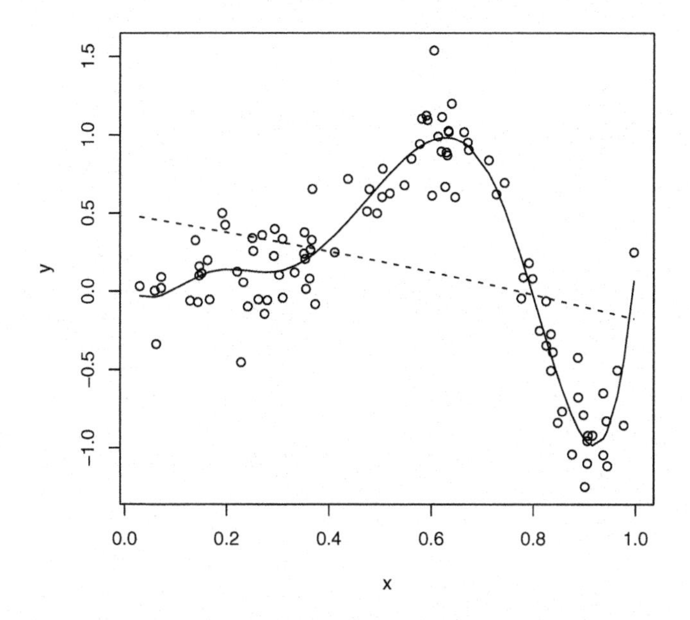

Fig. 6.6. Artificially generated data and estimated curves for $\lambda = 0$ (solid curve) and $\lambda = 80$ (dashed curve).

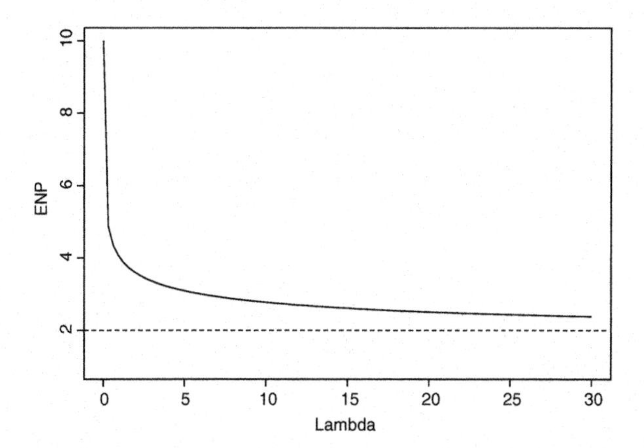

Fig. 6.7. Relationship between the value of smoothing parameter and the effective number of parameters.

7

Theoretical Development and Asymptotic Properties of the GIC

Information criteria have been constructed as estimators of the Kullback–Leibler information discrepancy between two probability distributions or, equivalently, the expected log-likelihood of a statistical model for prediction.

In this chapter, we introduce a general framework for constructing information criteria in the context of functional statistics and give technical arguments and a detailed derivation of the generalized information criterion (GIC) defined in (5.64). We also investigate the asymptotic properties of information criteria in the estimation of the expected log-likelihood of a statistical model.

7.1 Derivation of the GIC

7.1.1 Introduction

The GIC is a criterion for evaluating a statistical model $f(x|\hat{\boldsymbol{\theta}})$, in which the p-dimensional parameter vector $\boldsymbol{\theta}$ included in the density function $f(x|\boldsymbol{\theta})$ is replaced with a functional estimator $\hat{\boldsymbol{\theta}}$. The statistical model is a fitted model to the observed data $\boldsymbol{x}_n = \{x_1, x_2, \ldots, x_n\}$ drawn from the true distribution $G(x)$ having density $g(x)$.

The essential point in the derivation of the GIC is the bias correction of the log-likelihood

$$\sum_{\alpha=1}^{n} \log f(x_\alpha|\hat{\boldsymbol{\theta}}) \equiv \log f(\boldsymbol{x}_n|\hat{\boldsymbol{\theta}}) \tag{7.1}$$

in estimating the expected log-likelihood defined by

$$n \int \log f(z|\hat{\boldsymbol{\theta}}) dG(z). \tag{7.2}$$

In other words, the expectation of the difference between the log-likelihood and the expected log-likelihood,

$$D(\boldsymbol{X}_n; G) = \log f(\boldsymbol{X}_n|\hat{\boldsymbol{\theta}}) - n \int \log f(z|\hat{\boldsymbol{\theta}}) dG(z), \tag{7.3}$$

is evaluated.

In order to construct an information criterion that enables the evaluation of various types of statistical models, we employ a functional estimator with Fisher consistency. It is assumed that the ith element $\hat{\theta}_i$ of the estimator $\hat{\boldsymbol{\theta}} = (\hat{\theta}_1, \ldots, \hat{\theta}_p)^T$ is given by

$$\hat{\theta}_i = T_i(\hat{G}), \qquad i = 1, 2, \ldots, p, \tag{7.4}$$

where $T_i(\cdot)$ is a functional defined on the set of all distributions and $\hat{G}$ is the empirical distribution function based on the observed data. Writing the p-dimensional functional vector with $T_i(G)$ as the ith element as

$$\boldsymbol{T}(G) = (T_1(G), T_2(G), \ldots, T_p(G))^T, \tag{7.5}$$

the p-dimensional estimator can be expressed as

$$\hat{\boldsymbol{\theta}} = \boldsymbol{T}(\hat{G}) = \left(T_1(\hat{G}), T_2(\hat{G}), \ldots, T_p(\hat{G})\right)^T. \tag{7.6}$$

We can then see that

$$\lim_{n \to +\infty} \boldsymbol{T}(\hat{G}) = \boldsymbol{T}(G) \tag{7.7}$$

in probability.

We first decompose $D(\boldsymbol{X}_n; G)$ in (7.3) into three terms as follows (Figure 7.1):

$$\begin{aligned} D(\boldsymbol{X}_n; G) &= \log f(\boldsymbol{X}_n|\hat{\boldsymbol{\theta}}) - n \int \log f(z|\hat{\boldsymbol{\theta}}) dG(z), \\ &= D_1(\boldsymbol{X}_n; G) + D_2(\boldsymbol{X}_n; G) + D_3(\boldsymbol{X}_n; G), \end{aligned} \tag{7.8}$$

where

$$\begin{aligned} D_1(\boldsymbol{X}_n; G) &= \log f(\boldsymbol{X}_n|\hat{\boldsymbol{\theta}}) - \log f(\boldsymbol{X}_n|\boldsymbol{T}(G)), \\ D_2(\boldsymbol{X}_n; G) &= \log f(\boldsymbol{X}_n|\boldsymbol{T}(G)) - n \int \log f(z|\boldsymbol{T}(G)) dG(z), \\ D_3(\boldsymbol{X}_n; G) &= n \int \log f(z|\boldsymbol{T}(G)) dG(z) - n \int \log f(z|\hat{\boldsymbol{\theta}}) dG(z). \end{aligned} \tag{7.9}$$

Since the expectation of the second term in (7.8) is

$$\begin{aligned} E_G\left[D_2(\boldsymbol{X}_n; G)\right] &= E_G\left[\log f(\boldsymbol{X}_n|\boldsymbol{T}(G)) - n \int \log f(z|\boldsymbol{T}(G)) dG(z)\right] \\ &= \sum_{\alpha=1}^{n} E_G\left[\log f(X|\boldsymbol{T}(G))\right] - n \int \log f(z|\boldsymbol{T}(G)) dG(z) \\ &= 0, \end{aligned} \tag{7.10}$$

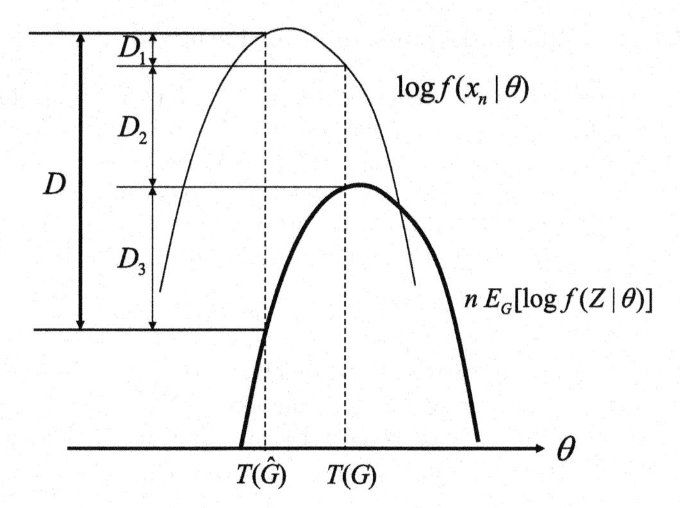

Fig. 7.1. Decomposition of the difference between the log-likelihood and the expected log-likelihood.

the bias calculation is reduced to

$$\begin{aligned} b(G) &\equiv E_G\left[D(\boldsymbol{X}_n; G)\right] \\ &= E_G\left[D_1(\boldsymbol{X}_n; G)\right] + E_G\left[D_3(\boldsymbol{X}_n; G)\right], \end{aligned} \tag{7.11}$$

where the expectation is taken with respect to the joint distribution of $\boldsymbol{X}_n$. Therefore, as for the derivation of the AIC, only two terms, $E_G\left[D_1(\boldsymbol{X}_n; G)\right]$ and $E_G\left[D_3(\boldsymbol{X}_n; G)\right]$, need to be evaluated.

Remark 1 The symbols O, O_p, o, and o_p, which are frequently used in this chapter, are defined as follows:

(i) O and o: Let $\{a_n\}, \{b_n\}$ be two sequences of real numbers. If $|a_n/b_n|$ is bounded when $n \to +\infty$, we write $a_n = O(b_n)$. Similarly, if $|a_n/b_n|$ converges to 0, we write $a_n = o(b_n)$.

(ii) O_p and o_p: Given a sequence of random variables $\{X_n\}$ and a sequence of real numbers $\{b_n\}$, if X_n/b_n is bounded in probability when $n \to +\infty$, then we write $X_n = O_p(b_n)$. If X_n/b_n converges in probability to 0, then we write $X_n = o_p(b_n)$. Note that a term bounded in probability means that there exist a constant c_ε and a natural number $n_0(\varepsilon)$ such that for any $\varepsilon > 0$,

$$\mathrm{P_r}\{|X_n| \le b_n c_\varepsilon\} \ge 1 - \varepsilon \tag{7.12}$$

if $n > n_0(\varepsilon)$.

In discussions of asymptotic theory with regard to the number of observations n, the quantity b_n becomes $n^{-1/2}$ or n^{-1}, which provides a good measure of the speed of convergence to a limit distribution or of the evaluation of the approximation accuracy.

7.1.2 Stochastic Expansion of an Estimator

To evaluate the bias correction term (7.11) for the log-likelihood, we employ the stochastic expansion of an estimator based on the functional Taylor series expansion. In this subsection, we drop the notational dependence on the estimator $\hat{\theta}_i$ and consider the stochastic expansion of $\hat{\theta}$.

Given a real-valued function $T(G)$ whose domain is the set of all distributions, for any distribution functions G and H, we write

$$h(\varepsilon) = T((1-\varepsilon)G + \varepsilon H), \qquad 0 \le \varepsilon \le 1. \tag{7.13}$$

The ith-order derivative of the functional $T(\cdot)$ at a point $(z_1, \ldots, z_i, G)$ is then defined as the symmetric function $T^{(i)}(z_1, \ldots, z_i; G)$ that satisfies the following equation with respect to any distribution function H [von Mises (1947), Withers (1983)]:

$$h^{(i)}(0) = \int \cdots \int T^{(i)}(z_1, \ldots, z_i; G) \prod_{j=1}^{i} d\left\{H(z_j) - G(z_j)\right\}. \tag{7.14}$$

Here, we impose the following condition to ensure the uniqueness of the derivative $T^{(i)}(z_1, \ldots, z_i; G)$:

$$\int T^{(i)}(z_1, \ldots, z_i; G) dG(z_k) = 0, \qquad 1 \le k \le i. \tag{7.15}$$

This permits the replacement of $d\left\{H(z_j) - G(z_j)\right\}$ in (7.14) with $dH(z_j)$.

In the next step, we expand $h(\varepsilon)$ in a Taylor series around $\varepsilon = 0$ in the form

$$h(\varepsilon) = h(0) + \varepsilon h'(0) + \frac{1}{2}\varepsilon^2 h''(0) + \cdots. \tag{7.16}$$

Since $h(1) = T(H)$ and $h(0) = T(G)$, by formally putting $\varepsilon = 1$ the above expansion is rewritten as

$$\begin{aligned} T(H) = T(G) &+ \int T^{(1)}(z_1; G) dH(z_1) \\ &+ \frac{1}{2} \int\int T^{(2)}(z_1, z_2; G) dH(z_1) dH(z_2) + \cdots. \end{aligned} \tag{7.17}$$

Since $\hat{G}$ is the empirical distribution function based on the observed data from the true distribution $G(x)$, $\hat{G}$ must converge to G as n tends to infinity. Thus, by replacing H in (7.17) with the empirical distribution function $\hat{G}$, we obtain the stochastic expansion for the estimator $\hat{\theta} = T(\hat{G})$ defined by the functional $T(\cdot)$ in the following:

$$T(\hat{G}) = T(G) + \frac{1}{n}\sum_{\alpha=1}^{n} T^{(1)}(x_\alpha; G) + \frac{1}{2n^2}\sum_{\alpha=1}^{n}\sum_{\beta=1}^{n} T^{(2)}(x_\alpha, x_\beta; G) + \cdots. \tag{7.18}$$

In addition, it follows from this stochastic expansion that we have

$$\sqrt{n}\left(T(\hat{G}) - T(G)\right) \approx \frac{1}{\sqrt{n}}\sum_{\alpha=1}^{n} T^{(1)}(x_\alpha; G). \tag{7.19}$$

Hence, it can be shown from the central limit theorem that $\sqrt{n}(T(\hat{G}) - T(G))$ is asymptotically distributed as a normal distribution with mean 0 and variance

$$\int \left\{T^{(1)}(x; G)\right\}^2 dG(x). \tag{7.20}$$

In the next subsection, we derive the GIC by using the stochastic expansion formula for an estimator $\hat{\theta}_i = T_i(\hat{G})$ $(i = 1, \ldots, p)$ defined by a statistical functional $T_i(\cdot)$.

For theoretical work on the functional Taylor series expansion, we refer to von Mises (1947), Filippova (1962), Reeds (1976), Serfling (1980), Fernholz (1983), Withers (1983), Konishi (1991), etc.

7.1.3 Derivation of the GIC

We recall that an estimator $\hat{\boldsymbol{\theta}} = (\hat{\theta}_1, \ldots, \hat{\theta}_p)^T$ is a functional, for which there exists a p-dimensional statistical functional $\boldsymbol{T}(\cdot)$ such that $\hat{\boldsymbol{\theta}} = \boldsymbol{T}(\hat{G}) = (T_1(\hat{G}), \ldots, T_p(\hat{G}))^T$. Here, as given in (7.18), the stochastic expansion of $\hat{\theta}_i = T_i(\hat{G})$ around $T_i(G)$ up to the term of order n^{-1} is

$$\hat{\theta}_i = T_i(G) + \frac{1}{n}\sum_{\alpha=1}^{n} T_i^{(1)}(X_\alpha; G) + \frac{1}{2n^2}\sum_{\alpha=1}^{n}\sum_{\beta=1}^{n} T_i^{(2)}(X_\alpha, X_\beta; G) + o_p(n^{-1}), \tag{7.21}$$

where $T_i^{(1)}(X_\alpha; G)$ and $T_i^{(2)}(X_\alpha, X_\beta; G)$ are respectively the first- and second-order derivatives defined in (7.14). We now express the stochastic expansion formula in vector form as follows:

$$\hat{\boldsymbol{\theta}} = \boldsymbol{T}(G) + \frac{1}{n}\sum_{\alpha=1}^{n} \boldsymbol{T}^{(1)}(X_\alpha; G) + \frac{1}{2n^2}\sum_{\alpha=1}^{n}\sum_{\beta=1}^{n} \boldsymbol{T}^{(2)}(X_\alpha, X_\beta; G) + o_p(n^{-1}), \tag{7.22}$$

where $\boldsymbol{T}^{(1)}(X_\alpha; G)$ and $\boldsymbol{T}^{(2)}(X_\alpha, X_\beta; G)$ are p-dimensional vectors given by

$$\boldsymbol{T}^{(1)}(X_\alpha; G) = \left(T_1^{(1)}(X_\alpha; G), \ldots, T_p^{(1)}(X_\alpha; G)\right)^T,$$
$$\boldsymbol{T}^{(2)}(X_\alpha, X_\beta; G) = \left(T_1^{(2)}(X_\alpha, X_\beta; G), \ldots, T_p^{(2)}(X_\alpha, X_\beta; G)\right)^T. \quad (7.23)$$

Noting that from the condition (7.15),

$$E_G\left[\boldsymbol{T}^{(1)}(X_\alpha; G)\right] = 0 \text{ and } E_G\left[\boldsymbol{T}^{(2)}(X_\alpha, X_\beta; G)\right] = 0, \ \alpha \neq \beta, \quad (7.24)$$

the expectation for the estimator $\hat{\boldsymbol{\theta}}$ in (7.22) can be calculated as

$$\begin{aligned} E_G\left[\hat{\boldsymbol{\theta}} - \boldsymbol{T}(G)\right] &= \frac{1}{2n^2}\sum_{\alpha=1}^{n}\sum_{\beta=1}^{n} E_G\left[\boldsymbol{T}^{(2)}(X_\alpha, X_\beta; G)\right] + o(n^{-1}) \\ &= \frac{1}{2n^2}\sum_{\alpha=1}^{n} E_G\left[\boldsymbol{T}^{(2)}(X_\alpha, X_\alpha; G)\right] + o(n^{-1}) \\ &= \frac{1}{n}\boldsymbol{b} + o(n^{-1}), \end{aligned} \quad (7.25)$$

where $\boldsymbol{b} = (b_1, b_2, \ldots, b_p)^T$ is an asymptotic bias of the estimator given by

$$\boldsymbol{b} = \frac{1}{2}\int \boldsymbol{T}^{(2)}(z, z; G)dG(z) \quad (7.26)$$

with i^{th} element

$$b_i = \frac{1}{2}\int T_i^{(2)}(z, z; G)dG(z). \quad (7.27)$$

The variance–covariance matrix of the estimator $\hat{\boldsymbol{\theta}}$ is asymptotically given by

$$\begin{aligned} &E_G\left[\left(\hat{\boldsymbol{\theta}} - \boldsymbol{T}(G)\right)\left(\hat{\boldsymbol{\theta}} - \boldsymbol{T}(G)\right)^T\right] \\ &= \frac{1}{n^2}\sum_{\alpha=1}^{n}\sum_{\beta=1}^{n} E_G\left[\boldsymbol{T}^{(1)}(X_\alpha; G)\boldsymbol{T}^{(1)}(X_\beta; G)^T\right] + o(n^{-1}) \\ &= \frac{1}{n^2}\sum_{\alpha=1}^{n} E_G\left[\boldsymbol{T}^{(1)}(X_\alpha; G)\boldsymbol{T}^{(1)}(X_\alpha; G)^T\right] + o(n^{-1}) \\ &= \frac{1}{n}\Sigma(G) + o(n^{-1}), \end{aligned} \quad (7.28)$$

where

$$\Sigma(G) = (\sigma_{ij}) = \int \boldsymbol{T}^{(1)}(z; G)\boldsymbol{T}^{(1)}(z; G)^T dG(z) \quad (7.29)$$

with $(i,j)^{th}$ element

$$\sigma_{ij} = \int T_i^{(1)}(z;G)T_j^{(1)}(z;G)dG(z). \tag{7.30}$$

Calculating the bias correction term $D_3(\boldsymbol{X}_n;G)$. Since $\hat{\boldsymbol{\theta}} = \boldsymbol{T}(\hat{G})$ converges to $\boldsymbol{T}(G)$ in probability as the sample size n tends to infinity, by expanding $\log f(z|\hat{\boldsymbol{\theta}})$ in a Taylor series around $\boldsymbol{T}(G)$, we obtain the stochastic expansion of the expected log-likelihood:

$$\begin{aligned}
&\int \log f(z|\hat{\boldsymbol{\theta}})dG(z) \\
&= \int \log f(z|\boldsymbol{T}(G))dG(z) + \left(\hat{\boldsymbol{\theta}} - \boldsymbol{T}(G)\right)^T \int \left.\frac{\partial \log f(z|\boldsymbol{\theta})}{\partial \boldsymbol{\theta}}\right|_{\boldsymbol{T}(G)} dG(z) \\
&\quad -\frac{1}{2}\left(\hat{\boldsymbol{\theta}} - \boldsymbol{T}(G)\right)^T J(G)\left(\hat{\boldsymbol{\theta}} - \boldsymbol{T}(G)\right) + \cdots,
\end{aligned} \tag{7.31}$$

where

$$J(G) = -\int \left.\frac{\partial^2 \log f(z|\boldsymbol{\theta})}{\partial \boldsymbol{\theta}\partial \boldsymbol{\theta}^T}\right|_{\boldsymbol{T}(G)} dG(z). \tag{7.32}$$

Then, by substituting the stochastic expansion formula for the estimator in (7.22) into (7.31), we have

$$\begin{aligned}
&\int \log f(z|\hat{\boldsymbol{\theta}})dG(z) - \int \log f(z|\boldsymbol{T}(G))dG(z) \\
&= \frac{1}{n}\sum_{\alpha=1}^{n} \boldsymbol{T}^{(1)}(X_\alpha;G)^T \int \left.\frac{\partial \log f(z|\boldsymbol{\theta})}{\partial \boldsymbol{\theta}}\right|_{\boldsymbol{T}(G)} dG(z) \\
&\quad + \frac{1}{2n^2}\sum_{\alpha=1}^{n}\sum_{\beta=1}^{n} \boldsymbol{T}^{(2)}(X_\alpha,X_\beta;G)^T \int \left.\frac{\partial \log f(z|\boldsymbol{\theta})}{\partial \boldsymbol{\theta}}\right|_{\boldsymbol{T}(G)} dG(z) \\
&\quad - \frac{1}{2n^2}\sum_{\alpha=1}^{n}\sum_{\beta=1}^{n} \boldsymbol{T}^{(1)}(X_\alpha;G)^T J(G)\boldsymbol{T}^{(1)}(X_\beta;G) + o_p(n^{-1}).
\end{aligned} \tag{7.33}$$

Taking the expectation term by term and using the results in (7.25) and (7.28), we obtain the expectation of $D_3(\boldsymbol{X}_n;G)$ in (7.9):

$$\begin{aligned}
&E_G\left[D_3(\boldsymbol{X}_n;G)\right] \\
&= E_G\left[n\int \log f(z|\boldsymbol{T}(G))dG(z) - n\int \log f(z|\hat{\boldsymbol{\theta}})dG(z)\right] \\
&= -\frac{1}{n}\sum_{\alpha=1}^{n} E_G\left[\frac{1}{2}\boldsymbol{T}^{(2)}(X_\alpha,X_\alpha;G)^T\right] \int \left.\frac{\partial \log f(z|\boldsymbol{\theta})}{\partial \boldsymbol{\theta}}\right|_{\boldsymbol{T}(G)} dG(z)
\end{aligned}$$

$$+\frac{1}{2n}\sum_{\alpha=1}^{n} E_G\left[\boldsymbol{T}^{(1)}(X_\alpha;G)^T J(G)\boldsymbol{T}^{(1)}(X_\alpha;G)\right]+o(1)$$
$$=-\boldsymbol{b}^T\int \left.\frac{\partial \log f(z|\boldsymbol{\theta})}{\partial \boldsymbol{\theta}}\right|_{\boldsymbol{T}(G)} dG(z)+\frac{1}{2}\mathrm{tr}\left\{J(G)\Sigma(G)\right\}+o(1). \quad (7.34)$$

Here, note that

$$E_G\left[\boldsymbol{T}^{(1)}(X_\alpha;G)^T J(G)\boldsymbol{T}^{(1)}(X_\alpha;G)\right]$$
$$=\mathrm{tr}\left\{J(G)E_G\left[\boldsymbol{T}^{(1)}(X_\alpha;G)\boldsymbol{T}^{(1)}(X_\alpha;G)^T\right]\right\}$$
$$=\mathrm{tr}\left\{J(G)\Sigma(G)\right\}. \quad (7.35)$$

Calculating the bias correction term $D_1(\boldsymbol{X}_n;G)$. Similarly, by expanding the log-likelihood $\log f(\boldsymbol{X}_n|\hat{\boldsymbol{\theta}})$ in a Taylor series around $\boldsymbol{T}(G)$, we obtain

$$\log f(\boldsymbol{X}_n|\hat{\boldsymbol{\theta}})$$
$$=\log f(\boldsymbol{X}_n|\boldsymbol{T}(G))+\left(\hat{\boldsymbol{\theta}}-\boldsymbol{T}(G)\right)^T\left.\frac{\partial \log f(\boldsymbol{X}_n|\boldsymbol{\theta})}{\partial \boldsymbol{\theta}}\right|_{\boldsymbol{T}(G)} \quad (7.36)$$
$$+\frac{1}{2}\left(\hat{\boldsymbol{\theta}}-\boldsymbol{T}(G)\right)^T\left.\frac{\partial^2 \log f(\boldsymbol{X}_n|\boldsymbol{\theta})}{\partial \boldsymbol{\theta}\partial \boldsymbol{\theta}^T}\right|_{\boldsymbol{T}(G)}\left(\hat{\boldsymbol{\theta}}-\boldsymbol{T}(G)\right)+o_p(1).$$

Then, by substituting the stochastic expansion formula of (7.22) for the estimator $\hat{\boldsymbol{\theta}}$, we obtain

$$\log f(\boldsymbol{X}_n|\hat{\boldsymbol{\theta}})-\log f(\boldsymbol{X}_n|\boldsymbol{T}(G))$$
$$=\frac{1}{n}\sum_{\alpha=1}^{n}\sum_{\beta=1}^{n}\boldsymbol{T}^{(1)}(X_\alpha;G)^T\left.\frac{\partial \log f(X_\beta|\boldsymbol{\theta})}{\partial \boldsymbol{\theta}}\right|_{\boldsymbol{T}(G)}$$
$$+\frac{1}{2n^2}\sum_{\alpha=1}^{n}\sum_{\beta=1}^{n}\sum_{\gamma=1}^{n}\boldsymbol{T}^{(2)}(X_\alpha,X_\beta;G)^T\left.\frac{\partial \log f(X_\gamma|\boldsymbol{\theta})}{\partial \boldsymbol{\theta}}\right|_{\boldsymbol{T}(G)} \quad (7.37)$$
$$+\frac{1}{2n^2}\sum_{\alpha=1}^{n}\sum_{\beta=1}^{n}\sum_{\gamma=1}^{n}\boldsymbol{T}^{(1)}(X_\alpha;G)^T\left.\frac{\partial^2 \log f(X_\gamma|\boldsymbol{\theta})}{\partial \boldsymbol{\theta}\partial \boldsymbol{\theta}^T}\right|_{\boldsymbol{T}(G)}\boldsymbol{T}^{(1)}(X_\beta;G)+o_p(1).$$

By using (7.25) and (7.28), the expectation of each term in this stochastic expansion formula can be calculated as follows:

$$\frac{1}{n}\sum_{\alpha=1}^{n}\sum_{\beta=1}^{n}E_G\left[\boldsymbol{T}^{(1)}(X_\alpha;G)^T\left.\frac{\partial \log f(X_\beta|\boldsymbol{\theta})}{\partial \boldsymbol{\theta}}\right|_{\boldsymbol{T}(G)}\right]$$
$$=\int \boldsymbol{T}^{(1)}(z;G)^T\left.\frac{\partial \log f(z|\boldsymbol{\theta})}{\partial \boldsymbol{\theta}}\right|_{\boldsymbol{T}(G)} dG(z)$$

$$= \mathrm{tr}\left\{ \int \boldsymbol{T}^{(1)}(z;G) \frac{\partial \log f(z|\boldsymbol{\theta})}{\partial \boldsymbol{\theta}^T}\bigg|_{\boldsymbol{T}(G)} dG(z) \right\}, \tag{7.38}$$

$$\frac{1}{2n^2}\sum_{\alpha=1}^{n}\sum_{\beta=1}^{n}\sum_{\gamma=1}^{n} E_G\left[\boldsymbol{T}^{(2)}(X_\alpha, X_\beta; G)^T \frac{\partial \log f(X_\gamma|\boldsymbol{\theta})}{\partial \boldsymbol{\theta}}\bigg|_{\boldsymbol{T}(G)} \right]$$
$$= \boldsymbol{b}^T \int \frac{\partial \log f(z|\boldsymbol{\theta})}{\partial \boldsymbol{\theta}}\bigg|_{\boldsymbol{T}(G)} dG(z) + o(1), \tag{7.39}$$

$$\frac{1}{2n^2}\sum_{\alpha=1}^{n}\sum_{\beta=1}^{n}\sum_{\gamma=1}^{n} E_G\left[\boldsymbol{T}^{(1)}(X_\alpha; G)^T \frac{\partial^2 \log f(X_\gamma|\boldsymbol{\theta})}{\partial \boldsymbol{\theta}\partial \boldsymbol{\theta}^T}\bigg|_{\boldsymbol{T}(G)} \boldsymbol{T}^{(1)}(X_\beta; G) \right]$$
$$= \frac{1}{2}\mathrm{tr}\left\{ E_G\left[\frac{\partial^2 \log f(Z|\boldsymbol{\theta})}{\partial \boldsymbol{\theta}\partial \boldsymbol{\theta}^T}\bigg|_{\boldsymbol{T}(G)} \right] E_G\left[\boldsymbol{T}^{(1)}(Z;G)\boldsymbol{T}^{(1)}(Z;G)^T \right] \right\} + o(1)$$
$$= -\frac{1}{2}\mathrm{tr}\left\{ J(G)\Sigma(G) \right\} + o(1). \tag{7.40}$$

Thus, the expectation of $D_1(\boldsymbol{X}_n; G)$ in (7.9) is given by

$$E_G\left[D_1(\boldsymbol{X}_n; G)\right]$$
$$= E_G\left[\log f(\boldsymbol{X}_n|\hat{\boldsymbol{\theta}}) - \log f(\boldsymbol{X}_n|\boldsymbol{T}(G)) \right]$$
$$= \mathrm{tr}\left\{ \int \boldsymbol{T}^{(1)}(z;G) \frac{\partial \log f(z|\boldsymbol{\theta})}{\partial \boldsymbol{\theta}^T}\bigg|_{\boldsymbol{T}(G)} dG(z) \right\} \tag{7.41}$$
$$+ \boldsymbol{b}^T \int \frac{\partial \log f(z|\boldsymbol{\theta})}{\partial \boldsymbol{\theta}}\bigg|_{\boldsymbol{T}(G)} dG(z) - \frac{1}{2}\mathrm{tr}\left\{ J(G)\Sigma(G) \right\} + o(1).$$

Calculating the bias correction term $D(\boldsymbol{X}_n; G)$. It follows from (7.34) and (7.41) that the asymptotic bias of the log-likelihood $\log f(\boldsymbol{X}_n|\hat{\boldsymbol{\theta}})$ in estimating the expected log-likelihood $E_G[\log f(z|\hat{\boldsymbol{\theta}})]$ is

$$b(G) = E_G\left[D(\boldsymbol{X}_n; G)\right]$$
$$= E_G\left[D_1(\boldsymbol{X}_n; G)\right] + E_G\left[D_3(\boldsymbol{X}_n; G)\right] \tag{7.42}$$
$$= \mathrm{tr}\left\{ \int \boldsymbol{T}^{(1)}(z;G) \frac{\partial \log f(z|\boldsymbol{\theta})}{\partial \boldsymbol{\theta}^T}\bigg|_{\boldsymbol{T}(G)} dG(z) \right\} + o(1).$$

The asymptotic bias correction term depends on the unknown distribution G, and hence by replacing G with the empirical distribution function $\hat{G}$, we obtain the bias estimate:

$$b(\hat{G}) = \frac{1}{n}\sum_{\alpha=1}^{n} \mathrm{tr}\left\{ \boldsymbol{T}^{(1)}(x_\alpha; G) \left. \frac{\partial \log f(x_\alpha|\boldsymbol{\theta})}{\partial \boldsymbol{\theta}^T} \right|_{\boldsymbol{\theta}=\hat{\boldsymbol{\theta}}} \right\}. \tag{7.43}$$

By subtracting the asymptotic bias estimate $b(\hat{G})$ from the log-likelihood, we obtain the GIC given in (5.64):

$$\begin{aligned} \mathrm{GIC} = & -2\sum_{\alpha=1}^{n} \log f(x_\alpha|\hat{\boldsymbol{\theta}}) \\ & + \frac{2}{n}\sum_{\alpha=1}^{n} \mathrm{tr}\left\{ \boldsymbol{T}^{(1)}(x_\alpha; \hat{G}) \left. \frac{\partial \log f(x_\alpha|\boldsymbol{\theta})}{\partial \boldsymbol{\theta}^T} \right|_{\boldsymbol{\theta}=\hat{\boldsymbol{\theta}}} \right\}. \end{aligned} \tag{7.44}$$

Information criteria for stochastic processes were investigated by Uchida and Yoshida (2001, 2004) [see also Yoshida (1997)].

7.2 Asymptotic Properties and Higher-Order Bias Correction

7.2.1 Asymptotic Properties of Information Criteria

Information criteria were constructed as estimators of the Kullback–Leibler information discrepancy between the true distribution $g(z)$ and the statistical model $f(z|\hat{\boldsymbol{\theta}})$ or, equivalently, the expected log-likelihood $E_{G(z)}[\log f(Z|\hat{\boldsymbol{\theta}})]$. We estimate the expected log-likelihood by the log-likelihood $f(\boldsymbol{x}_n|\hat{\boldsymbol{\theta}})$. The bias correction for the log-likelihood of a statistical model in the estimation of the expected log-likelihood is essential for constructing an information criterion. The bias correction term is generally given as an asymptotic bias. According to the assumptions made for model estimation and the relationship between the specified model and the true distribution, the asymptotic bias takes a different form, and consequently we can obtain the information criteria introduced previously, including the AIC.

In this subsection, we discuss, within a general framework, the theoretical evaluation of the asymptotic accuracy of an information criterion as an estimator of the expected log-likelihood. In the following, we assume that the p-dimensional parameter vector for a model $f(x|\boldsymbol{\theta})$ is estimated by $\hat{\boldsymbol{\theta}} = \boldsymbol{T}(\hat{G})$ for a suitable p-dimensional functional $\boldsymbol{T}(G)$. The aim is to estimate the expected log-likelihood of the statistical model $f(x|\hat{\boldsymbol{\theta}})$ defined by

$$\eta(G; \hat{\boldsymbol{\theta}}) \equiv E_{G(z)}\left[\log f(Z|\hat{\boldsymbol{\theta}})\right] = \int \log f(z|\hat{\boldsymbol{\theta}}) dG(z). \tag{7.45}$$

The expected log-likelihood is conditional on the observed data $\boldsymbol{x}_n$ and also depends on the unknown distribution G generating the data.

We suppose that under certain regularity conditions, the expectation of $\eta(G; \hat{\boldsymbol{\theta}})$ over the sampling distribution G of $\boldsymbol{X}_n$ can be expanded in the form

$$
\begin{aligned}
&E_{G(\boldsymbol{x})}\left[\eta(G;\hat{\boldsymbol{\theta}})\right] \\
&\quad = E_{G(\boldsymbol{x})}\left[E_{G(z)}\left[\log f(Z|\hat{\boldsymbol{\theta}})\right]\right] \\
&\quad = \int \log f(z|\boldsymbol{T}(G))dG(z) + \frac{1}{n}\eta_1(G) + \frac{1}{n^2}\eta_2(G) + O(n^{-3}).
\end{aligned} \tag{7.46}
$$

The objective is to estimate this quantity from observed data as accurately as possible. In other words, we want to obtain an estimator $\hat{\eta}(\hat{G};\hat{\boldsymbol{\theta}})$ of $\eta(G;\hat{\boldsymbol{\theta}})$ that satisfies the condition

$$
E_{G(\boldsymbol{x})}\left[\hat{\eta}(\hat{G};\hat{\boldsymbol{\theta}}) - \eta(G;\hat{\boldsymbol{\theta}})\right] = O(n^{-j}) \tag{7.47}
$$

for j as large as possible. For example, if $j = 2$, (7.46) indicates that the estimator agrees up to a term of order $1/n$.

An obvious estimator is the log-likelihood ($\times\ 1/n$)

$$
\eta(\hat{G};\hat{\boldsymbol{\theta}}) \equiv \frac{1}{n}\sum_{\alpha=1}^{n} \log f(x_\alpha|\hat{\boldsymbol{\theta}}), \tag{7.48}
$$

which is obtained by replacing the unknown probability distribution G of the expected log-likelihood $\eta(G;\hat{\boldsymbol{\theta}})$ with the empirical distribution function $\hat{G}$. In this subsection, because of the order of the expansion formula, we refer to the above equation divided by n as the log-likelihood.

By using the stochastic expansion of a statistical functional, the expectation of the log-likelihood gives a valid expansion of the following form:

$$
\begin{aligned}
&E_{G(\boldsymbol{x})}\left[\eta(\hat{G};\hat{\boldsymbol{\theta}})\right] \\
&\quad = \int \log f(z|\boldsymbol{T}(G))dG(z) + \frac{1}{n}L_1(G) + \frac{1}{n^2}L_2(G) + O(n^{-3}).
\end{aligned} \tag{7.49}
$$

Therefore, the log-likelihood as an estimator of the expected log-likelihood (7.46) only agrees in the first term, and the term of order $1/n$ remains as a bias. Specifically, the asymptotic expansions in (7.46) and (7.49) differ in the term of order n^{-1}, namely,

$$
E_{G(\boldsymbol{x})}\left[\eta(\hat{G};\hat{\boldsymbol{\theta}}) - \eta(G;\hat{\boldsymbol{\theta}})\right] = \frac{1}{n}\left\{L_1(G) - \eta_1(G)\right\} + O(n^{-2}). \tag{7.50}
$$

In (7.42) in the preceding section, we showed that this bias is given by

$$
\begin{aligned}
b(G) &= L_1(G) - \eta_1(G) \\
&= \mathrm{tr}\left\{\int \boldsymbol{T}^{(1)}(z;G)\left.\frac{\partial \log f(z|\boldsymbol{\theta})}{\partial \boldsymbol{\theta}^T}\right|_{\boldsymbol{T}(G)} dG(z)\right\},
\end{aligned} \tag{7.51}
$$

within the framework of a regular functional.

The asymptotic bias of the log-likelihood given by $\{L_1(G) - \eta_1(G)\}/n$ (= $b_1(G)/n$) may be estimated by $b_1(\hat{G})/n = \{L_1(\hat{G}) - \eta_1(\hat{G})\}/n$, and the bias-corrected version of the log-likelihood is

$$\eta_{\mathrm{IC}}(\hat{G};\hat{\boldsymbol{\theta}}) = \eta(\hat{G};\hat{\boldsymbol{\theta}}) - \frac{1}{n}b_1(\hat{G}). \tag{7.52}$$

Noting that the difference between $E_G[b_1(\hat{G})]$ and $b_1(G)$ is usually of order n^{-1}, that is, $E_{G(\boldsymbol{x})}[b_1(\hat{G})] = b_1(G) + O(n^{-1})$, we have

$$E_G\left[\eta(\hat{G};\hat{\boldsymbol{\theta}}) - \frac{1}{n}b_1(\hat{G}) - \eta(G;\hat{\boldsymbol{\theta}})\right] = O(n^{-2}). \tag{7.53}$$

Hence, the bias-corrected log-likelihood $\eta_{bc}(\hat{G};\hat{\boldsymbol{\theta}})$ is *second-order correct* or *accurate* for $\eta(G;\hat{\boldsymbol{\theta}})$ in the sense that the expectations of two quantities are in agreement up to and including the term of order n^{-1} and that the order of the remainder is n^{-2}.

It can be readily seen that the $-(2n)^{-1}$ times information criteria AIC, TIC, and GIC are all second-order correct for the corresponding expected log-likelihood. In contrast, the log-likelihood itself is only first-order correct.

If the specified parametric family of densities includes the true distribution and the maximum likelihood estimate is used to estimate the underlying density, then the asymptotic bias of the log-likelihood is given by the number of estimated parameters, giving AIC $= -2n\{\eta(\hat{F};\hat{\boldsymbol{\theta}}_{\mathrm{ML}}) - p/n\}$. In this case, the bias-corrected version of the log-likelihood is given by

$$\eta_{\mathrm{ML}}(\hat{F};\hat{\boldsymbol{\theta}}_{\mathrm{ML}}) = \eta(\hat{F};\hat{\boldsymbol{\theta}}_{\mathrm{ML}}) - \frac{1}{n}p - \frac{1}{n^2}\{L_2(\hat{F}) - \eta_2(\hat{F})\}. \tag{7.54}$$

It can be readily checked that

$$E_F\left[\eta_{\mathrm{ML}}(\hat{F};\hat{\boldsymbol{\theta}}_{\mathrm{ML}}) - \eta(F;\hat{\boldsymbol{\theta}}_{\mathrm{ML}})\right] = O(n^{-3}), \tag{7.55}$$

which implies that $\eta_{\mathrm{ML}}(\hat{F};\hat{\boldsymbol{\theta}}_{\mathrm{ML}})$ is third-order correct for $\eta(F;\hat{\boldsymbol{\theta}}_{\mathrm{ML}})$.

In practice, we need to derive the second-order bias-corrected term $L_2(\hat{F}) - \eta_2(\hat{F})$ analytically for each estimator, though it seems to be of no practical use. In such cases, bootstrap methods may be applied to estimate the bias of the log-likelihood, and the same asymptotic order as for $\eta_{\mathrm{ML}}(\hat{F};\hat{\boldsymbol{\theta}}_{\mathrm{ML}})$ can be achieved by bootstrapping $\eta(\hat{F};\hat{\boldsymbol{\theta}}_{\mathrm{ML}}) - p/n$ or equivalently $\eta(\hat{F};\hat{\boldsymbol{\theta}}_{\mathrm{ML}})$ (see Section 8.2).

7.2.2 Higher-Order Bias Correction

The information criteria are derived by correcting the asymptotic bias of the log-likelihood in the estimation of the expected log-likelihood of a statistical model. Obtaining information criteria, as estimators for the expected log-likelihood, that have higher orders of accuracy remains a problem. For particular situations when distributional and structural assumptions of the models are made, Sugiura (1978), Hurvich and Tsai (1989, 1991, 1993), Fujikoshi and Satoh (1997), Satoh et al. (1997), Hurvich et al. (1998), and McQuarrie and Tsai (1998) have investigated the asymptotic properties of the AIC and demonstrated the effectiveness of bias reduction in autoregressive time series models and parametric and nonparametric regression models, both theoretically and numerically. Most of these studies employed the normality assumption, and the proposed criteria were relatively simple and easy to apply in practical situations.

Here we develop a general theory for bias reduction in evaluating the bias of a log-likelihood in the context of smooth functional statistics and introduce an information criterion that yields more refined results.

We showed in (7.53) that information criteria based on the asymptotic bias-corrected log-likelihood are second-order correct for the expected log-likelihood $\eta(G;\hat{\boldsymbol{\theta}})$ in the sense that the expectations of $\eta(\hat{G};\hat{\boldsymbol{\theta}})$ $-b_1(\hat{G})/n$ and $\eta(G;\hat{\boldsymbol{\theta}})$ are in agreement up to and including the term of order n^{-1}, while the expectations of $\eta(\hat{G};\hat{\boldsymbol{\theta}})$ and $\eta(G;\hat{\boldsymbol{\theta}})$ differ in the term of order n^{-1}. We now consider higher-order bias correction for information criteriaD

The bias of the asymptotic bias-corrected log-likelihood as an estimate of the expected log-likelihood is given by

$$\begin{aligned} E_{G(\boldsymbol{x})} & \left[\eta_{\mathrm{IC}}(\hat{G};\hat{\boldsymbol{\theta}}) - \eta(G;\hat{\boldsymbol{\theta}})\right] \\ &= E_{G(\boldsymbol{x})}\left[\eta(\hat{G};\hat{\boldsymbol{\theta}}) - \frac{1}{n}b_1(\hat{G}) - \eta(G;\hat{\boldsymbol{\theta}})\right] \\ &= E_{G(\boldsymbol{x})}\left[\eta(\hat{G};\hat{\boldsymbol{\theta}}) - \eta(G;\hat{\boldsymbol{\theta}})\right] - \frac{1}{n}E_G\left[b_1(\hat{G})\right]. \end{aligned} \tag{7.56}$$

The first term in the right-hand side is the bias of the log-likelihood and may be expanded as

$$E_{G(\boldsymbol{x})}\left[\eta(\hat{G};\hat{\boldsymbol{\theta}}) - \eta(G;\hat{\boldsymbol{\theta}})\right] = \frac{1}{n}b_1(G) + \frac{1}{n^2}b_2(G) + O(n^{-3}), \tag{7.57}$$

where $b_1(G)$ is the first-order or asymptotic bias correction term. The expected value of $b_1(\hat{G})$ can also be expanded as

$$E_{G(\boldsymbol{x})}\left[b_1(\hat{G})\right] = b_1(G) + \frac{1}{n}\Delta b_1(G) + O(n^{-2}). \tag{7.58}$$

Hence, the bias of the asymptotic bias-corrected log-likelihood is given by

$$\begin{aligned} E_{G(\boldsymbol{x})} & \left[\eta(\hat{G};\hat{\boldsymbol{\theta}}) - \frac{1}{n}b_1(\hat{G}) - \eta(G;\hat{\boldsymbol{\theta}})\right] \\ &= \frac{1}{n^2}\left\{b_2(G) - \Delta b_1(G)\right\} + O(n^{-3}). \end{aligned} \tag{7.59}$$

Konishi and Kitagawa (2003) have developed a general theory for bias reduction in evaluating the bias of a log-likelihood in the context of smooth functional estimators and derived the second-order bias correction term given by

$$\begin{aligned} &b_2(G) - \Delta b_1(G) \\ &\quad = b_1(G) + \frac{1}{2}\left\{\sum_{i=1}^{p}\int T_i^{(2)}(z,z;G)dG(z)\int \frac{\partial \log f(z|\boldsymbol{T}(G))}{\partial\theta_i}dG(z)\right. \\ &\qquad - \sum_{i=1}^{p}\int T_i^{(2)}(z,z;G)\frac{\partial \log f(z|\boldsymbol{T}(G))}{\partial\theta_i}dG(z) \\ &\qquad + \sum_{i=1}^{p}\sum_{j=1}^{p}\int T_i^{(1)}(z;G)T_j^{(1)}(z;G)dG(z)\int \frac{\partial^2 \log f(z|\boldsymbol{T}(G))}{\partial\theta_i\partial\theta_j}dG(z) \\ &\qquad \left. - \sum_{i=1}^{p}\sum_{j=1}^{p}\int T_i^{(1)}(z;G)T_j^{(1)}(z;G)\frac{\partial^2 \log f(z|\boldsymbol{T}(G))}{\partial\theta_i\partial\theta_j}dG(z)\right\}. \end{aligned} \tag{7.60}$$

This second-order bias correction term is estimated by $b_2(\hat{G}) - \Delta b_1(\hat{G})$, in which the unknown probability distribution G is replaced by the empirical distribution function $\hat{G}$. Then, by further correcting the bias in the information criterion $\eta_{\mathrm{IC}}(G;\hat{\boldsymbol{\theta}})$ with the first-order bias correction, we obtain the following theorem:

GIC with a second-order bias correction. Assume that the statistical model $f(x|\hat{\boldsymbol{\theta}})$ is estimated with $\hat{\boldsymbol{\theta}} = \boldsymbol{T}(\hat{G}) = (T_1(\hat{G}), T_2(\hat{G}), \ldots, T_p(\hat{G}))^T$, using the regular functional $\boldsymbol{T}(\cdot)$. Then the generalized information criterion with a second-order bias correction is given by

$$\mathrm{SGIC} \equiv -2\sum_{\alpha=1}^{n}\log f(X_\alpha|\hat{\boldsymbol{\theta}}) + 2\left\{b_1(\hat{G}) + \frac{1}{n}\left(b_2(\hat{G}) - \Delta b_1(\hat{G})\right)\right\}, \tag{7.61}$$

where $b_1(\hat{G})$ is the asymptotic bias term given in (7.43), and $b_2(\hat{G}) - \Delta b_1(\hat{G})$ is the second-order bias correction term given in (7.60) with $\hat{G}$.

It can be shown that the information criterion SGIC with a second-order bias correction is *third-order correct* or *accurate* in the sense that the order of (7.47) is $O(n^{-3})$, that is, the expectations are in agreement up to the term of order n^{-2} and that the order of the remainder is n^{-3}.

Example 1 (Gaussian linear regression model) Suppose that we have n observations $\{(y_\alpha, \boldsymbol{x}_\alpha);\ \alpha = 1, \ldots, n\}$ of a response variable y and a

p-dimensional vector of explanatory variables $\boldsymbol{x}$. The Gaussian linear regression model is

$$\boldsymbol{y} = X\boldsymbol{\beta} + \boldsymbol{\varepsilon}, \qquad \boldsymbol{\varepsilon} \sim N(\mathbf{0}, \sigma^2 I_n), \tag{7.62}$$

where $\boldsymbol{y} = (y_1, y_2, \ldots, y_n)^T$, X is an $n \times p$ design matrix, and $\boldsymbol{\beta}$ is a p-dimensional parameter vector. The maximum likelihood estimates of the parameters $\boldsymbol{\theta} = (\boldsymbol{\beta}^T, \sigma^2)^T$ $(\in \Theta \subset R^{p+1})$ are given by

$$\hat{\boldsymbol{\beta}} = (X^T X)^{-1} X^T \boldsymbol{y} \quad \text{and} \quad \hat{\sigma}^2 = \frac{1}{n}(\boldsymbol{y} - X\hat{\boldsymbol{\beta}})^T(\boldsymbol{y} - X\hat{\boldsymbol{\beta}}). \tag{7.63}$$

Here we assume that the true distribution that generates the data is contained in the specified parametric model. In other words, the true distribution is given as an n-dimensional normal distribution with mean $X\boldsymbol{\beta}_0$ and variance covariance matrix $\sigma_0^2 I_n$ for some $\boldsymbol{\beta}_0$ and σ_0^2. Then, for an n-dimensional observation vector $\boldsymbol{z}$ obtained randomly independent of $\boldsymbol{y}$, the statistical model can be expressed as

$$f(\boldsymbol{z}|\hat{\boldsymbol{\theta}}) = \left(2\pi\hat{\sigma}^2\right)^{-n/2} \exp\left\{-\frac{(\boldsymbol{z} - X\hat{\boldsymbol{\beta}})^T(\boldsymbol{z} - X\hat{\boldsymbol{\beta}})}{2\hat{\sigma}^2}\right\}. \tag{7.64}$$

The log-likelihood and the expected log-likelihood of this model are, respectively, given by

$$\begin{aligned} \log f(\boldsymbol{y}|\hat{\boldsymbol{\theta}}) &= -\frac{n}{2}\left\{\log(2\pi\hat{\sigma}^2) + 1\right\}, \\ \int \log f(\boldsymbol{z}|\hat{\boldsymbol{\theta}}) dG(\boldsymbol{z}) &= -\frac{n}{2}\left\{\log(2\pi\hat{\sigma}^2) + \frac{\sigma_0^2}{\hat{\sigma}^2}\right. \\ &\quad \left. + \frac{(X\boldsymbol{\beta}_0 - X\hat{\boldsymbol{\beta}})^T(X\boldsymbol{\beta}_0 - X\hat{\boldsymbol{\beta}})}{n\hat{\sigma}^2}\right\}. \end{aligned} \tag{7.65}$$

In this case, the bias of the log-likelihood can be evaluated exactly using the properties of the normal distribution, as discussed in Subsection 3.5.1, and it is given by

$$E_G\left[\log f(\boldsymbol{y}|\hat{\boldsymbol{\theta}}) - \int \log f(\boldsymbol{z}|\hat{\boldsymbol{\theta}}) dG(\boldsymbol{z})\right] = \frac{n(p+1)}{n-p-2} \tag{7.66}$$

[Sugiura (1978)]. Hence, under the assumption that the true model that generates the data is contained in the specified Gaussian linear regression model, we obtain the following information criterion for which the bias of the log-likelihood is exactly corrected:

$$\begin{aligned} \text{AIC}_C &= -2\log f(\boldsymbol{y}|\hat{\boldsymbol{\theta}}) + 2\frac{n(p+1)}{n-p-2} \\ &= n\left\{\log(2\pi\hat{\sigma}^2) + 1\right\} + 2\frac{n(p+1)}{n-p-2}. \end{aligned} \tag{7.67}$$

On the other hand, recall that the AIC is

$$\text{AIC} = -2\log f(\boldsymbol{y}|\hat{\boldsymbol{\theta}}) + 2(p+1), \tag{7.68}$$

in which the number of free parameters for the model was adjusted to the log-likelihood.

The exact bias correction term in (7.67) can be expanded as

$$\frac{n(p+1)}{n-p-2} = (p+1)\left\{1 + \frac{1}{n}(p+2) + \frac{1}{n^2}(p+2)^2 + \cdots\right\}. \tag{7.69}$$

The $p+1$ factor in the first term on the right-hand side is the asymptotic bias. Hence, for the AIC, the asymptotic bias for the log-likelihood of the model is corrected. Although accurate bias corrections are thus possible for specific models and estimation methods, they are difficult to discuss within a general framework.

Example 2 (Normal model) Although the second-order bias correction term $b_2(G)$ for the log-likelihood and bias $\Delta b_1(G)$ take quite complex forms, such as in (7.60), when determined within the framework of functionals, the results of $b_2(G) - \Delta b_1(G)$ can be simplified substantially for specific models. Here we give these correction terms for the normal model, $N(\mu, \sigma^2)$. First, the derivatives of statistical functionals $T_\mu(G)$ and $T_{\sigma^2}(G)$ are given as follows:

$$\begin{aligned}
T_\mu^{(1)}(x;G) &= x-\mu, \quad T_\mu^{(j)}(x_1,\ldots,x_j;G) = 0 \quad (j \geq 2),\\
T_{\sigma^2}^{(1)}(x;G) &= (x-\mu)^2 - \sigma^2,\\
T_{\sigma^2}^{(2)}(x,y;G) &= -2(x-\mu)(y-\mu),\\
T_{\sigma^2}^{(j)}(x_1,\ldots,x_j;G) &= 0 \quad (j \geq 3).
\end{aligned} \tag{7.70}$$

Using these results, we can obtain the second-order bias correction terms:

$$\begin{aligned}
b_2(G) &= 3 - \frac{\mu_4}{\sigma^4} - \frac{1}{2}\frac{\mu_6}{\sigma^6} + 4\frac{\mu_3^2}{\sigma^6} + \frac{3}{2}\frac{\mu_4^2}{\sigma^8},\\
\Delta b_1(G) &= 3 - \frac{3}{2}\frac{\mu_4}{\sigma^4} - \frac{\mu_6}{\sigma^6} + 4\frac{\mu_3^2}{\sigma^6} + \frac{3}{2}\frac{\mu_4^2}{\sigma^8},\\
b_2(G) - \Delta b_1(G) &= \frac{1}{2}\left(\frac{\mu_4}{\sigma^4} + \frac{\mu_6}{\sigma^6}\right),
\end{aligned} \tag{7.71}$$

where μ_j is the jth-order central moment of the true distribution G. These results indicate that although $b_2(G)$ and $\Delta b_1(G)$ are somewhat complex, $b_2(G) - \Delta b_1(G)$ has a relatively simple form. Consequently, the bias correction term with third-order accuracy is given by

$$\begin{aligned}
&b_1(G) - \frac{1}{n}\Delta b_1(G) + \frac{1}{n}b_2(G)\\
&\quad = \frac{1}{2}\left(1 + \frac{\mu_4}{\sigma^4}\right) + \frac{1}{2n}\left(\frac{\mu_4}{\sigma^4} + \frac{\mu_6}{\sigma^6}\right).
\end{aligned} \tag{7.72}$$

Table 7.1. Bias correction terms for the normal distribution model and Laplace distribution model.

True bias	$b_1(G)$	$b_1(G) + \frac{1}{n} b_2(G)$	$\frac{1}{n} \Delta b_1(G)$
Normal distribution	2	$2 + \frac{6}{n}$	$-\frac{3}{n}$
Laplace distribution	3.5	$3.5 + \frac{6}{n}$	$-\frac{42}{n}$

Example 3 (Numerical results) We now show the results of Monte Carlo experiments for two cases, the normal distribution and the Laplace distribution (two-sided exponential distribution):

$$
\begin{aligned}
g(x) &= \frac{1}{\sqrt{2\pi}} \exp\left(-\frac{x^2}{2}\right), \\
g(x) &= \frac{1}{2} \exp(-|x|).
\end{aligned}
\tag{7.73}
$$

The specified model is a normal distribution $\{f(x|\mu, \sigma^2); (\mu, \sigma^2) \in \Theta\}$, and the unknown parameters μ and σ^2 are estimated by the maximum likelihood method. The central moments are $\mu_3 = 0$C$\mu_4 = 3$, and $\mu_6 = 15$ if the true model is a normal distribution and $\mu_3 = 0$C $\mu_4 = 6$, and $\mu_6 = 90$ if the true model is a Laplace distribution.

Table 7.1 shows the asymptotic bias $b_1(G)$ of the log-likelihood, the second-order correction term $b_1(G) + \frac{1}{n} b_2(G)$, and the bias $\frac{1}{n} \Delta b_1(G)$ for the asymptotic bias of the maximum likelihood model $f(x|\hat{\mu}, \hat{\sigma}^2)$ calculated using the results in Example 2. If the true distribution is a normal distribution, then the absolute value of $\Delta b_1(G)$ is half $b_2(G)$. However, for a Laplace distribution, it is more than seven times greater than $b_2(G)$. Therefore, in general, it would be meaningless to correct for only $b_2(G)$. One of the advantages of the AIC is that the bias correction term does not depend on the distribution G, and, therefore, $\Delta b_{\mathrm{AIC}}(\hat{G}) = 0$.

Tables 7.2 and 7.3 show the values of $b_1(\hat{G})$C $b_1(\hat{G}) + \frac{1}{n} b_2(\hat{G})$, and $b_1(\hat{G}) + \frac{1}{n}(b_2(\hat{G}) - \Delta b_1(\hat{G}))$ obtained by substituting the empirical distribution function $\hat{G}$ into the true bias $b(G)$, asymptotic bias $b_1(G)$, and second-order correction terms $b_1(G) + \frac{1}{n} b_2(G)$ by assuming that the true distribution is a normal distribution (Table 7.2) and a Laplace distribution (Table 7.3). These values were obtained by conducting 10,000 Monte Carlo iterations.

For $n = 200$ or higher, the bias correction term yields substantially good estimators, not only for the case in which the true distribution G is used, but also for the case in which the empirical distribution function $\hat{G}$ is used. In contrast, for $n = 25$, the asymptotic bias is substantially underevaluated, indicating the effectiveness of the second-order correction.

Table 7.2. Bias correction terms and their estimates for normal distribution models.

Sample size n	25	50	100	200	400	800
True bias $b(G)$	2.27	2.13	2.06	2.03	2.02	2.01
$b_1(G)$	2.00	2.00	2.00	2.00	2.00	2.00
$b_1(G) + \frac{1}{n}b_2(G)$	2.24	2.12	2.06	2.03	2.02	2.01
$b_1(\hat{G})$	1.89	1.94	1.97	1.99	1.99	2.00
$b_1(\hat{G}) + \frac{1}{n}b_2(\hat{G})$	2.18	2.08	2.04	2.02	2.01	2.00
$b_1(\hat{G}) + \frac{1}{n}(b_2(\hat{G}) - \Delta b_1(\hat{G}))$	2.18	2.10	2.06	2.03	2.01	2.01

In practical situations, G is unknown and we have to estimate the first- and second-order bias correction terms. When the true distribution is assumed to be a normal distribution, the estimator $b_1(\hat{G})$ of the asymptotic bias takes a smaller value, 1.89, than the value corrected by the AIC, i.e., 2. The difference -0.11 is in close agreement with the bias $\Delta b_1(G)/n = -3/25 = -0.12$. In contrast, the second-order bias correction gives a good approximation to the true bias.

If the true distribution is assumed to be a Laplace distribution, then the correction terms $b_1(G)$ and $b_1(G) + \frac{1}{n}b_2(G)$ yield relatively good approximations to $b(G)$. However, their estimates $b_1(\hat{G})$ and $b_1(\hat{G}) + \frac{1}{n}b_2(\hat{G})$ have significant biases because of the large value of the bias of the asymptotic bias estimate $b_1(\hat{G})$, $\Delta b_1(G)/n = -42/n$. In fact, the bias correction $b_1(\hat{G})+(b_2(\hat{G}) - \Delta b_1(\hat{G}))/n$ gives a remarkably accurate approximation to the true bias.

We notice that while the correction with $\Delta b_1(G)/n$ works well when $n = 50$ or higher, it is virtually useless when $n = 25$. This is due to the poor estimation accuracy of the first-order corrected bias and seems to indicate a limitation of high-order correction techniques. A possible solution to this problem is the bootstrap method shown in the next chapter.

Table 7.3. Bias correction terms and their estimates for Laplace distribution models.

Sample size n	25	50	100	200	400	800
True bias $b(G)$	3.88	3.66	3.57	3.53	3.52	3.51
$b_1(G)$	3.50	3.50	3.50	3.50	3.50	3.50
$b_1(G) + \frac{1}{n} b_2(G)$	3.74	3.62	3.56	3.53	3.52	3.51
$b_1(\hat{G})$	2.59	2.93	3.17	3.31	3.40	3.45
$b_1(\hat{G}) + \frac{1}{n} b_2(\hat{G})$	3.30	3.31	3.34	3.39	3.43	3.46
$b_1(\hat{G}) + \frac{1}{n}(b_2(\hat{G}) - \Delta b_1(\hat{G}))$	3.28	3.43	3.49	3.51	3.51	3.51

8

Bootstrap Information Criterion

Advances in computing now allow numerical methods to be used for modeling complex systems, instead of analytic methods. Complex Bayesian models can now be used for practical applications by using numerical methods such as the Markov chain Monte Carlo (MCMC) technique. Also, when the maximum likelihood estimator cannot be obtained analytically, it is possible to obtain it by a numerical optimization method. In conjunction with the development of numerical methods, model evaluation must now deal with extremely complex and increasingly diverse models. The bootstrap information criterion [Efron (1983), Wong (1983), Konishi and Kitagawa (1996), Ishiguro et al. (1997), Cavanaugh and Shumway (1997), and Shibata (1997)], obtained by applying the bootstrap methods originally proposed by Efron (1979), permits the evaluation of models estimated through complex processes.

8.1 Bootstrap Method

The bootstrap method has received considerable interest due to its ability to provide effective solutions to problems that cannot be solved by analytic approaches based on theories or formulas. A salient feature of the bootstrap method is that it uses massive iterative computer calculations rather than analytic expressions. This makes the bootstrap method a flexible statistical method that can be applied to complex problems employing very weak assumptions.

As a solution to the problem of nonparametric estimation of the bias and variance (or standard error) of an estimator, Efron (1979) introduced the bootstrap method as a more effective technique than the traditional jackknife method. As Efron showed, the bootstrap method can address the problems of variance estimation for sample medians and the estimation of the prediction error in discriminant analysis. Subsequently, the bootstrap method has been applied to the estimation of percentile points in probability distributions of estimators and to the construction of confidence intervals of parameters.

Studies on improving the approximation accuracy of confidence intervals have clarified the theoretical structure of the bootstrap method, and the bootstrap method has become an established practical statistical technique for a variety of applications.

Example books focusing on applications and practical aspects of the bootstrap method to statistical problems are those of Efron and Tibshirani (1993) and Davison and Hinkley (1997). Works addressing the theoretical aspects of the bootstrap method are those of Efron (1982), Hall (1992), and Shao and Tu (1995). In addition, Diaconis and Efron (1983), Efron and Gong (1983), and Efron and Tibshirani (1986) provide introductions to the basic concepts underlying the bootstrap method. In this section, we introduce the basic concepts and procedures for the bootstrap method through the evaluation of the bias and variance of an estimator.

Let $\boldsymbol{X}_n = \{X_1, X_2, \ldots, X_n\}$ be a random sample of size n drawn from an unknown probability distribution $G(x)$. We estimate a parameter θ with respect to the probability distribution $G(x)$ by using an estimator $\hat{\theta} = \hat{\theta}(\boldsymbol{X}_n)$. When observed data $\boldsymbol{x}_n = \{x_1, x_2, \ldots, x_n\}$ are obtained, critical statistical analysis tasks are estimating the parameter θ by the estimator $\hat{\theta} = \hat{\theta}(\boldsymbol{x}_n)$ and evaluating the reliability of the estimation.

The basic quantities used to assess the error in the estimation are the following bias and variance of the estimator:

$$b(G) = E_G[\,\hat{\theta}\,] - \theta, \qquad \sigma^2(G) = E_G\left[\left\{\hat{\theta} - E_G[\hat{\theta}]\right\}^2\right]. \tag{8.1}$$

Both the bias and variance express the statistical error of an estimator and depend on the true probability distribution $G(x)$. The task is to estimate them from the data. Instead of attempting to estimate these quantities analytically for each estimator, the bootstrap method provides an algorithm for estimating them numerically with a computer. Basically, the procedure of the bootstrap method is executed through the following steps:

(1) Estimate the unknown probability distribution $G(x)$ from an empirical distribution function $\hat{G}(x)$, where $\hat{G}(x)$ is a probability distribution function with an equal probability $1/n$ at each point of the n observations $\{x_1, x_2, \ldots, x_n\}$. (See Subsection 5.1.1 for a description of empirical distribution functions.)

(2) Random samples from the empirical distribution function $\hat{G}(x)$ are referred to as *bootstrap samples* and are denoted as $\boldsymbol{X}_n^* = \{X_1^*, X_2^*, \ldots, X_n^*\}$. Similarly, the estimator based on a bootstrap sample is denoted as $\hat{\theta}^* = \hat{\theta}(\boldsymbol{X}_n^*)$. The bias and variance of the estimator in (8.1) are then estimated as

$$b(\hat{G}) = E_{\hat{G}}[\hat{\theta}^*] - \hat{\theta}, \qquad \sigma^2(\hat{G}) = E_{\hat{G}}\left[\left\{\hat{\theta}^* - E_{\hat{G}}[\hat{\theta}^*]\right\}^2\right], \tag{8.2}$$

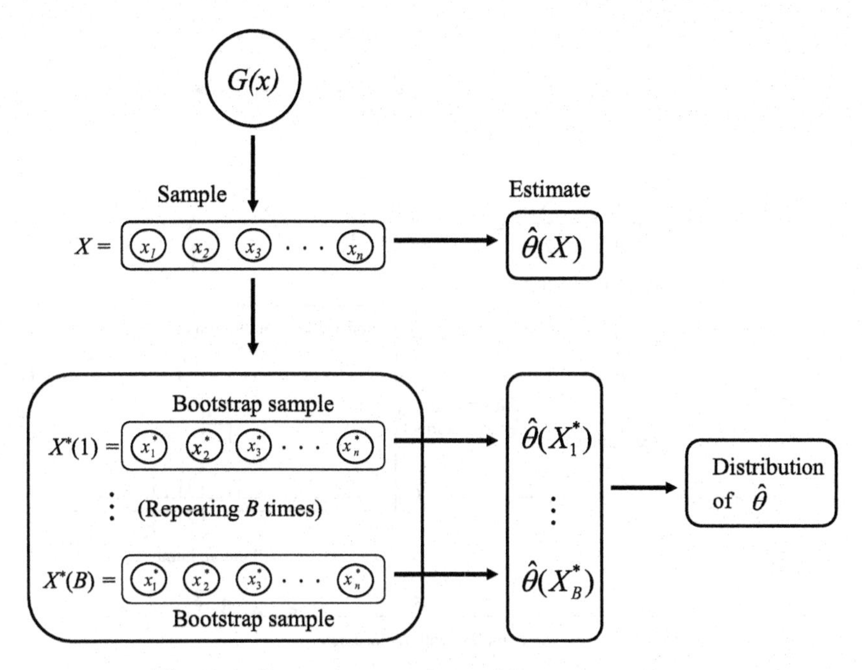

Fig. 8.1. Bootstrap samples and bootstrap estimate.

respectively, where $E_{\hat{G}}$ denotes the expectation with respect to the empirical distribution function $\hat{G}(x)$. The expressions $b(\hat{G})$ and $\sigma^2(\hat{G})$ are referred to as the *bootstrap estimates* of $b(G)$ and $\sigma^2(G)$, respectively.

(3) Exploiting the fact that a bootstrap sample $\boldsymbol{X}_n^*(i) = \{x_1^*(i), \ldots, x_n^*(i)\}$ is obtained by n repeated samples with replacement from the observed data, the bootstrap estimates in (8.2) are numerically approximated by using the Monte Carlo method (see Remark 1). Specifically, bootstrap samples of size n are extracted repeatedly B times, i.e., $\{\boldsymbol{X}_n^*(i); i = 1, \ldots, B\}$, and the corresponding B estimators are denoted as $\{\hat{\theta}^*(i) = \hat{\theta}(\boldsymbol{X}_n^*(i)); i = 1, \ldots, B\}$. Then the bootstrap estimates of the bias and variance in (8.2) are respectively approximated as

$$b(\hat{G}) \approx \frac{1}{B}\sum_{i=1}^{B} \hat{\theta}^*(i) - \hat{\theta}, \qquad \sigma^2(\hat{G}) \approx \frac{1}{B-1}\sum_{i=1}^{B} \left\{\hat{\theta}^*(i) - \hat{\theta}^*(\cdot)\right\}^2,$$

where $\hat{\theta}^*(\cdot) = \sum_{i=1}^{B} \hat{\theta}^*(i)/B$ (see Figure 8.1).

Remark 1 (Bootstrap sample) The following is a brief explanation of a bootstrap sample. Generally, given any distribution function $G(x)$, random numbers that follow the distribution $G(x)$ can be obtained by generating uniform random numbers u over the interval $[0, 1)$ and substituting them

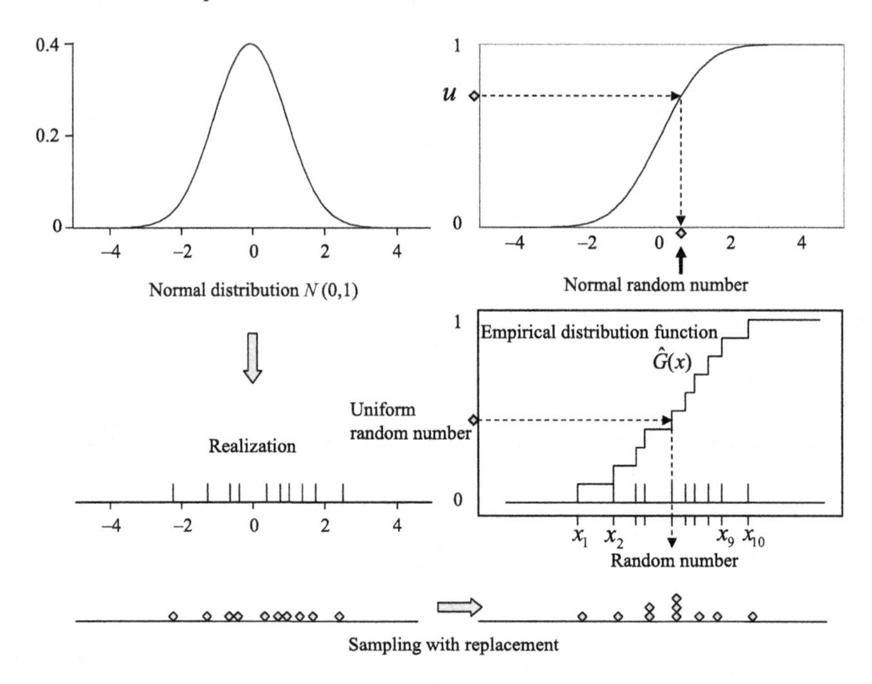

Fig. 8.2. Generation of random numbers (bootstrap samples) from an empirical distribution function.

into the inverse function $G^{-1}(u)$ of $G(x)$. This principle can be applied to empirical distribution functions. Since an empirical distribution function is a discrete distribution with an equal probability $1/n$ at each of the n data points $x_1, x_2, \ldots, x_n$, it follows that

$$\hat{G}^{-1}(u) = \{\text{one of the observations } x_1, \ldots, x_n\}. \tag{8.3}$$

It is clear that the bootstrap sample obtained by repeating this process is simply a set of n data points that are sampled with replacement from n observations.

Figure 8.2 shows the relationship among density functions, distribution functions, empirical distribution functions, and bootstrap samples. The upper left plot shows a normal density function. The upper right plot shows the distribution function obtained by integrating the normal density function. In this plot, a normal random number can be obtained by generating a uniform random number u over the interval [0,1) on the ordinate, determining the intersection between the line drawn horizontally from the number and the distribution function, tracing a line perpendicularly downward from the intersection, and determining the point at which the line crosses the x-axis. The lower left plot shows data generated from a standard normal distribution $N(0, 1)$. The plot on the lower right shows an empirical distribution function determined by these data points and an example in which random numbers are generated in a similar manner using the distribution function. From the figure, it is clear that random numbers can be obtained in equal probabilities. The figure also shows that bootstrap samples can be obtained by sampling with replacement of the observed data used in the construction of the empirical distribution function.

Remark 2 (Bootstrap simulation) The bootstrap method can be applied to a broad range of complex inference problems because the Monte Carlo method employed in step (3) above permits the numerical approximation of bootstrap estimates. For a parameter θ and a function determined by the estimator $\hat{\theta}$, such as $\hat{\theta} - \theta$ and $\{\hat{\theta} - E_G[\hat{\theta}]\}^2$, we write $r(\hat{\theta}, \theta)$. The bias and variance of the estimator can be expressed as

$$E_G\left[r(\hat{\theta}, \theta)\right] = \int \cdots \int r(\hat{\theta}, \theta) \prod_{\alpha=1}^{n} dG(x_\alpha), \tag{8.4}$$

that is, the expectation of $r(\hat{\theta}, \theta)$, appropriately defined [see Subsection 3.1.1 for a description of $dG(x)$]. The bootstrap method estimates this quantity by using

$$E_{\hat{G}}\left[r(\hat{\theta}^*, \hat{\theta})\right] = \int \cdots \int r(\hat{\theta}^*, \hat{\theta}) \prod_{\alpha=1}^{n} d\hat{G}(x_\alpha^*). \tag{8.5}$$

In other words, the bootstrap method performs an inference process based on $\{G, \theta, \hat{\theta}\}$ by replacing it with $\{\hat{G}, \hat{\theta}, \hat{\theta}^*\}$.

The expectation in (8.4) cannot be computed since the probability distribution $G(x)$ is unknown. In contrast, since the expectation in (8.5) is taken with respect to the joint distribution $\prod_{\alpha=1}^{n} d\hat{G}(x_\alpha^*)$ of the empirical distribution function, which is a known probability distribution, it can be numerically approximated using a Monte Carlo simulation. Specifically, a set of n random numbers (bootstrap sample) that follows the empirical distribution function is generated repeatedly, and the expectation is numerically approximated as

$$\begin{aligned} E_{\hat{G}}\left[r(\hat{\theta}^*, \hat{\theta})\right] &= \int \cdots \int r(\hat{\theta}^*, \hat{\theta}) \prod_{\alpha=1}^{n} d\hat{G}(x_\alpha^*) \\ &\approx \frac{1}{B} \sum_{i=1}^{B} r(\hat{\theta}^*(i), \hat{\theta}), \end{aligned} \tag{8.6}$$

where $\hat{\theta}^*(i)$ denotes an estimate based on the ith set of random numbers obtained by repeatedly generating random numbers of size n B times from $\hat{G}(x)$.

This method exploits the fact that a set of random numbers of size n from an empirical distribution function, that is, a bootstrap sample of size n, is equivalent to the sampling with replacement of a sample of size n from the observed data $\{x_1, x_2, \ldots, x_n\}$. It is clear, therefore, that this sampling process cannot be performed unless n observations are obtained independently from the same distribution.

Remark 3 (Number of bootstrap samples) Errors in the approximation by Monte Carlo simulation can be ignored if the number B of bootstrap repetitions becomes infinitely large. In practice, however, the number of bootstrap repetitions for estimating a bias or variance (standard error) is usually $B = 50 \sim 200$. In contrast, the estimation of percentage points of the probability distribution of an estimator requires $B = 1000 \sim 2000$.

8.2 Bootstrap Information Criterion

8.2.1 Bootstrap Estimation of Bias

Recall that the information criterion is obtained by correcting the bias,

$$b(G) = E_{G(\boldsymbol{x})}\left[\sum_{\alpha=1}^{n} \log f(X_\alpha|\hat{\boldsymbol{\theta}}(\boldsymbol{X}_n)) - nE_{G(z)}\left[\log f(Z|\hat{\boldsymbol{\theta}}(\boldsymbol{X}_n))\right]\right], \quad (8.7)$$

when the expected log-likelihood of a model is estimated by the log-likelihood, where $E_{G(\boldsymbol{x})}$ denotes the expectation with respect to the joint distribution of a random sample $\boldsymbol{X}_n$, and $E_{G(z)}$ represents the expectation with respect to the probability distribution G.

The second term $E_{G(z)}\left[\log f(Z|\hat{\boldsymbol{\theta}}(\boldsymbol{X}_n))\right]$ on the right-hand side of (8.7) can be expressed as

$$E_{G(z)}\left[\log f(Z|\hat{\boldsymbol{\theta}}(\boldsymbol{X}_n))\right] = \int \log f(z|\hat{\boldsymbol{\theta}}(\boldsymbol{X}_n))dG(z). \quad (8.8)$$

This represents the expectation with respect to the distribution $G(z)$ of the future data z that is independent of the random sample $\boldsymbol{X}_n$. In addition, the first term on the right-hand side is the log-likelihood, which can be expressed as an integral by using the empirical distribution function $\hat{G}(x)$,

$$\sum_{\alpha=1}^{n} \log f(X_\alpha|\hat{\boldsymbol{\theta}}(\boldsymbol{X}_n)) = n\int \log f(z|\hat{\boldsymbol{\theta}}(\boldsymbol{X}_n))d\hat{G}(z). \quad (8.9)$$

Recall here that the information criteria AIC, TIC, and GIC were obtained analytically based on asymptotic theory for the terms in (8.7) under suitable conditions. In contrast, the bootstrap information criterion is obtained through a numerical approximation by using the bootstrap method, instead of analytically deriving the bias of the log-likelihood for each statistical model.

In constructing the bootstrap information criterion, the true distribution $G(x)$ is replaced with an empirical distribution function $\hat{G}(x)$. In connection with this replacement, the random variable and estimator contained in (8.7) are substituted as follows:

$$\begin{aligned} G(x) &\longrightarrow \hat{G}(x), \\ X_\alpha \sim G(x) &\longrightarrow X_\alpha^* \sim \hat{G}(x), \\ Z \sim G(z) &\longrightarrow Z^* \sim \hat{G}(z), \\ E_{G(\boldsymbol{x})},\ E_{G(z)} &\longrightarrow E_{\hat{G}(\boldsymbol{x}^*)},\ E_{\hat{G}(z^*)}, \\ \hat{\boldsymbol{\theta}} = \hat{\boldsymbol{\theta}}(\boldsymbol{X}) &\longrightarrow \hat{\boldsymbol{\theta}}^* = \hat{\boldsymbol{\theta}}(\boldsymbol{X}^*). \end{aligned}$$

Therefore, the bootstrap bias estimate of (8.7) becomes

$$b^*(\hat{G}) = E_{\hat{G}(\boldsymbol{x}^*)}\left[\sum_{\alpha=1}^{n} \log(X_\alpha^* | \hat{\boldsymbol{\theta}}(\boldsymbol{X}_n^*)) - nE_{\hat{G}(z^*)}\left[\log f(Z^*|\hat{\boldsymbol{\theta}}(\boldsymbol{X}_n^*))\right]\right]. \tag{8.10}$$

In the following, we describe in detail how the terms are replaced in the framework of the bootstrap method.

Given a set of data $\boldsymbol{x}_n = \{x_1, x_2, \ldots, x_n\}$, in the bootstrap method, the true distribution function $G(x)$ is first substituted by an empirical distribution function $\hat{G}(x)$. A statistical model $f(x|\hat{\boldsymbol{\theta}}(\boldsymbol{X}_n^*))$ is constructed based on a bootstrap sample $\boldsymbol{X}_n^*$ from the empirical distribution function. Then, the expected log-likelihood of the model $f(x|\hat{\boldsymbol{\theta}}(\boldsymbol{X}_n^*))$ when the empirical distribution function is considered as the true distribution is calculated as

$$\begin{aligned} E_{\hat{G}(z)}\left[\log f(Z|\hat{\boldsymbol{\theta}}(\boldsymbol{X}_n^*))\right] &= \int \log f(z|\hat{\boldsymbol{\theta}}(\boldsymbol{X}_n^*)) d\hat{G}(z) \\ &= \frac{1}{n}\sum_{\alpha=1}^{n} \log f(x_\alpha|\hat{\boldsymbol{\theta}}(\boldsymbol{X}_n^*)) \\ &\equiv \frac{1}{n}\ell(\boldsymbol{x}_n|\hat{\boldsymbol{\theta}}(\boldsymbol{X}_n^*)). \end{aligned} \tag{8.11}$$

Thus, if $\hat{G}(x)$ is considered as the true distribution, the expected log-likelihood is simply the log-likelihood.

On the other hand, since the log-likelihood, which is an estimator of the expected log-likelihood, is constructed by reusing the bootstrap sample $\boldsymbol{X}_n^*$, it can be represented as

$$\begin{aligned} E_{\hat{G}^*(z)}\left[\log f(Z|\hat{\boldsymbol{\theta}}(\boldsymbol{X}_n^*))\right] &= \int \log f(z|\hat{\boldsymbol{\theta}}(\boldsymbol{X}^*)) d\hat{G}^*(z) \\ &= \frac{1}{n}\sum_{\alpha=1}^{n} \log f(X_\alpha^*|\hat{\boldsymbol{\theta}}(\boldsymbol{X}_n^*)) \\ &\equiv \frac{1}{n}\ell(\boldsymbol{X}_n^*|\hat{\boldsymbol{\theta}}(\boldsymbol{X}_n^*)), \end{aligned} \tag{8.12}$$

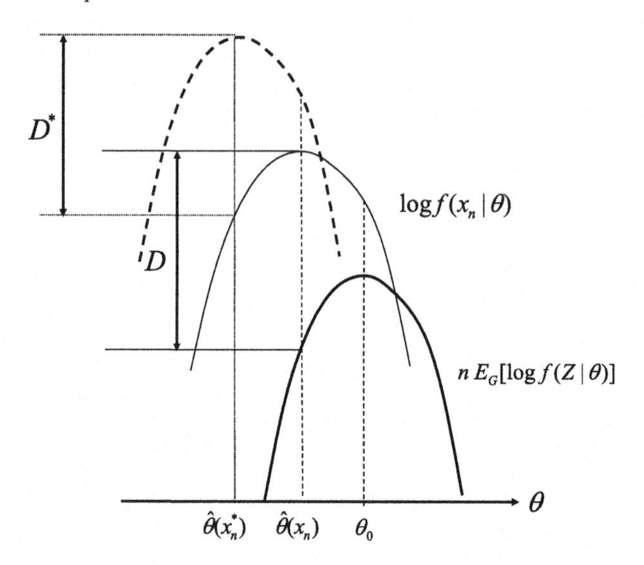

Fig. 8.3. Estimation of the bias of the log-likelihood by the bootstrap method. The bold curve shows the expected log-likelihood, the thin curve the log-likelihood, and the dashed curve the log-likelihood based on a bootstrap sample.

where $\hat{G}^*$ is an empirical distribution function based on the bootstrap sample $\boldsymbol{X}_n^*$. Consequently, using the bootstrap method, the bootstrap bias estimate in (8.7) can be written as

$$\begin{aligned} b^*(\hat{G}) &= E_{\hat{G}(\boldsymbol{x}^*)}\left[\ell(\boldsymbol{X}_n^*|\hat{\boldsymbol{\theta}}(\boldsymbol{X}_n^*)) - \ell(\boldsymbol{X}_n|\hat{\boldsymbol{\theta}}(\boldsymbol{X}_n^*))\right] \\ &= \int \cdots \int \left\{\ell(\boldsymbol{X}_n^*|\hat{\boldsymbol{\theta}}(\boldsymbol{X}_n^*)) - \ell(\boldsymbol{X}_n|\hat{\boldsymbol{\theta}}(\boldsymbol{X}_n^*))\right\} \prod_{\alpha=1}^{n} d\hat{G}(x_\alpha^*). \end{aligned} \tag{8.13}$$

As noted in the preceding section, the most significant feature of the bootstrap information criterion is that this integral can be approximated numerically by the Monte Carlo method by using the fact that $\hat{G}$ is a known probability distribution (the empirical distribution function).

In the bootstrap information criterion, we use D^* instead of D in Figure 8.3, which is equivalent to determining the expectation of the difference between $E_{\hat{G}(z)}\left[\log f(Z|\hat{\boldsymbol{\theta}}(\boldsymbol{X}_n^*))\right]$ and $E_{\hat{G}^*(z)}\left[\log f(Z|\hat{\boldsymbol{\theta}}(\boldsymbol{X}_n^*))\right]$ instead of determining the expectation of the difference between the expected log-likelihood $E_{G(z)}\left[\log f(Z|\hat{\boldsymbol{\theta}}(\boldsymbol{X}_n))\right]$ and the log-likelihood $nE_{\hat{G}(z)}\left[\log f(Z|\hat{\boldsymbol{\theta}}(\boldsymbol{X}_n))\right] = \log f(\boldsymbol{X}_n|\hat{\boldsymbol{\theta}}(\boldsymbol{X}_n))$.

8.2.2 Bootstrap Information Criterion, EIC

Let us extract B sets of bootstrap samples of size n and write the ith bootstrap sample as $\boldsymbol{X}_n^*(i) = \{X_1^*(i), X_2^*(i), \ldots, X_n^*(i)\}$. We denote the difference between (8.12) and (8.11) with respect to the sample $\boldsymbol{X}_n^*(i)$ as

$$D^*(i) = \ell(\boldsymbol{X}_n^*(i)|\hat{\boldsymbol{\theta}}(\boldsymbol{X}_n^*(i))) - \ell(\boldsymbol{x}_n|\hat{\boldsymbol{\theta}}(\boldsymbol{X}_n^*(i))), \tag{8.14}$$

where $\hat{\boldsymbol{\theta}}(\boldsymbol{X}_n^*(i))$ is an estimate of $\boldsymbol{\theta}$ obtained from the ith bootstrap sample. Then the expectation in (8.13) based on B bootstrap samples can be numerically approximated as

$$b^*(\hat{G}) \approx \frac{1}{B}\sum_{i=1}^{B} D^*(i) \equiv b_B(\hat{G}). \tag{8.15}$$

The quantity $b_B(\hat{G})$ is the bootstrap estimate of the bias $b(G)$ of the log-likelihood. Consequently, the bootstrap methods yield an information criterion as follows:

Bootstrap information criterion, EIC. Let $f(x|\hat{\boldsymbol{\theta}})$ be a statistical model estimated by a procedure such as the maximum likelihood, and let $b_B(\hat{G})$ be the bootstrap bias estimate of the log-likelihood. The bootstrap information criterion is given by

$$\text{EIC} = -2\sum_{\alpha=1}^{n} \log f(X_\alpha|\hat{\boldsymbol{\theta}}) + 2b_B(\hat{G}). \tag{8.16}$$

This quantity was referred to as the extended information criterion (EIC) by Ishiguro et al. (1997). Konishi and Kitagawa (1996) have given a theoretical justification for the use of the bootstrap method in the bias estimate of a log-likelihood. For the use of the bootstrap for model uncertainty, we refer to Kishino and Hasegawa (1989), Shimodaira and Hasegawa (1999), Burnham and Anderson (2002, Chapter 6), and Shimodaira (2004).

8.3 Variance Reduction Method

8.3.1 Sampling Fluctuation by the Bootstrap Method

The bootstrap method can be applied without analytically cumbersome procedures under very weak assumptions, that is, the estimator is invariant with respect to the reordering of the sample. In applying the bootstrap method, however, care should be paid to the magnitude of the fluctuations due to bootstrap simulations and approximation errors, in addition to the sample fluctuations of the bias estimate itself.

Table 8.1. True bias $b(G)$ and the means and variances of the bootstrap estimate.

n	25	100	400	1,600
$b(G)$	2.27	2.06	2.02	2.00
$\hat{\mathrm{E}}(b_B(\hat{G}))$	2.23	2.04	2.01	2.00
$\mathrm{Var}(b_B(\hat{G}))$	0.51	0.61	2.07	8.04
$\mathrm{Var}(D^*)$	24.26	56.06	203.63	797.66

For a set of given observations, the approximation $b_B(\hat{G})$ in (8.15) converges to the bootstrap estimate $b^*(\hat{G})$ of the bias in (8.13), with probability one, if the number of bootstrap resampling B goes to infinity. However, because simulation errors occur for finite B, procedures must be devised to reduce the error. This can be considered a reduction of simulation error for $b_B(\hat{G})$ for a given sample. The variance reduction method described in the next section, called the *efficient bootstrap simulation method* or the *efficient resampling method*, provides an effective, yet extremely simple method of reducing any fluctuation in the bootstrap bias estimation of log-likelihood.

Example 1 (Variance of bootstrap bias estimate) Table 8.1 shows the true bias $b(G)$ and bootstrap estimates of $b(G)$ when the true distribution $G(x)$ is assumed to be the standard normal distribution $N(0, 1)$ and the parameters of the normal distribution model $N(\mu, \sigma^2)$ are estimated by the maximum likelihood method. The table shows the average of $b_B(\hat{G})$, variance of $b_B(\hat{G})$, and variance of $D^*(i)$, obtained by setting the number of bootstrap replications to $B = 100$ and repeating the Monte Carlo simulation 10,000 times. The table shows that the variance of $b_B(\hat{G})$ grows as n increases and, when the sample size n is large, an accurate estimate cannot be obtained if the number of bootstrap replications is moderate, e.g., $B = 100$. It is clear that the variance of $D^*(i)$ is approximately B times $b_B(\hat{G})$ and is approximately half the sample size n. The variance of $D^*(i)$ divided by B (i.e., 100 in this example) is attributable to the bootstrap approximation error due to the fluctuation of $b_B(\hat{G})$. Therefore, it can be seen that reducing the variance caused by the bootstrap simulation is essential, especially when the sample size n is large.

8.3.2 Efficient Bootstrap Simulation

We set the difference between the log-likelihood of the model in (8.7) and (n times) the expected log-likelihood as

$$D(\boldsymbol{X}_n; G) = \log f(\boldsymbol{X}_n|\hat{\boldsymbol{\theta}}) - n \int \log f(z|\hat{\boldsymbol{\theta}}) dG(z), \tag{8.17}$$

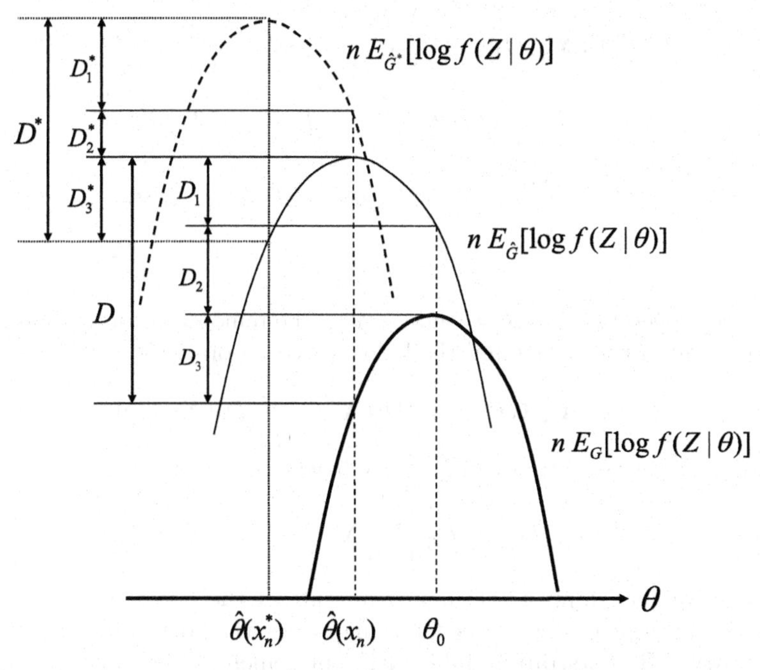

Fig. 8.4. Decomposition of the bias term for variance reduction. For simplicity, a maximum likelihood estimator is shown for which $\hat{\theta}(X)$ attains the maximum of the function.

where $\log f(\boldsymbol{X}_n|\hat{\boldsymbol{\theta}}) = \sum_{\alpha=1}^{n} \log f(X_\alpha|\hat{\boldsymbol{\theta}})$. In this case, $D(\boldsymbol{X}_n; G)$ can be decomposed into three terms (Figure 8.4):

$$D(\boldsymbol{X}_n; G) = D_1(\boldsymbol{X}_n; G) + D_2(\boldsymbol{X}_n; G) + D_3(\boldsymbol{X}_n; G), \tag{8.18}$$

where

$$\begin{aligned} D_1(\boldsymbol{X}_n; G) &= \log f(\boldsymbol{X}_n|\hat{\boldsymbol{\theta}}) - \log f(\boldsymbol{X}_n|\boldsymbol{\theta}), \\ D_2(\boldsymbol{X}_n; G) &= \log f(\boldsymbol{X}_n|\boldsymbol{\theta}) - n\int \log f(z|\boldsymbol{\theta})dG(z), \\ D_3(\boldsymbol{X}_n; G) &= n\int \log f(z|\boldsymbol{\theta})dG(z) - n\int \log f(z|\hat{\boldsymbol{\theta}})dG(z). \end{aligned} \tag{8.19}$$

In the derivation of the information criterion, the bias represents the expected value of $D(\boldsymbol{X}_n; G)$ with respect to the joint distribution of a random sample $\boldsymbol{X}_n$. By taking the expectation term by term on the right-hand side of (8.18), we obtain the second term as

$$
\begin{aligned}
&E_G\left[D_2(\boldsymbol{X}_n;G)\right]\\
&\qquad = E_G\left[\log f(\boldsymbol{X}_n|\boldsymbol{\theta}) - n\int \log f(z|\boldsymbol{\theta})dG(z)\right]\\
&\qquad = \sum_{\alpha=1}^{n} E_G\left[\log f(X_\alpha|\boldsymbol{\theta})\right] - nE_G[\log f(Z|\boldsymbol{\theta})]\\
&\qquad = 0.
\end{aligned} \tag{8.20}
$$

Thus, the expectation in the second term can be removed from the bias of the log-likelihood of the model and the following equation holds:

$$
E_G\left[D(\boldsymbol{X}_n;G)\right] = E_G\left[D_1(\boldsymbol{X}_n;G) + D_3(\boldsymbol{X}_n;G)\right]. \tag{8.21}
$$

Similarly, for the bootstrap estimate, we have

$$
E_{\hat{G}}\left[D(\boldsymbol{X}_n^*;\hat{G})\right] = E_{\hat{G}}\left[D_1(\boldsymbol{X}_n^*;\hat{G}) + D_3(\boldsymbol{X}_n^*;\hat{G})\right]. \tag{8.22}
$$

Therefore, in the Monte Carlo approximation of the bootstrap estimate, it suffices to take the average of the following values as a bootstrap bias estimate after drawing B bootstrap samples with replacement:

$$
\begin{aligned}
&D_1(\boldsymbol{X}_n^*(i);\hat{G}) + D_3(\boldsymbol{X}_n^*(i);\hat{G})\\
&\qquad = \log f(\boldsymbol{X}_n^*(i)|\hat{\boldsymbol{\theta}}^*(i)) - \log f(\boldsymbol{X}_n^*(i)|\hat{\boldsymbol{\theta}})\\
&\qquad\quad + \log f(\boldsymbol{X}_n|\hat{\boldsymbol{\theta}}) - \log f(\boldsymbol{X}_n|\hat{\boldsymbol{\theta}}^*(i)).
\end{aligned} \tag{8.23}
$$

This implies that we may use

$$
b_B(\hat{G}) = \frac{1}{B}\sum_{i=1}^{B}\left\{D_1(\boldsymbol{X}_n^*(i);\hat{G}) + D_3(\boldsymbol{X}_n^*(i);\hat{G})\right\} \tag{8.24}
$$

as a bootstrap bias estimate.

In fact, conditional on the observed data, it can be shown that the orders of asymptotic conditional variances of two bootstrap estimates are

$$
\mathrm{Var}\left[\frac{1}{B}\sum_{i=1}^{B}\left\{D(\boldsymbol{X}_n^*;\hat{G})\right\}\right] = \frac{1}{B}O(n), \tag{8.25}
$$

$$
\mathrm{Var}\left[\frac{1}{B}\sum_{i=1}^{B}\left\{D_1(\boldsymbol{X}_n^*;\hat{G}) + D_3(\boldsymbol{X}_n^*;\hat{G})\right\}\right] = \frac{1}{B}O(1). \tag{8.26}
$$

The difference between these orders can be explained by noting that, whereas the order of the asymptotic variance of the terms $B^{-1}\sum_{i=1}^{B} D_1(X_n^*;\hat{G})$ and

$B^{-1}\sum_{i=1}^{B} D_3(X_n^*;\hat{G})$ in (8.26) is $O(1)$ if $\hat{\boldsymbol{\theta}}$ is the maximum likelihood estimator, the order of the asymptotic variance of $B^{-1}\sum_{i=1}^{B} D_2(X_n^*;\hat{G})$ is $O(n)$. The theoretical justification for using the simple variance reduction technique mentioned above is as follows.

If there exists a function $\mathrm{IF}(X;G)$ such that its expectation is 0, then the expectation of $D(\boldsymbol{X}_n;G)$ and the expectation of $D(\boldsymbol{X}_n;G)-\sum_{\alpha=1}^{n}\mathrm{IF}(X_\alpha;G)$ in (8.17) are equal. Satisfying such a property is

$$\mathrm{IF}(X;G) \equiv \log f(X|\boldsymbol{\theta}) - \int \log f(z|\boldsymbol{\theta})dG(z). \tag{8.27}$$

This is the influence function of $D(\boldsymbol{X}_n;G)$, which indicates that, while the expectation remains unchanged, the order of the asymptotic variance of $D(\boldsymbol{X}_n;G)$ is $O(n)$, whereas the order of the asymptotic variance of $D(\boldsymbol{X}_n;G)-\sum_{\alpha=1}^{n}\mathrm{IF}(X_\alpha;G)$ is $O(1)$. Therefore, by using

$$\begin{aligned}E_{\hat{G}}\left[D(\boldsymbol{X}_n^*;\hat{G})\right] &= E_{\hat{G}}\left[D(\boldsymbol{X}_n^*;\hat{G}) - \sum_{\alpha=1}^{n}\mathrm{IF}(X_\alpha^*;\hat{G})\right]\\ &= E_{\hat{G}}\left[\log f(\boldsymbol{X}_n^*|\hat{\boldsymbol{\theta}}^*) - \log f(\boldsymbol{X}_n^*|\hat{\boldsymbol{\theta}})\right.\\ &\qquad \left. + \log f(\boldsymbol{X}_n|\hat{\boldsymbol{\theta}}) - \log f(\boldsymbol{X}_n|\hat{\boldsymbol{\theta}}^*)\right]\end{aligned} \tag{8.28}$$

as a bootstrap bias estimate instead of (8.17), the variance due to bootstrap resampling can be reduced significantly.

This variance reduction technique was originally proposed by Konishi and Kitagawa (1996) and Ishiguro et al. (1997), who verified the effectiveness of this method both theoretically and numerically. Other studies on information criteria based on the bootstrap method include those of Cavanaugh and Shumway (1997) and Shibata (1997).

Example 2 (Variance reduction in bootstrap bias estimates) We show the effect of the variance reduction method for normal distribution models with unknown mean μ and variance σ^2 by assuming a standard normal distribution $N(0,1)$ for the true distribution.

Table 8.2 shows the bias terms D, D_1+D_3, D_1, D_2, and D_3 for sample sizes $n=25$, 100, 400, and 1,600, respectively. For each n, the first and the second rows show the exact bias term and an estimate obtained by putting $B=100$, namely by using 100 bootstrap resamples. The table shows the values obtained by taking an average over 1,000,000 different samples $\boldsymbol{x}$. The values in brackets show the variances of bootstrap bias estimates. The table shows the merit of using the variance reduction method for large sample size n.

Figure 8.5 shows box plots of the distributions of bootstrap estimates $DCD_1+D_3CD_1CD_2$, and D_3 for $n=25$, 100, 400, and 1,600. The figure

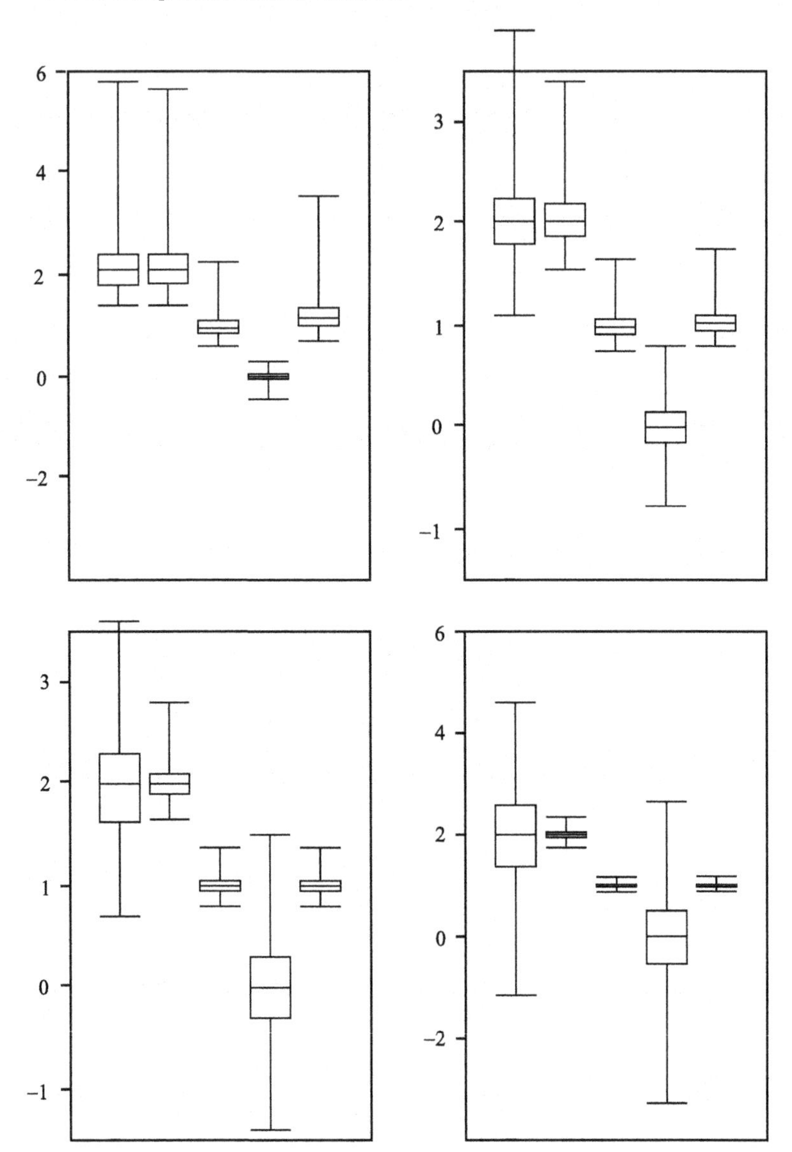

Fig. 8.5. Box–plots of the bootstrap distributions of D, $D_1 + D_3$, D_1, D_2, and D_3 for $n = 25$ (top)C 100 (top right), 400 (bottom left), and 1,600 (bottom right).

Table 8.2. Bias correction terms for normal distribution models when the true distribution is a normal distribution. n is the sample size.

n		D	$D_1 + D_3$	D_1	D_2	D_3
25	Exact	2.27	2.27	1.04	0.00	1.23
	Bootstrap	2.23(0.51)	2.23(0.35)	1.00(0.05)	0.00(0.11)	1.23(0.14)
100	Exact	2.06	2.06	1.01	0.00	1.05
	Bootstrap	2.04(0.61)	2.04(0.10)	1.00(0.02)	0.00(0.49)	1.04(0.03)
400	Exact	2.02	2.02	1.00	0.00	1.01
	Bootstrap	2.01(2.04)	2.01(0.06)	1.00(0.01)	0.00(1.98)	1.01(0.01)
1600	Exact	2.00	2.00	1.00	0.00	1.00
	Bootstrap	2.00(7.98)	2.00(0.04)	1.00(0.01)	0.00(7.97)	1.00(0.01)

Table 8.3. Effect of variance reduction method.

n	25		100		400	
D	0.023	0.237	0.057	0.113	0.206	0.223
$D_1 + D_3$	0.008	0.231	0.005	0.061	0.004	0.019

clearly shows that as n increases, D and $D_1 + D_3$ fluctuate in a different manner because of the spreading of the distribution of D_2. For small n, such as 25, the fluctuations of D_1 and D_3 are large compared with that of D_2, and, as a result, the fluctuations of D and $D_1 + D_3$ are not very different. However, when n increases, the fluctuation of D_2 becomes dominant and that of $D_1 + D_3$ becomes significantly smaller than that of D.

Table 8.3 shows the variances of bootstrap estimates for $n = 25$, 100, and 400. For each n, the left-hand values show the changes in $\hat{G}$ of the bootstrap bias correction terms of (8.15) and (8.24), that is, the variance due to differences in the data. The right-hand values indicate the variance in bootstrap bias estimates obtained by (8.15) and (8.24). The values in the table were obtained for $B = 100$ and are considered to be inversely proportional to B. The table shows that the method of decomposing the difference between the log-likelihood and the expected log-likelihood into D_1 and D_3, respectively, can have a dramatic effect, especially when the sample size n is large. Furthermore, it is shown that since there exists a variance due to fluctuations of the sample, as indicated by the left-hand values, simply increasing the number of bootstrap replications would be meaningless.

8.3.3 Accuracy of Bias Correction

Under certain conditions, the bias of the log-likelihood in (8.7) can be expanded in the form

$$b(G) = b_1(G) + \frac{1}{n}b_2(G) + \frac{1}{n^2}b_3(G) + \cdots. \tag{8.29}$$

In this case, the expectation of the bootstrap estimate of the bias becomes

$$\begin{aligned} E_G\left[b^*(\hat{G})\right] &= E_G\left[b_1(\hat{G}) + \frac{1}{n}b_2(\hat{G})\right] + o(n^{-1}) \\ &= b_1(G) + \frac{1}{n}\Delta b_1(G) + \frac{1}{n}b_2(G) + o(n^{-1}), \end{aligned} \tag{8.30}$$

where $\Delta b_1(G)$ denotes the bias of the first-order bias estimate $b_1(\hat{G})$. Therefore, it follows that when $\Delta b_1(G) = 0$, the bootstrap bias estimate automatically yields the second-order bias correction.

In contrast, the GIC with asymptotic bias estimate $b_1(\hat{G})$ obtained analytically gives

$$E_G\left[b_1(\hat{G})\right] = b_1(G) + \frac{1}{n}\Delta b_1(G) + O(n^{-1}). \tag{8.31}$$

Consequently, even when $\Delta b_1(G) = 0$, a second-order bias correction does not occur, since the second-order bias correction term is given by

$$b_1(G) + \frac{1}{n}\{b_2(G) - \Delta b_1(G)\}. \tag{8.32}$$

Although in the preceding section we derived a second-order bias correction term analytically, in practical situations the bootstrap method offers an alternative approach for estimating it numerically. If $b_1(G)$ is evaluated analytically, then the bootstrap estimate of the second-order bias correction term can be obtained by using

$$\begin{aligned} \frac{1}{n}b_2^*(\hat{G}) = E_{\hat{G}(\boldsymbol{x}^*)} &\left[\log f(\boldsymbol{X}_n^*|\hat{\boldsymbol{\theta}}(\boldsymbol{X}_n^*)) - b_1(\hat{G}) \right. \\ &\left. - nE_{\hat{G}(z)}\left[\log f(Z|\hat{\boldsymbol{\theta}}(\boldsymbol{X}_n^*))\right]\right]. \end{aligned} \tag{8.33}$$

On the other hand, in situations where it is difficult to analytically determine the first-order correction term $b_1(G)$, an estimate of the second-order correction term can be obtained by employing the following two-step bootstrap method:

$$\begin{aligned} \frac{1}{n}b_2^{**}(\hat{G}) = E_{\hat{G}(\boldsymbol{x}^*)} &\left[\log f(\boldsymbol{X}_n^*|\hat{\boldsymbol{\theta}}(\boldsymbol{X}_n^*)) - b_B^*(\hat{G}) \right. \\ &\left. - nE_{\hat{G}(z^*)}[\log f(Z^*|\hat{\boldsymbol{\theta}}(\boldsymbol{X}_n^*))]\right], \end{aligned} \tag{8.34}$$

Table 8.4. Bias correction terms for normal distribution models when the true distribution is normal. n is the sample size.

n	$b(G)$	B_1	B_2	$\hat{B}_1$	$\hat{B}_2$	B_1^*	B_2^*	B_2^{**}
25	2.27	2.00	2.24	1.89	2.18	2.20	2.24	2.33
100	2.06	2.00	2.06	1.97	2.06	2.04	2.06	2.06
400	2.02	2.00	2.02	1.99	2.02	2.01	2.02	2.02

where $b_B^*(\hat{G})$ is the bootstrap estimate of the first-order correction term obtained by (8.15).

Example 3 (Bootstrap higher-order bias correction: normal distribution) We show the effect of the second-order correction for normal distribution models with unknown mean μ and variance σ^2. The true distribution is assumed to be the standard normal distribution $N(0,1)$.

The centered moments of the normal distribution are $\mu_3 = 0$, $\mu_4 = 3$, and $\mu_6 = 15$. As shown in Table 7.1 in Section 7.2, the first-order bias correction term $b_1(G)$ is a function only of the number of observations.

Table 8.4 shows the bias correction terms obtained by running 10,000 Monte Carlo trials for three sample sizes, $n = 25$, 100, and 400, under the assumption that the true distribution is a standard normal distribution. Here, $b(G)$ represents the exact bias, which can be evaluated analytically and can be given as $2n/(n-3)$. In Table 8.4, B_1 and B_2 represent respectively the following first- and second-order correction terms, as indicated in (5.73) and (7.71):

$$B_1 = b_1(G), \quad B_2 = b_1(G) + \frac{1}{n}(b_2(G) - \Delta b_1(G)).$$

In the table, the hat symbol ($\hat{}$) denotes the case in which the empirical distribution function $\hat{G}$ is substituted for the true distribution G, and the symbols * and ** represent estimates obtained by performing 1,000 bootstrap repetitions and the two-stage bootstrap method of (8.34), respectively.

In this case, since the model contains the true distribution, B_1 agrees with the bias correction term (the number of estimated parameters) in the AIC. For $n = 400$, the asymptotic bias B_1 and all other bias estimates are close to the true value, resulting in good approximations. For $n = 25$, however, B_1 substantially underestimates the true value, whereas B_2 gives a good approximation. In practice, however, the true distribution G is unknown, and it should be noted that the quantities $\hat{B}_1$ and $\hat{B}_2$ are used in place of B_1 and B_2, respectively. In this case, $\hat{B}_1 = 1.89$ is substantially smaller than B_1, but the difference of 0.11 is equal to the bias of the first-order bias correction term $\Delta b_1/n = -3/25 = -0.12$. Although the second-order correction term $\hat{B}_2$ yields a considerable underestimate for $n = 25$, it gives accurate values for $n = 100$ and 400. The first-order bootstrap estimate B_1^* gives a value close to the second-order analytical correction term B_2, due to the fact that the model in this example contains the true distribution, in which case B_1 becomes a constant, as discussed in Section 7.2, and consequently reverts to $\Delta b_1 = 0$, and the bootstrap estimate automatically performs second-order corrections.

Example 4 (Bootstrap higher-order bias correction: Laplace distribution) We consider now the bias correction terms for the normal distribution model when the true distribution is a Laplace distribution:

$$g(x) = \frac{1}{\sqrt{2}} \exp\left\{-\sqrt{2}|x|\right\}. \tag{8.35}$$

The centered moments for the Laplace distribution are $\mu_3 = 0$, $\mu_4 = 6$, and $\mu_6 = 90$. Table 8.5 shows the first- and second-order bias correction terms. In this case, compared with correction term 2 of the AIC, B_1 and B_2 yield substantially good estimates of the true value. $\hat{B}_1$ and $\hat{B}_2$ estimated using $\hat{G}$, however, contain significantly large biases. The bias of the first-order bias correction term is $\Delta b_1/n = -42/n$, which may account for some of the large bias. In this case, the bootstrap estimate B_1^* gives a better approximation to the bias $b(G)$ than does $\hat{B}_1$. B_2^* and B_2^{**} are second-order bootstrap bias correction terms by (8.33) and (8.34). For $n = 25$, B_2^{**} yields a better approximation than $\hat{B}_2$ or B_2^*, which may be due to the fact that B_1^* produces a better approximation than $\hat{B}_1$.

Example 5 (Bootstrap bias correction for robust estimation) As an example of evaluating a model whose parameters are estimated using a technique other than the maximum likelihood method, Table 8.6 shows the parameters μ and σ estimated using a median $\hat{\mu}_m = \text{med}_i\{X_i\}$ and a median absolute deviation $\hat{\sigma}_m = c^{-1}\text{med}_i\{|X_i - \text{med}_j\{X_j\}|\}$, respectively, where $c = \Phi^{-1}(0.75)$. The bootstrap method can also be applied to such estimates. In this case, Table 8.6 shows that the averages of D_1 and D_3 take entirely different values and that the bootstrap method produces appropriate estimates. Although the asymptotic bias $b_1(G)$ is the same as that for the maximum likelihood estimate, it is noteworthy that for $n = 100$ or 400, the AIC gives an appropriate approximation for models estimated by a robust procedure (see Subsection 5.2.3).

Table 8.5. Bias of a normal distribution model. The true distribution is assumed to be a Laplace distribution.

n	$b(G)$	B_1	B_2	$\hat{B}_1$	$\hat{B}_2$	B_1^*	B_2^*	B_2^{**}
25	3.87	3.50	3.74	2.60	3.28	3.09	3.30	3.52
100	3.57	3.50	3.56	3.16	3.49	3.33	3.50	3.50
400	3.56	3.50	3.52	3.40	3.51	3.43	3.51	3.50

Table 8.6. Bias correction terms of the normal distribution model when the parameters are estimated using the median. The true model is assumed to be a normal distribution.

n		$\hat{b}(G)$	B_1	B_1^*	B_2^{**}
25	$D_1 + D_3$	2.58	1.89	2.57	2.63
	D_1	−0.47	0.94	−0.56	−0.54
	D_3	3.04	0.94	3.14	3.16
100	$D_1 + D_3$	2.12	1.97	2.25	2.27
	D_1	−0.18	0.98	−0.37	−0.35
	D_3	2.30	0.98	2.61	2.62
400	$D_1 + D_3$	2.02	1.99	2.06	2.06
	D_1	−0.16	0.99	−0.19	−0.19
	D_3	2.18	0.99	2.25	2.26

8.3.4 Relation Between Bootstrap Bias Correction Terms

It is appropriate at this point to comment on the relation between the bootstrap bias correction terms proposed in literature.

For Gaussian state-space model selection, Cavanaugh and Shumway (1997) proposed a criterion by bootstrapping $(\boldsymbol{\theta}_0 - \hat{\boldsymbol{\theta}})^T J(\boldsymbol{\theta}_0)(\boldsymbol{\theta}_0 - \hat{\boldsymbol{\theta}})$, where $J(\theta_0)$ is the Fisher information matrix. The bias correction term in this criterion is $2D_3$ in our notation. As can be seen in Table 8.2, $2D_3$ overestimate the true bias $b(G)$ even for a simple normal distribution model, particularly for small sample sizes. However, although $2D_3$ may work well as an order selection criterion in practice, this criterion cannot be applied as a general estimation procedure. As shown in Table 8.6, D_1 and D_3 for models estimated using a method other than the maximum likelihood method take different values even for large n, and thus $2D_3$ cannot yield a reasonable estimate of $b(G)$.

Shibata (1997) presented six candidate bias correction terms, $b_1, \ldots, b_6$. These bias correction terms can be clearly explained by the decomposition shown in Figure 8.4 and can be expressed as $b_1 = D_1 + D_2 + D_3$, $b_2 = D_3$, $b_3 = D_1$, $b_4 = D_2 + D_3$, $b_5 = D_1 + D_2$, and $b_6 = D_2$. The difference between the bootstrap variances of these estimates can be clearly explained by our decomposition. However, the most efficient bias correction term $D_1 + D_3$ was not included. This is probably because only a small sample size $n = 50$ was used in the Monte Calro simulation, and thus the necessity of removing the middle term D_2 did not become apparent.

8.4 Applications of Bootstrap Information Criterion

8.4.1 Change Point Model

Let x_α denote the data observed at time α, where the data points are ordered either temporally or spatially. For n observations $x_1, \ldots, x_n$, we refer to $[1, n]$ as the *total interval*. In the following, we assume that the data do not follow a distribution in the total interval, but if the total interval is partitioned into several intervals, the data in each interval follow a certain distribution. However, the appropriate partition of the interval and the distribution in each subinterval are unknown. As the simplest model that represents such a situation, we consider the following change point model.

We assume that the interval $[1, n]$ is partitioned into k subintervals $[1, n_1], [n_1 + 1, n_2], \ldots, [n_{k-1} + 1, n]$, and that in each subinterval the data x_α follow a normal distribution with mean μ_j and variance σ_j^2. In other words, for $j = 1, \ldots, k$, we assume

$$x_\alpha \sim N(\mu_j, \sigma_j^2) \qquad \alpha = n_{j-1} + 1, \ldots, n_j, \tag{8.36}$$

where the number of subintervals k is unknown. If we write $\boldsymbol{\theta}_k = (\mu_1, \ldots, \mu_k, \sigma_1^2, \ldots, \sigma_k^2)^T$, then the density function for a model with k subinterval is

$$f(\boldsymbol{x}|\boldsymbol{\theta}_k) = \prod_{j=1}^{k} \prod_{\alpha=n_{j-1}+1}^{n_j} \frac{1}{\sqrt{2\pi\sigma_j^2}} \exp\left\{ -\frac{(x_\alpha - \mu_j)^2}{2\sigma_j^2} \right\}. \tag{8.37}$$

Consequently, the log-likelihood function is

$$\begin{aligned} \ell_k(\boldsymbol{\theta}_k) &= -\frac{n}{2}\log 2\pi - \frac{1}{2}\sum_{j=1}^{k}(n_j - n_{j-1})\log\sigma_j^2 \\ &\quad -\frac{1}{2}\sum_{j=1}^{k}\frac{1}{\sigma_j^2}\sum_{\alpha=n_{j-1}+1}^{n_j}(x_\alpha - \mu_j)^2, \end{aligned} \tag{8.38}$$

and the maximum likelihood estimators for μ_j and σ_j^2 $(j = 1, \ldots, k)$ are given by

$$\begin{aligned} \hat{\mu}_j &= \frac{1}{n_j - n_{j-1}} \sum_{\alpha=n_{j-1}+1}^{n_j} x_\alpha, \\ \hat{\sigma}_j^2 &= \frac{1}{n_j - n_{j-1}} \sum_{\alpha=n_{j-1}+1}^{n_j} (x_\alpha - \hat{\mu}_j)^2. \end{aligned} \tag{8.39}$$

In this case, the maximum log-likelihood is

$$\ell_k(\hat{\boldsymbol{\theta}}_k) = -\frac{n}{2}(\log 2\pi + 1) - \frac{1}{2}\sum_{j=1}^{k}(n_j - n_{j-1})\log\hat{\sigma}_j^2. \tag{8.40}$$

Therefore, since the number of unknown parameters in the model is $2k$, corresponding to μ_j and σ_j^2, the AIC is given by

$$\mathrm{AIC}_k = n(\log 2\pi + 1) + \sum_{j=1}^{k}(n_j - n_{j-1}) \log \hat{\sigma}_j^2 + 4k. \tag{8.41}$$

In practice, however, the partition points are also unknown, and it is unclear as to whether they should be added to the number of parameters in the information criterion. Therefore, we attempt to evaluate the bias correction term by the bootstrap method through the following procedure:

For a number of intervals $k = 1, \ldots, K$, the following steps are repeated:

(1) Estimate the endpoint of the subintervals $n_1, \ldots, n_{k-1}$ and the parameters $\{(\mu_j, \sigma_j^2); j = 1, \ldots, k\}$ of the models.
(2) Calculate the residual by $\hat{\varepsilon}_\alpha = x_\alpha - \hat{\mu}_j$ $(\alpha = n_{j-1} + 1, \ldots, n_j)$.
(3) By resampling the residual, generate $\hat{\varepsilon}_\alpha^*$ $(\alpha = 1, \ldots, n)$ and a bootstrap sample $x_\alpha^* = \hat{\varepsilon}_\alpha^* + \hat{\mu}_j$ $(\alpha = n_{j-1} + 1, \ldots, n_j)$.
(4) Assuming that the number of intervals k is known, estimate $n_1^*, \ldots, n_{k-1}^*$ and the parameters $\mu_1^*, \ldots, \mu_k^*$, $\sigma_1^{2*}, \ldots, \sigma_k^{2*}$ by the maximum likelihood method.
(5) Repeat steps 3 to 4 B times and estimate the bias:

$$b_B(\hat{G}) = \frac{1}{B}\sum_{i=1}^{B}\left\{\log f(\boldsymbol{x}^*(i)|\hat{\boldsymbol{\theta}}_k^*) - \log f(\boldsymbol{x}|\hat{\boldsymbol{\theta}}_k^*)\right\}, \tag{8.42}$$

where $\boldsymbol{x}^*(i)$ is the bootstrap sample obtained in step 3.

In the next example, we use this algorithm to examine the relationship between the number of parameters and the bias correction term.

Example 6 (Numerical result) We assume that the data $x_1, \ldots, x_n$ with $n = 100$ are generated from two normal distributions:

$$\begin{aligned} x_1, \ldots, x_{50} &\sim N(0, 1), \\ x_{51}, \ldots, x_{100} &\sim N(c, 1). \end{aligned}$$

Namely, the true model is specified by $k = 2$ and $n_1 = 50$. Table 8.7 shows the maximum log-likelihood, AIC bias correction term b_{AIC}, and bootstrap bias correction term $b_B(\hat{G})$ obtained by fitting the models with $k = 1$, 2, and 3. Whereas $\ell(\hat{\boldsymbol{\theta}}_k)$ represents only the case of $c = 1$, the bias correction term $b_B(\hat{G})$ represents six cases, $c = 0, 0.5, 1, 2, 4$, and 8. Whereas $k = 3$ is selected by the AIC for $c = 1$, $k = 2$ is selected by the bootstrap information criterion EIC.

Note that the closer c is to 0, the greater the bias correction term $b_B(\hat{G})$. This can easily be understood by considering the true number of intervals,

Table 8.7. Bias correction terms for the change point model. k denotes the number of subintervals and c is the amount of level shift.

k	$\ell(\hat{\boldsymbol{\theta}}_k)$	b_{AIC}	Amount of level shift c					
			0	0.5	1	2	4	8
1	−157.16	2	1.9	1.9	1.9	1.7	1.4	1.2
2	−142.55	4	10.3	8.7	6.5	5.6	4.4	4.0
3	−138.62	6	22.6	19.6	15.5	12.3	14.6	14.2

$k = 2$. If $c = \infty$, then the endpoint n_1 can be detected with probability one. Therefore, this case is equivalent to estimating two normal distribution models independently, and from the means and variances for two subintervals, it follows that $b_B(\hat{G}) = 4$. In contrast, if $c \to 0$, n_1 fluctuates randomly between 1 and n. Consequently, contrary to the apparent goodness of fit, n_1 deviates greatly from the true distribution, resulting in a large bias. The bias associated with the model with $k = 3$ is extremely large, indicating that the log-likelihood without bias correction significantly overestimates the expected log-likelihood.

8.4.2 Subset Selection in a Regression Model

We now fit the regression model

$$y_\alpha = \sum_{j=1}^{k} \beta_j x_{\alpha j} + \varepsilon_\alpha, \quad \varepsilon_\alpha \sim N(0, \sigma^2) \tag{8.43}$$

to n data points $\{(y_\alpha, x_{\alpha 1}, \ldots, x_{\alpha k});\ \alpha = 1, 2, \ldots, n\}$ observed for a response variable Y and k explanatory variables $x_1, \ldots, x_k$. Except for certain models such as an autoregressive model or polynomial regression model for which the order of the explanatory variables in the model is predetermined naturally, the priority by which the k explanatory variables are selected is generally not predetermined. Therefore, we have to consider ${}_kC_m$ candidate models in fitting regression models with m explanatory variables. In particular, in the extreme case in which all the coefficients are zero, i.e., $\beta_j = 0$, the apparent best model obtained by maximizing the log-likelihood actually yields the worst model. This suggests that bias correction by the AIC, that is, by the number of free parameters, is inadequate for the selection of variables in a regression model.

Table 8.8 shows the bootstrap estimate of the bias correction terms when subset regression models of order m ($m = 0, 1, \ldots, 20$) are fitted to the data generated for a model with $k = 20$ with all the coefficients assumed to be $\beta_j = 0$. For the sake of simplicity, we assume that the explanatory variable $x_{\alpha j}$ is an orthogonal variable. In the table, EIC$_1$ represents the bias correction term for the bootstrap information criterion EIC of the regression model

Table 8.8. Comparison of bias correction terms in subset regression models.

m	AIC	EIC_1	EIC_2	m	AIC	EIC_1	EIC_2
0	1	0.96	0.80	11	12	12.29	20.10
1	2	1.79	3.36	12	13	13.49	21.17
2	3	2.65	5.60	13	14	14.85	22.13
3	4	3.55	7.71	14	15	16.29	23.02
4	5	4.48	9.63	15	16	17.78	23.80
5	6	5.44	11.44	16	17	19.35	24.51
6	7	6.46	13.15	17	18	21.02	25.12
7	8	7.51	14.77	18	19	22.76	25.71
8	9	8.60	16.25	19	20	24.58	27.29
9	10	9.74	17.65	20	21	26.51	26.67
10	11	10.93	18.94				

for which explanatory variables are incorporated into the model in the order $x_1, x_2, \ldots, x_k$. In contrast, EIC_2 represents the bias estimates in the EIC for the case in which a subset regression model is selected by the maximum likelihood criterion for each m, the number of explanatory variables. In each case, for $n = 100$, the number of bootstrap replication was set to $B = 100$ and the computations were repeated 1,000 times. Since EIC_1 and EIC_2 must agree when $k = 0$ and $k = 20$, the difference between these quantities can be considered to be the error due to the bootstrap approximation.

Here, EIC_1, which corresponds to an ordinary regression model, is more or less equal to the AIC bias correction term for order 14 or less, but increases rapidly at higher orders. This can be attributed to an increase in the number of parameters relative to the number of data points. In contrast, EIC_2 for the subset regression model at first increases rapidly as m is large, but at the maximum order $m = 20$, it becomes approximately the same as that for the ordinary regression model. This indicates that subset regression models are easy to adopt from models of apparently good fit and that, consequently, their bias is not uniform and the value of m tends to be skewed toward small values. In the estimation of subset regression models, the use of the EIC can prevent the problem of overfitting by incorporating too many variables that appear to improve the goodness of fit.

9

Bayesian Information Criteria

This chapter considers model selection and evaluation criteria from a Bayesian point of view. A general framework for constructing the Bayesian information criterion (BIC) is described. The BIC is also extended such that it can be applied to the evaluation of models estimated by regularization. Section 9.2 presents Akaike's Bayesian information criterion (ABIC) developed for the evaluation of Bayesian models having prior distributions with hyperparameters. In the latter half of this chapter, we consider information criteria for the evaluation of predictive distributions of Bayesian models. In particular, Section 9.3 gives examples of analytical evaluations of bias correction for linear Gaussian Bayes models. Section 9.4 describes, for general Bayesian models, how to estimate the asymptotic biases and how to perform the second-order bias correction by means of Laplace's method for integrals.

9.1 Bayesian Model Evaluation Criterion (BIC)

9.1.1 Definition of BIC

The Bayesian information criterion (BIC) or Schwarz's information criterion (SIC) proposed by Schwarz (1978) is an evaluation criterion for models defined in terms of their posterior probability [see also Akaike (1977)]. It is derived as follows.

Let $M_1, M_2, \ldots, M_r$ be r candidate models, and assume that each model M_i is characterized by a parametric distribution $f_i(x|\boldsymbol{\theta}_i)$ ($\boldsymbol{\theta}_i \in \Theta_i \subset R^{k_i}$) and the prior distribution $\pi_i(\boldsymbol{\theta}_i)$ of the k_i-dimensional parameter vector $\boldsymbol{\theta}_i$. When n observations $\boldsymbol{x}_n = \{x_1, \ldots, x_n\}$ are given, then, for the ith model M_i, the marginal distribution or probability of $\boldsymbol{x}_n$ is given by

$$p_i(\boldsymbol{x}_n) = \int f_i(\boldsymbol{x}_n|\boldsymbol{\theta}_i)\pi_i(\boldsymbol{\theta}_i)d\boldsymbol{\theta}_i. \tag{9.1}$$

This quantity can be considered as the likelihood of the ith model and is referred to as the *marginal likelihood* of the data.

According to Bayes' theorem, if we suppose that the prior probability of the ith model is $P(M_i)$, the posterior probability of the ith model is given by

$$P(M_i|\boldsymbol{x}_n) = \frac{p_i(\boldsymbol{x}_n)P(M_i)}{\sum_{j=1}^{r} p_j(\boldsymbol{x}_n)P(M_j)}, \qquad i = 1, 2, \ldots, r. \tag{9.2}$$

This posterior probability indicates the probability of the data being generated from the ith model when data $\boldsymbol{x}_n$ are observed. Therefore, if one model is to be selected from r models, it would be natural to adopt the model that has the largest posterior probability. This principle means that the model that maximizes the numerator $p_i(\boldsymbol{x}_n)P(M_i)$ must be selected, since all models share the same denominator in (9.2).

If we further assume that the prior probabilities $P(M_i)$ are equal in all models, it follows that the model that maximizes the marginal likelihood $p_i(\boldsymbol{x}_n)$ of the data must be selected. Therefore, if an approximation to the marginal likelihood expressed in terms of an integral in (9.1) can readily be obtained, the need to compute the integral on a problem-by-problem basis will vanish, thus making the BIC suitable for use as a general model selection criterion.

The BIC is actually defined as the natural logarithm of the integral multiplied by -2, and we have

$$\begin{aligned} -2\log p_i(\boldsymbol{x}_n) &= -2\log\left\{\int f_i(\boldsymbol{x}_n|\boldsymbol{\theta}_i)\pi_i(\boldsymbol{\theta}_i)d\boldsymbol{\theta}_i\right\} \\ &\approx -2\log f_i(\boldsymbol{x}_n|\hat{\boldsymbol{\theta}}_i) + k_i \log n, \end{aligned} \tag{9.3}$$

where $\hat{\boldsymbol{\theta}}_i$ is the maximum likelihood estimator of the k_i-dimensional parameter vector $\boldsymbol{\theta}_i$ of the model $f_i(x|\boldsymbol{\theta}_i)$. Consequently, from the r models that are to be evaluated using the maximum likelihood method, the model that minimizes the value of BIC can be selected as the optimal model for the data.

Thus, even under the assumption that all models have equal prior probabilities, the posterior probability obtained by using the information from the data serves to contrast the models and helps to identify the model that generated the data. We see in the next section that the BIC can be obtained by approximating the integral using Laplace's method.

Bayes factors. For simplicity, let us compare two models, say M_1 and M_2. When the data produce the posterior probabilities $P(M_i|\boldsymbol{x}_n)$ $(i = 1, 2)$, the posterior odds in favor of model M_1 against model M_2 are

$$\frac{P(M_1|\boldsymbol{x}_n)}{P(M_2|\boldsymbol{x}_n)} = \frac{p_1(\boldsymbol{x}_n)}{p_2(\boldsymbol{x}_n)} \frac{P(M_1)}{P(M_2)}. \tag{9.4}$$

Then the ratio

$$B_{12} = \frac{p_1(\boldsymbol{x}_n)}{p_2(\boldsymbol{x}_n)} = \frac{\int f_1(\boldsymbol{x}_n|\boldsymbol{\theta}_1)\pi_1(\boldsymbol{\theta}_1)d\boldsymbol{\theta}_1}{\int f_2(\boldsymbol{x}_n|\boldsymbol{\theta}_2)\pi_2(\boldsymbol{\theta}_2)d\boldsymbol{\theta}_2} \tag{9.5}$$

is defined as the *Bayes factor.*

Akaike (1983a) showed that model comparisons based on the AIC are asymptotically equivalent to those based on Bayes factors. Kass and Raftery (1995) commented that from a Bayesian viewpoint this is true only if the precision of the prior is comparable to that of the likelihood, but not in the more usual situation where prior information is limited relative to the information provided by the data. For Bayes factors, we refer to Kass and Raftery (1995), O'Hagan (1995), and Berger and Pericchi (2001) and references given therein.

9.1.2 Laplace Approximation for Integrals

In order to explain the Laplace approximation method [Tierney and Kadane (1986), Davison (1986), and Barndorff-Nielsen and Cox (1989, p. 169)], we consider the approximation of a simple integral given by

$$\int \exp\{nq(\boldsymbol{\theta})\}d\boldsymbol{\theta}, \tag{9.6}$$

where $\boldsymbol{\theta}$ is a p-dimensional parameter vector. Notice that in the Laplace approximation of an actual likelihood function, the form of $q(\boldsymbol{\theta})$ also changes as the number n of observations increases.

The basic concept underlying the Laplace approximation takes advantage of the fact that when the number n of observations is large, the integrand is concentrated in a neighborhood of the mode $\hat{\boldsymbol{\theta}}$ of $q(\boldsymbol{\theta})$, and consequently, the value of the integral depends solely on the behavior of the integrand in that neighborhood of $\hat{\boldsymbol{\theta}}$.

It follows from $\partial q(\boldsymbol{\theta})/\partial\boldsymbol{\theta}|_{\boldsymbol{\theta}=\hat{\boldsymbol{\theta}}} = \mathbf{0}$ that the Taylor expansion of $q(\boldsymbol{\theta})$ around $\hat{\boldsymbol{\theta}}$ yields the following:

$$q(\boldsymbol{\theta}) = q(\hat{\boldsymbol{\theta}}) - \frac{1}{2}(\boldsymbol{\theta} - \hat{\boldsymbol{\theta}})^T J_q(\hat{\boldsymbol{\theta}})(\boldsymbol{\theta} - \hat{\boldsymbol{\theta}}) + \cdots, \tag{9.7}$$

where

$$J_q(\hat{\boldsymbol{\theta}}) = -\left.\frac{\partial^2 q(\boldsymbol{\theta})}{\partial\boldsymbol{\theta}\partial\boldsymbol{\theta}^T}\right|_{\boldsymbol{\theta}=\hat{\boldsymbol{\theta}}}. \tag{9.8}$$

Substituting the Taylor expansion of $q(\boldsymbol{\theta})$ into (9.6) gives

$$\begin{aligned}&\int \exp\left[n\left\{q(\hat{\boldsymbol{\theta}}) - \frac{1}{2}(\boldsymbol{\theta} - \hat{\boldsymbol{\theta}})^T J_q(\hat{\boldsymbol{\theta}})(\boldsymbol{\theta} - \hat{\boldsymbol{\theta}}) + \cdots\right\}\right] d\boldsymbol{\theta} \\ &\qquad \approx \exp\left\{nq(\hat{\boldsymbol{\theta}})\right\}\int \exp\left\{-\frac{n}{2}(\boldsymbol{\theta} - \hat{\boldsymbol{\theta}})^T J_q(\hat{\boldsymbol{\theta}})(\boldsymbol{\theta} - \hat{\boldsymbol{\theta}})\right\} d\boldsymbol{\theta}.\end{aligned} \tag{9.9}$$

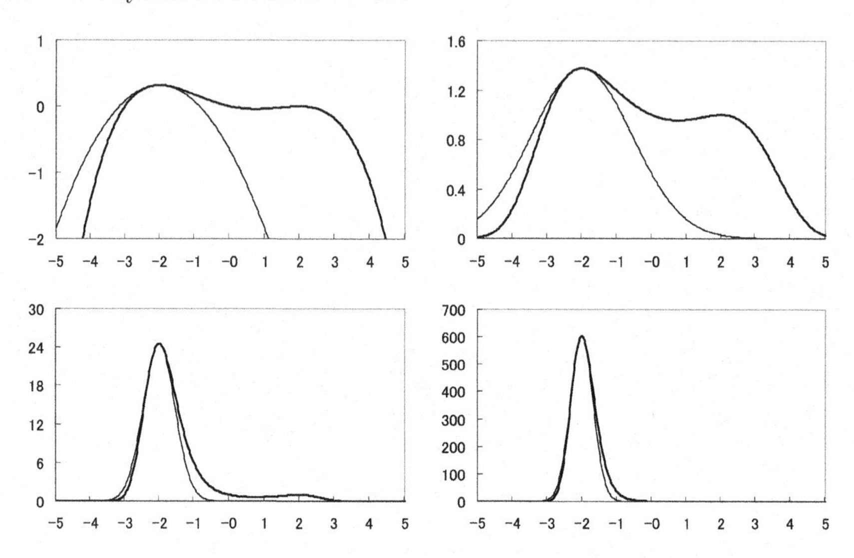

Fig. 9.1. Laplace approximation. Top left: $q(\theta)$ and its quadratic function approximation. Top right, bottom left, and bottom right: $\exp\{nq(\theta)\}$ and Laplace approximations with n=1, 10, and 20, respectively.

By noting the fact that the p-dimensional random vector $\boldsymbol{\theta}$ follows the p-variate normal distribution with mean vector $\hat{\boldsymbol{\theta}}$ and variance covariance matrix $n^{-1}J_q(\hat{\boldsymbol{\theta}})^{-1}$, calculation of the integral on the right-hand side of (9.9) yields

$$\int \exp\left\{-\frac{n}{2}(\boldsymbol{\theta}-\hat{\boldsymbol{\theta}})^T J_q(\hat{\boldsymbol{\theta}})(\boldsymbol{\theta}-\hat{\boldsymbol{\theta}})\right\} d\boldsymbol{\theta} = \frac{(2\pi)^{p/2}}{n^{p/2}|J_q(\hat{\boldsymbol{\theta}})|^{1/2}}. \tag{9.10}$$

Therefore, we obtain the following Laplace approximation of the integral (9.6).

Laplace approximation of integrals. Let $q(\boldsymbol{\theta})$ be a real-valued function of a p-dimensional parameter vector $\boldsymbol{\theta}$, and let $\hat{\boldsymbol{\theta}}$ be the mode of $q(\boldsymbol{\theta})$. Then the Laplace approximation of the integral is given by

$$\int \exp\{nq(\boldsymbol{\theta})\}d\boldsymbol{\theta} \approx \frac{(2\pi)^{p/2}}{n^{p/2}|J_q(\hat{\boldsymbol{\theta}})|^{1/2}} \exp\left\{nq(\hat{\boldsymbol{\theta}})\right\}, \tag{9.11}$$

where $J_q(\hat{\boldsymbol{\theta}})$ is defined by (9.8).

Example 1 (Laplace approximation)@ Figure 9.1 shows how Laplace's method for integrals works. The upper left graph illustrates a suitably defined function $q(\boldsymbol{\theta})$ and its approximation in terms of its Taylor expansion. The curve with two peaks shown in bold lines represents the function $q(\boldsymbol{\theta})$, and the thin line indicates its approximation by the Taylor series expansion up

Table 9.1. The integral of the function given in Figure 9.1 and its Laplace approximation.

n	1	10	20	50
Integral	398.05	1678.76	26378.39	240282578
Laplace approximation	244.51	1403.40	24344.96	240282578
Relative errors	0.386	0.164	0.077	0

to the second term. In this graph, only the left peak of the two peaks is approximated, and it can hardly be considered a good approximation. The other three graphs show the integrand $\exp\{nq(\boldsymbol{\theta})\}$ and approximations to it. The upper right, lower left, and lower right graphs represent the cases $n = 1$, 10, and 20, in the indicated order. The graph for $n = 1$ fails to describe the peak on the right side. However, as n increases to $n = 10$ and $n = 20$, the right peak vanishes rapidly, indicating that making use of the Taylor series expansion yields a good approximation. Therefore, it is clear that, when the value of n is large, this method provides a good approximation to the integral.

Table 9.1 shows the integral of the function $\exp\{nq(\boldsymbol{\theta})\}$ given in Figure 9.1, its Laplace approximation, and the relative error (= |true value − approximation|/|true value|). In this case, the relative error is as large as 0.386 when $n = 1$, but it diminishes as n increases, and the relative error becomes 0 when $n = 50$.

9.1.3 Derivation of the BIC

The marginal likelihood or the marginal distribution of data $\boldsymbol{x}_n$ can be approximated by using Laplace's method for integrals. In this section, we drop the notational dependence on the model M_i and represent the marginal likelihood of (9.1) as

$$p(\boldsymbol{x}_n) = \int f(\boldsymbol{x}_n|\boldsymbol{\theta})\pi(\boldsymbol{\theta})d\boldsymbol{\theta}, \tag{9.12}$$

where $\boldsymbol{\theta}$ is a p-dimensional parameter vector. This equation may be rewritten as

$$\begin{aligned} p(\boldsymbol{x}_n) &= \int \exp\left\{\log f(\boldsymbol{x}_n|\boldsymbol{\theta})\right\}\pi(\boldsymbol{\theta})d\boldsymbol{\theta} \\ &= \int \exp\left\{\ell(\boldsymbol{\theta})\right\}\pi(\boldsymbol{\theta})d\boldsymbol{\theta}, \end{aligned} \tag{9.13}$$

where $\ell(\boldsymbol{\theta})$ is the log-likelihood function $\ell(\boldsymbol{\theta}) = \log f(\boldsymbol{x}_n|\boldsymbol{\theta})$.

The Laplace approximation takes advantage of the fact that when the number n of observations is sufficiently large, the integrand is concentrated in a neighborhood of the mode of $\ell(\boldsymbol{\theta})$ or, in this case, in a neighborhood of the maximum likelihood estimator $\hat{\boldsymbol{\theta}}$, and that the value of the integral depends on the behavior of the function in this neighborhood. Since $\partial \ell(\boldsymbol{\theta})/\partial \boldsymbol{\theta}|_{\boldsymbol{\theta}=\hat{\boldsymbol{\theta}}} = 0$ holds for the maximum likelihood estimator $\hat{\boldsymbol{\theta}}$ of the parameter $\boldsymbol{\theta}$, the Taylor expansion of the log-likelihood function $\ell(\boldsymbol{\theta})$ around $\hat{\boldsymbol{\theta}}$ yields

$$\ell(\boldsymbol{\theta}) = \ell(\hat{\boldsymbol{\theta}}) - \frac{n}{2}(\boldsymbol{\theta} - \hat{\boldsymbol{\theta}})^T J(\hat{\boldsymbol{\theta}})(\boldsymbol{\theta} - \hat{\boldsymbol{\theta}}) + \cdots, \tag{9.14}$$

where

$$J(\hat{\boldsymbol{\theta}}) = -\frac{1}{n}\frac{\partial^2 \ell(\boldsymbol{\theta})}{\partial \boldsymbol{\theta} \partial \boldsymbol{\theta}^T}\bigg|_{\boldsymbol{\theta}=\hat{\boldsymbol{\theta}}} = -\frac{1}{n}\frac{\partial^2 \log f(\boldsymbol{x}_n|\boldsymbol{\theta})}{\partial \boldsymbol{\theta} \partial \boldsymbol{\theta}^T}\bigg|_{\boldsymbol{\theta}=\hat{\boldsymbol{\theta}}}. \tag{9.15}$$

Similarly, we can expand the prior distribution $\pi(\boldsymbol{\theta})$ in a Taylor series around the maximum likelihood estimator $\hat{\boldsymbol{\theta}}$ as

$$\pi(\boldsymbol{\theta}) = \pi(\hat{\boldsymbol{\theta}}) + (\boldsymbol{\theta} - \hat{\boldsymbol{\theta}})^T \frac{\partial \pi(\boldsymbol{\theta})}{\partial \boldsymbol{\theta}}\bigg|_{\boldsymbol{\theta}=\hat{\boldsymbol{\theta}}} + \cdots. \tag{9.16}$$

Substituting (9.14) and (9.16) into (9.13) and simplifying the results lead to the approximation of the marginal likelihood as follows:

$$\begin{aligned} p(\boldsymbol{x}_n) &= \int \exp\left\{\ell(\hat{\boldsymbol{\theta}}) - \frac{n}{2}(\boldsymbol{\theta} - \hat{\boldsymbol{\theta}})^T J(\hat{\boldsymbol{\theta}})(\boldsymbol{\theta} - \hat{\boldsymbol{\theta}}) + \cdots\right\} \\ &\quad \times \left\{\pi(\hat{\boldsymbol{\theta}}) + (\boldsymbol{\theta} - \hat{\boldsymbol{\theta}})^T \frac{\partial \pi(\boldsymbol{\theta})}{\partial \boldsymbol{\theta}}\bigg|_{\boldsymbol{\theta}=\hat{\boldsymbol{\theta}}} + \cdots\right\} d\boldsymbol{\theta} \\ &\approx \exp\left\{\ell(\hat{\boldsymbol{\theta}})\right\} \pi(\hat{\boldsymbol{\theta}}) \int \exp\left\{-\frac{n}{2}(\boldsymbol{\theta} - \hat{\boldsymbol{\theta}})^T J(\hat{\boldsymbol{\theta}})(\boldsymbol{\theta} - \hat{\boldsymbol{\theta}})\right\} d\boldsymbol{\theta}. \end{aligned} \tag{9.17}$$

Here we used the fact that $\hat{\boldsymbol{\theta}}$ converges to $\boldsymbol{\theta}$ in probability with order $\hat{\boldsymbol{\theta}} - \boldsymbol{\theta} = O_p(n^{-1/2})$ and also that the following equation holds:

$$\int (\boldsymbol{\theta} - \hat{\boldsymbol{\theta}}) \exp\left\{-\frac{n}{2}(\boldsymbol{\theta} - \hat{\boldsymbol{\theta}})^T J(\hat{\boldsymbol{\theta}})(\boldsymbol{\theta} - \hat{\boldsymbol{\theta}})\right\} d\boldsymbol{\theta} = \mathbf{0}. \tag{9.18}$$

In (9.17), integrating with respect to the parameter vector $\boldsymbol{\theta}$ yields

$$\int \exp\left\{-\frac{n}{2}(\boldsymbol{\theta} - \hat{\boldsymbol{\theta}})^T J(\hat{\boldsymbol{\theta}})(\boldsymbol{\theta} - \hat{\boldsymbol{\theta}})\right\} d\boldsymbol{\theta} = (2\pi)^{p/2} n^{-p/2} |J(\hat{\boldsymbol{\theta}})|^{-1/2}, \tag{9.19}$$

since the integrand is the density function of the p-dimensional normal distribution with mean vector $\hat{\boldsymbol{\theta}}$ and variance covariance matrix $J^{-1}(\hat{\boldsymbol{\theta}})/n$. Consequently, when the sample size n becomes large, it is clear that the marginal likelihood can be approximated as

$$p(\boldsymbol{x}_n) \approx \exp\left\{\ell(\hat{\boldsymbol{\theta}})\right\} \pi(\hat{\boldsymbol{\theta}}) (2\pi)^{p/2} n^{-p/2} |J(\hat{\boldsymbol{\theta}})|^{-1/2}. \tag{9.20}$$

Taking the logarithm of this expression and multiplying it by -2, we obtain

$$\begin{aligned} -2\log p(\boldsymbol{x}_n) &= -2\log\left\{\int f(\boldsymbol{x}_n|\boldsymbol{\theta})\pi(\boldsymbol{\theta})d\boldsymbol{\theta}\right\} \qquad (9.21) \\ &\approx -2\ell(\hat{\boldsymbol{\theta}}) + p\log n + \log|J(\hat{\boldsymbol{\theta}})| - p\log(2\pi) - 2\log\pi(\hat{\boldsymbol{\theta}}). \end{aligned}$$

Then the following model evaluation criterion BIC can be obtained by ignoring terms with order less than $O(1)$ with respect to the sample size n.

Bayesian information criterion (BIC). Let $f(\boldsymbol{x}_n|\hat{\boldsymbol{\theta}})$ be a statistical model estimated by the maximum likelihood method. Then the Bayesian information criterion BIC is given by

$$\text{BIC} = -2\log f(\boldsymbol{x}_n|\hat{\boldsymbol{\theta}}) + p\log n. \qquad (9.22)$$

From the above argument, it can be seen that, BIC is an evaluation criterion for models estimated by using the maximum likelihood method and that the criterion is obtained under the condition that the sample size n is made sufficiently large. We also see that it was obtained by approximating the marginal likelihood associated with the posterior probability of the model by Laplace's method for integrals and that it is not an information criterion, leading to an unbiased estimation of the K-L information.

We shall now consider how to extend the BIC to an evaluation criterion that permits the evaluation of models estimated by the regularization method described in Subsection 5.2.4. In the next section, we derive a model evaluation criterion that represents an extension of the BIC through the application of Laplace approximation.

Minimum description length (MDL). Rissanen (1978, 1989) proposed a model evaluation criterion (MDL) based on the concept of minimum description length in transmitting a set of data by coding using a family of probability models $\{f(x|\boldsymbol{\theta}); \boldsymbol{\theta} \in \Theta \subset \boldsymbol{R}^p\}$.

Assume that the data $\boldsymbol{x}_n = \{x_1, x_2, \ldots, x_n\}$ are obtained from $f(x|\boldsymbol{\theta})$. Since the parameter vector $\boldsymbol{\theta}$ of the model is unknown, we first encode $\boldsymbol{\theta}$ and send it to the receiver, and then encode and send the data $\boldsymbol{x}_n$ by using the probability distribution $f(x|\boldsymbol{\theta})$ specified by $\boldsymbol{\theta}$. Then, given the parameter vector $\boldsymbol{\theta}$, the description length necessary for encoding the data is $-\log f(\boldsymbol{x}_n|\boldsymbol{\theta})$ and the total description length is defined by $-\log f(\boldsymbol{x}_n|\boldsymbol{\theta})$ plus the description length of the probability distribution model. The probability distribution model that minimizes this total description length is such a model that can encode the data $\boldsymbol{x}_n$ in minimum length.

If the parameter is a real number, an infinite description length is necessary for exact coding. Therefore, we consider encoding the parameter by discretizing through segmentation of the parameter space $\Theta \in \boldsymbol{R}^p$ into infinitesimal cubes of size δ. Then the total description length depends on the value of δ, and its minimum can be approximated as

$$\ell(\boldsymbol{x}_n) = -\log f(\boldsymbol{x}_n|\hat{\boldsymbol{\theta}}) + \frac{p}{2}\log n - \frac{p}{2}\log 2\pi$$
$$+ \log \int \sqrt{|J(\boldsymbol{\theta})|}d\boldsymbol{\theta} + O(n^{-1/2}), \tag{9.23}$$

where $J(\boldsymbol{\theta})$ is Fisher's information matrix. By considering terms up to order $O(\log n)$, the minimum description length is defined as

$$\text{MDL} = -\log f(\boldsymbol{x}_n|\boldsymbol{\theta}) + \frac{p}{2}\log n. \tag{9.24}$$

The first term on the right-hand side is the description length in sending the data $\boldsymbol{x}_n$ by using the probability distribution $f(x|\hat{\boldsymbol{\theta}})$ specified by the maximum likelihood estimator $\hat{\boldsymbol{\theta}}$ as the encoding function, and the second term is the description length for encoding the maximum likelihood estimate $\hat{\boldsymbol{\theta}}$ with accuracy $\delta = O(n^{-1/2})$. In any case, it is interesting that the minimum description length MDL coincides with the BIC that was derived in terms of the posterior probability of the model within the Bayesian framework.

9.1.4 Extension of the BIC

Let $f(x|\hat{\boldsymbol{\theta}}_P)$ be a statistical model estimated by the regularization method for the parametric model $f(x|\boldsymbol{\theta})$ $(\boldsymbol{\theta} \in \Theta \subset R^p)$, where $\hat{\boldsymbol{\theta}}_P$ is an estimator of dimension p obtained by maximizing the penalized log-likelihood function

$$\ell_\lambda(\boldsymbol{\theta}) = \log f(\boldsymbol{x}_n|\boldsymbol{\theta}) - \frac{n\lambda}{2}\boldsymbol{\theta}^T K\boldsymbol{\theta}, \tag{9.25}$$

and where K is a $p\times p$ specified matrix with rank $d = p-k$ [for the typical form of K, see (5.135)]. Our objective here is to obtain a criterion for evaluation and selection of a statistical model $f(x|\hat{\boldsymbol{\theta}}_P)$, from a Bayesian perspective.

The penalized log-likelihood function in (9.25) can be rewritten as

$$\ell_\lambda(\boldsymbol{\theta}) = \log f(\boldsymbol{x}_n|\boldsymbol{\theta}) + \log\left\{\exp\left(-\frac{n\lambda}{2}\boldsymbol{\theta}^T K\boldsymbol{\theta}\right)\right\}$$
$$= \log\left\{f(\boldsymbol{x}_n|\boldsymbol{\theta})\exp\left(-\frac{n\lambda}{2}\boldsymbol{\theta}^T K\boldsymbol{\theta}\right)\right\}. \tag{9.26}$$

By considering the exponential term on the right-hand side as a p-dimensional degenerate normal distribution with mean vector $\mathbf{0}$ and singular variance covariance matrix $(n\lambda K)^-$ and adding a constant term to yield a density function, we obtain

$$\pi(\boldsymbol{\theta}|\lambda) = (2\pi)^{-d/2}(n\lambda)^{d/2}|K|_+^{1/2}\exp\left(-\frac{n\lambda}{2}\boldsymbol{\theta}^T K\boldsymbol{\theta}\right), \tag{9.27}$$

where $|K|_+$ denotes the product of nonzero eigenvalues of the specified matrix K with rank d. This distribution can be thought of as a prior distribution in which the smoothing parameter λ is a hyperparameter.

Given the data distribution $f(\boldsymbol{x}_n|\boldsymbol{\theta})$ and the prior distribution $\pi(\boldsymbol{\theta}|\lambda)$ with hyperparameter λ, the marginal likelihood of the model is defined by

$$p(\boldsymbol{x}_n|\lambda) = \int f(\boldsymbol{x}_n|\boldsymbol{\theta})\pi(\boldsymbol{\theta}|\lambda)d\boldsymbol{\theta}. \tag{9.28}$$

When the prior distribution of $\boldsymbol{\theta}$ is given by the p-dimensional normal distribution in (9.27), this marginal likelihood can be rewritten as

$$\begin{aligned} p(\boldsymbol{x}_n|\lambda) &= \int f(\boldsymbol{x}_n|\boldsymbol{\theta})\pi(\boldsymbol{\theta}|\lambda)d\boldsymbol{\theta} \\ &= \int \exp\left[n \times \frac{1}{n}\log\left\{f(\boldsymbol{x}_n|\boldsymbol{\theta})\pi(\boldsymbol{\theta}|\lambda)\right\}\right] d\boldsymbol{\theta} \\ &= \int \exp\left\{nq(\boldsymbol{\theta}|\lambda)\right\} d\boldsymbol{\theta}, \end{aligned} \tag{9.29}$$

where

$$\begin{aligned} q(\boldsymbol{\theta}|\lambda) &= \frac{1}{n}\log\left\{f(\boldsymbol{x}_n|\boldsymbol{\theta})\pi(\boldsymbol{\theta}|\lambda)\right\} \\ &= \frac{1}{n}\left\{\log f(\boldsymbol{x}_n|\boldsymbol{\theta}) + \log \pi(\boldsymbol{\theta}|\lambda)\right\} \\ &= \frac{1}{n}\left\{\log f(\boldsymbol{x}_n|\boldsymbol{\theta}) - \frac{n\lambda}{2}\boldsymbol{\theta}^T K\boldsymbol{\theta}\right\} \\ &\quad - \frac{1}{2n}\left\{d\log(2\pi) - d\log(n\lambda) - \log|K|_+\right\}. \end{aligned} \tag{9.30}$$

We note here that the mode, $\hat{\boldsymbol{\theta}}_P$, of $q(\boldsymbol{\theta}|\lambda)$ in the above equation coincides with a solution obtained by maximizing the penalized log-likelihood function (9.25). By approximating it using Laplace's method for integrals in (9.11), we have

$$\int \exp\{nq(\boldsymbol{\theta})\}d\boldsymbol{\theta} \approx \frac{(2\pi)^{p/2}}{n^{p/2}|J_\lambda(\hat{\boldsymbol{\theta}}_P)|^{1/2}} \exp\left\{nq(\hat{\boldsymbol{\theta}}_P)\right\}. \tag{9.31}$$

Taking the logarithm of this expression and multiplying it by -2, we obtain the following model evaluation criterion [Konishi et al. (2004)]:

Generalized Bayesian information criterion (GBIC). Suppose that the model $f(\boldsymbol{x}_n|\boldsymbol{\theta}_P)$ is constructed by maximizing the penalized log-likelihood function (9.25). Then the model evaluation criterion based on a Bayesian approach is given by

$$\begin{aligned} \mathrm{GBIC} &= -2\log f(\boldsymbol{x}_n|\hat{\boldsymbol{\theta}}_P) + n\lambda\hat{\boldsymbol{\theta}}_P^T K\hat{\boldsymbol{\theta}}_P + (p-d)\log n \\ &\quad + \log|J_\lambda(\hat{\boldsymbol{\theta}}_P)| - d\log\lambda - \log|K|_+ - (p-d)\log(2\pi), \end{aligned} \tag{9.32}$$

where K is a $p \times p$ specified matrix of rank d, $|K|_+$ is the product of the d nonzero eigenvalues of K, and

$$J_\lambda(\hat{\boldsymbol{\theta}}_P) = -\frac{1}{n} \frac{\partial^2 \log f(\boldsymbol{x}_n|\boldsymbol{\theta})}{\partial \boldsymbol{\theta} \partial \boldsymbol{\theta}^T}\bigg|_{\hat{\boldsymbol{\theta}}_P} + \lambda K. \tag{9.33}$$

Since the model evaluation criterion GBIC can be used for the selection of a smoothing parameter λ, we select λ that minimizes the GBIC as the optimal smoothing parameter. This results in the selection of an optimal model from a family of models characterized by smoothing parameters.

By interpreting the regularization method based on the above argument from a Bayesian point of view, it can be seen that the regularized estimator agrees with the estimate that is obtained through maximization (mode) of the following posterior probability, depending on the value of the smoothing parameter:

$$\pi(\boldsymbol{\theta}|\boldsymbol{x}_n;\lambda) = \frac{f(\boldsymbol{x}_n|\boldsymbol{\theta})\pi(\boldsymbol{\theta}|\lambda)}{\displaystyle\int f(\boldsymbol{x}_n|\boldsymbol{\theta})\pi(\boldsymbol{\theta}|\lambda)d\boldsymbol{\theta}}, \tag{9.34}$$

where $\pi(\boldsymbol{\theta}|\lambda)$ is the density function resulting from (9.27) as a prior probability of the p-dimensional parameter $\boldsymbol{\theta}$ for the model $f(\boldsymbol{x}_n|\boldsymbol{\theta})$. For the Bayesian justification of the maximum penalized likelihood approach, we refer to Silverman (1985) and Wahba (1990, Chapter 1).

The use of Laplace's method for integrals has been extensively investigated as a useful tool for approximating Bayesian predictive distributions, Bayes factors, and Bayesian model selection criteria [Davison (1986), Clarke and Barron (1994), Kass and Wasserman (1995), Kass and Raftery (1995), O'Hagan (1995), Konishi and Kitagawa (1996), Neath and Cavanaugh (1997), Pauler (1998), Lanterman (2001), and Konishi et al. (2004)].

Example 2 (Nonlinear regression models) Suppose that n observations $\{(\boldsymbol{x}_\alpha, y_\alpha);\ \alpha = 1, 2, \ldots, n\}$ are obtained in terms of a p-dimensional vector of explanatory variables $\boldsymbol{x}$ and a response variable Y. We assume the regression model based on the basis expansion described in Section 6.1 as follows:

$$\begin{aligned} y_\alpha &= \sum_{i=1}^{m} w_i b_i(\boldsymbol{x}_\alpha) + \varepsilon_\alpha \\ &= \boldsymbol{w}^T \boldsymbol{b}(\boldsymbol{x}_\alpha) + \varepsilon_\alpha, \qquad \alpha = 1, 2, \ldots, n, \end{aligned} \tag{9.35}$$

where $\boldsymbol{b}(\boldsymbol{x}_\alpha) = (b_1(\boldsymbol{x}_\alpha), \ldots, b_m(\boldsymbol{x}_\alpha))^T$ and ε_α, $\alpha = 1, 2, \ldots, n$, are independently and normally distributed with mean zero and variance σ^2. Then the regression model based on the basis expansion can be expressed in terms of the probability density function

$$f(y_\alpha|\boldsymbol{x}_\alpha;\boldsymbol{\theta}) = \frac{1}{\sqrt{2\pi\sigma^2}}\exp\left[-\frac{\{y_\alpha - \boldsymbol{w}^T\boldsymbol{b}(\boldsymbol{x}_\alpha)\}^2}{2\sigma^2}\right], \tag{9.36}$$

where $\boldsymbol{\theta} = (\boldsymbol{w}^T, \sigma^2)^T$.

If we estimate the parameter vector $\boldsymbol{\theta}$ of the model by maximizing the penalized log-likelihood function (9.25), the estimators for $\boldsymbol{w}$ and σ^2 are respectively given by

$$\hat{\boldsymbol{w}} = (B^T B + n\lambda\hat{\sigma}^2 K)^{-1}B^T\boldsymbol{y}, \quad \hat{\sigma}^2 = \frac{1}{n}(\boldsymbol{y} - B\hat{\boldsymbol{w}})^T(\boldsymbol{y} - B\hat{\boldsymbol{w}}), \tag{9.37}$$

where B is an $n \times m$ basis function matrix given by $B = (\boldsymbol{b}(\boldsymbol{x}_1), \boldsymbol{b}(\boldsymbol{x}_2), \cdots, \boldsymbol{b}(\boldsymbol{x}_n))^T$ (see Section 6.1). Then the probability density function $f(y_\alpha|\boldsymbol{x}_\alpha;\hat{\boldsymbol{\theta}}_P)$ in which the parameters $\boldsymbol{\theta} = (\boldsymbol{w}^T, \sigma^2)^T$ in (9.36) are replaced with their estimators $\hat{\boldsymbol{\theta}}_P = (\hat{\boldsymbol{w}}^T, \hat{\sigma}^2)^T$ is the resulting statistical model.

By applying the GBIC in (9.32), the model evaluation criterion for the statistical model $f(y_\alpha|\boldsymbol{x}_\alpha;\hat{\boldsymbol{\theta}}_P)$ estimated by the regularization method is given by

$$\begin{aligned}\text{GBIC} = &\ n\log\hat{\sigma}^2 + n\lambda\hat{\boldsymbol{w}}^T K\hat{\boldsymbol{w}} + n + n\log(2\pi)\\ &+ (m+1-d)\log n + \log|J_\lambda(\hat{\boldsymbol{\theta}}_P)| - \log|K|_+\\ &- d\log\lambda - (m+1-d)\log(2\pi),\end{aligned} \tag{9.38}$$

where the $(m+1)\times(m+1)$ matrix $J_\lambda(\hat{\boldsymbol{\theta}}_P)$ is

$$J_\lambda(\hat{\boldsymbol{\theta}}_P) = \frac{1}{n\hat{\sigma}^2}\begin{bmatrix} B^T B + n\lambda\hat{\sigma}^2 K & \dfrac{1}{\hat{\sigma}^2}B^T\boldsymbol{e} \\ \dfrac{1}{\hat{\sigma}^2}\boldsymbol{e}^T B & \dfrac{n}{2\hat{\sigma}^2}\end{bmatrix} \tag{9.39}$$

with the n-dimensional residual vector

$$\boldsymbol{e} = \Big(y_1 - \hat{\boldsymbol{w}}^T\boldsymbol{b}(\boldsymbol{x}_1), y_2 - \hat{\boldsymbol{w}}^T\boldsymbol{b}(\boldsymbol{x}_2), \cdots, y_n - \hat{\boldsymbol{w}}^T\boldsymbol{b}(\boldsymbol{x}_n)\Big)^T, \tag{9.40}$$

and K is an $m \times m$ specified matrix of rank d and $|K|_+$ is the product of the d nonzero eigenvalues of K.

Example 3 (Nonlinear logistic regression models) Let $y_1, \ldots, y_n$ be independent binary random variables with

$$\Pr(Y_\alpha = 1|\boldsymbol{x}_\alpha) = \pi(\boldsymbol{x}_\alpha) \quad \text{and} \quad \Pr(Y_\alpha = 0|\boldsymbol{x}_\alpha) = 1 - \pi(\boldsymbol{x}_\alpha), \tag{9.41}$$

where $\boldsymbol{x}_\alpha$ are p-dimensional explanatory variables. We model $\pi(\boldsymbol{x}_\alpha)$ by

$$\log\left\{\frac{\pi(\boldsymbol{x}_\alpha)}{1-\pi(\boldsymbol{x}_\alpha)}\right\} = w_0 + \sum_{i=1}^m w_i b_i(\boldsymbol{x}_\alpha), \tag{9.42}$$

where $\{b_1(\boldsymbol{x}_\alpha), \ldots, b_m(\boldsymbol{x}_\alpha)\}$ are basis functions. Estimating the $(m+1)$-dimensional parameter vector $\boldsymbol{w} = (w_0, w_1, \ldots, w_m)^T$ by maximization of the penalized log-likelihood function (9.25) yields the model

$$f(y_\alpha|\boldsymbol{x}_\alpha; \hat{\boldsymbol{w}}) = \hat{\pi}(\boldsymbol{x}_\alpha)^{y_\alpha}\{1 - \hat{\pi}(\boldsymbol{x}_\alpha)\}^{1-y_\alpha}, \qquad \alpha = 1, \ldots, n, \tag{9.43}$$

where $\hat{\pi}(\boldsymbol{x}_\alpha)$ is the estimated conditional probability given by

$$\hat{\pi}(\boldsymbol{x}_\alpha) = \frac{\exp\left\{\hat{\boldsymbol{w}}^T \boldsymbol{b}(\boldsymbol{x}_\alpha)\right\}}{1 + \exp\left\{\hat{\boldsymbol{w}}^T \boldsymbol{b}(\boldsymbol{x}_\alpha)\right\}}. \tag{9.44}$$

By using the GBIC in (9.32), we obtain the model evaluation criterion for the model $f(y_\alpha|\boldsymbol{x}_\alpha; \hat{\boldsymbol{w}})$ estimated by the regularization method as follows:

$$\begin{aligned} \text{GBIC} = {} & 2\sum_{\alpha=1}^{n} \left[\log\left\{1 + \exp\left(\hat{\boldsymbol{w}}^T b(\boldsymbol{x}_\alpha)\right)\right\} - y_\alpha \hat{\boldsymbol{w}}^T b(\boldsymbol{x}_\alpha)\right] + n\lambda\hat{\boldsymbol{w}}^T K\hat{\boldsymbol{w}} \\ & - (m+1-d)\log(2\pi/n) + \log|Q_\lambda^{(L)}(\hat{\boldsymbol{w}})| - \log|K|_+ - d\log\lambda, \end{aligned} \tag{9.45}$$

where $Q_\lambda^{(\mathrm{L})}(\hat{\boldsymbol{w}}) = B^{\mathrm{T}}\Gamma^{(\mathrm{L})}B/n + \lambda K$ with

$$\Gamma_{\alpha\alpha}^{(\mathrm{L})} = \frac{\exp\{\hat{\boldsymbol{w}}^T \boldsymbol{b}(\boldsymbol{x}_\alpha)\}}{[1 + \exp\{\hat{\boldsymbol{w}}^T \boldsymbol{b}(\boldsymbol{x}_\alpha)\}]^2} \tag{9.46}$$

as the αth diagonal element of $\Gamma^{(\mathrm{L})}$.

Example 4 (Numerical results) For illustration, binary observations $y_1, \ldots, y_{100}$ were generated from the true models

$$\begin{aligned} &(1) \quad \Pr(Y = 1|x) = \frac{1}{1 + \exp\{-\cos(1.5\pi x)\}}, \\ &(2) \quad \Pr(Y = 1|x) = \frac{1}{1 + \exp\{-\exp(-3x)\cos(3\pi x)\}}, \end{aligned} \tag{9.47}$$

where the design points are uniformly distributed in $[0, 1]$. We fitted the nonlinear logistic regression model based on B-splines discussed in Subsection 6.2.1 to the simulated data. The number of basis functions and the value of a smoothing parameter were selected as $m = 17$ and $\lambda = 0.251$ for case (1), and $m = 6$ and $\lambda = 6.31 \times 10^{-5}$ for case (2). Figure 9.2 shows the true and estimated conditional probability functions; the circles indicate the data.

9.2 Akaike's Bayesian Information Criterion (ABIC)

Let $f(\boldsymbol{x}_n|\boldsymbol{\theta})$ be the data distribution of $\boldsymbol{x}_n$ with respect to a parametric model $\{f(x|\boldsymbol{\theta});\ \boldsymbol{\theta} \in \Theta \subset \mathrm{R}^p\}$, and let $\pi(\boldsymbol{\theta}|\boldsymbol{\lambda})$ be the prior distribution of the p-dimensional parameter vector $\boldsymbol{\theta}$ with q-dimensional hyperparameter vector $\boldsymbol{\lambda}$

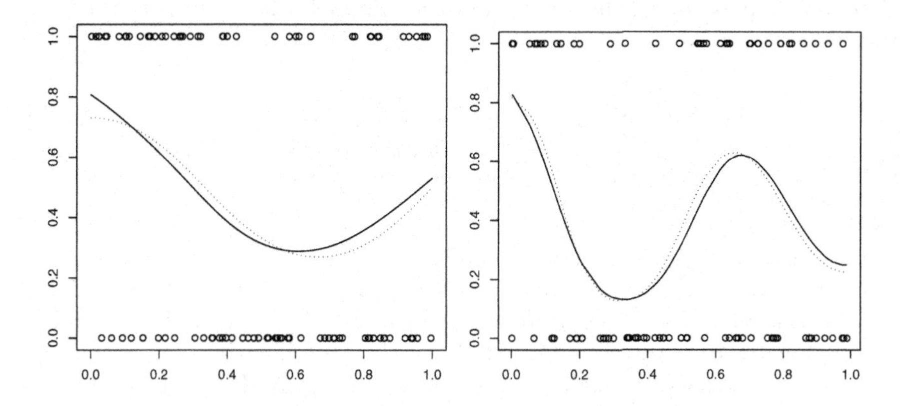

Fig. 9.2. B-spline logistic regression; the true (dashed line) and estimated (solid line) conditional probability functions. Case (1): left, case (2): right.

$(\in \Lambda \subset \mathrm{R}^q)$. Then the marginal distribution or marginal likelihood of the data $\boldsymbol{x}_n$ is given by

$$p(\boldsymbol{x}_n|\boldsymbol{\lambda}) = \int f(\boldsymbol{x}_n|\boldsymbol{\theta})\pi(\boldsymbol{\theta}|\boldsymbol{\lambda})d\boldsymbol{\theta}. \tag{9.48}$$

If the marginal distribution $p(\boldsymbol{x}_n|\boldsymbol{\lambda})$ of the Bayes model is considered to be a parametric model with hyperparameter $\boldsymbol{\lambda}$, then evaluation of the model can be considered within the framework of the AIC, and the criterion is given by

$$\begin{aligned} \mathrm{ABIC} &= -2\log\left\{\max_{\boldsymbol{\lambda}} p(\boldsymbol{x}_n|\boldsymbol{\lambda})\right\} + 2q \\ &= -2\max_{\boldsymbol{\lambda}}\log\left\{\int f(\boldsymbol{x}_n|\boldsymbol{\theta})\pi(\boldsymbol{\theta}|\boldsymbol{\lambda})d\boldsymbol{\theta}\right\} + 2q. \end{aligned} \tag{9.49}$$

This criterion for model evaluation, originally proposed by Akaike (1980b), is referred to as *Akaike's Bayesian information criterion* (ABIC).

According to the Bayesian approach based on the ABIC, the value of the hyperparameter $\boldsymbol{\lambda}$ of a Bayes model can be estimated by maximizing either the marginal likelihood $p(\boldsymbol{x}_n|\boldsymbol{\lambda})$ or the marginal log-likelihood $\log p(\boldsymbol{x}|\boldsymbol{\lambda})$. In other words, the hyperparameter $\boldsymbol{\lambda}$ can be regarded as being estimated using the maximum likelihood method in terms of $p(\boldsymbol{x}_n|\boldsymbol{\lambda})$. If there are two or more Bayes models characterized by a hyperparameter and if it is necessary to compare their goodness of fit, it suffices to select the model that minimizes the ABIC.

If the hyperparameter estimated in this way is denoted by $\hat{\boldsymbol{\lambda}}$, then we can determine the posterior distribution of the parameter $\boldsymbol{\theta}$ in terms of the prior distribution $\pi(\boldsymbol{\theta}|\hat{\boldsymbol{\lambda}})$ as

$$\pi(\boldsymbol{\theta}|\boldsymbol{x}_n;\hat{\boldsymbol{\lambda}}) = \frac{f(\boldsymbol{x}_n|\boldsymbol{\theta})\pi(\boldsymbol{\theta}|\hat{\boldsymbol{\lambda}})}{\displaystyle\int f(\boldsymbol{x}_n|\boldsymbol{\theta})\pi(\boldsymbol{\theta}|\hat{\boldsymbol{\lambda}})d\boldsymbol{\theta}}. \tag{9.50}$$

In general, the mode of the posterior distribution (9.50) is used in practical applications, i.e., the value $\hat{\boldsymbol{\theta}}$ that maximizes $\pi(\boldsymbol{\theta}|\boldsymbol{x}_n;\hat{\boldsymbol{\lambda}}) \propto f(\boldsymbol{x}_n|\boldsymbol{\theta})\pi(\boldsymbol{\theta}|\hat{\boldsymbol{\lambda}})$.

The ultimate objective of modeling using the information criterion ABIC is not to estimate the hyperparameter $\boldsymbol{\lambda}$. Rather, the objective is to estimate the parameter $\boldsymbol{\theta}$ or the distribution of data $\boldsymbol{x}_n$ specified by the parameters. Inferences performed through the minimization of the ABIC can be thought of as a two-step estimation process consisting first of the estimation of a hyperparameter and the selection of a model using the maximum likelihood method on the data distribution $p(\boldsymbol{x}_n|\boldsymbol{\lambda})$, which is given as a marginal distribution, and second, the determination of an estimate of $\boldsymbol{\theta}$ by maximizing the posterior distribution $\pi(\boldsymbol{\theta}|\boldsymbol{x}_n;\hat{\boldsymbol{\lambda}})$ of the parameter $\boldsymbol{\theta}$.

The ABIC minimization method was originally used for the development of seasonal adjustments of econometric data [Akaike (1980b, 1980c) and Akaike and Ishiguro (1980a, 1980b, 1980c)]. Subsequently, it has been used for the development of a variety of new models, including cohort analyses [Nakamura (1986)], binary regression models [Sakamoto and Ishiguro (1988)], and earth tide analyses [Ishiguro and Sakamoto (1984)].

Akaike (1987) showed the relationship between, AIC and ABIC by introducing the Bayesian approach to control the occurrence of improper solutions in normal theory maximum likelihood factor analysis [see also Martin and McDonald (1975)].

9.3 Bayesian Predictive Distributions

Predictive distributions based on a Bayesian approach are constructed using a parametric model $\{f(x|\boldsymbol{\theta}); \boldsymbol{\theta} \in \Theta \subset \mathrm{R}^p\}$ that defines the data distribution and a prior distribution $\pi(\boldsymbol{\theta})$ for the parameter vector $\boldsymbol{\theta}$. If the prior distribution, in turn, has a hyperparameter $\boldsymbol{\lambda}$, its distribution is denoted by $\pi(\boldsymbol{\theta}|\boldsymbol{\lambda})$ ($\boldsymbol{\lambda} \in \Theta_\lambda \subset \mathrm{R}^q$; $q < p$).

9.3.1 Predictive Distributions and Predictive Likelihood

Let $\boldsymbol{x}_n = \{x_1, \ldots, x_n\}$ be n observations that are generated from an unknown probability distribution $G(x)$ having density function $g(x)$. Let $f(x|\boldsymbol{\theta})$ denote a parametric model having a p-dimensional parameter $\boldsymbol{\theta}$, and let us consider a Bayes model for which the prior distribution of the parameter $\boldsymbol{\theta}$ is $\pi(\boldsymbol{\theta})$.

Given data $\boldsymbol{x}_n$ and the distribution $f(\boldsymbol{x}_n|\boldsymbol{\theta})$, it follows from Bayes' theorem that the *posterior distribution* of $\boldsymbol{\theta}$ is defined by

$$\pi(\boldsymbol{\theta}|\boldsymbol{x}_n) = \frac{f(\boldsymbol{x}_n|\boldsymbol{\theta})\pi(\boldsymbol{\theta})}{\displaystyle\int f(\boldsymbol{x}_n|\boldsymbol{\theta})\pi(\boldsymbol{\theta})d\boldsymbol{\theta}}. \tag{9.51}$$

Let $\boldsymbol{z} = \{z_1, \cdots, z_n\}$ be future data generated independently of the observed data $\boldsymbol{x}_n$. Using the posterior distribution (9.51), we approximate the distribution $g(\boldsymbol{z})$ of the future data by

$$\begin{aligned} h(\boldsymbol{z}|\boldsymbol{x}_n) &= \int f(\boldsymbol{z}|\boldsymbol{\theta})\pi(\boldsymbol{\theta}|\boldsymbol{x}_n)d\boldsymbol{\theta} \\ &= \frac{\displaystyle\int f(\boldsymbol{z}|\boldsymbol{\theta})f(\boldsymbol{x}_n|\boldsymbol{\theta})\pi(\boldsymbol{\theta})d\boldsymbol{\theta}}{\displaystyle\int f(\boldsymbol{x}_n|\boldsymbol{\theta})\pi(\boldsymbol{\theta})d\boldsymbol{\theta}}. \end{aligned} \tag{9.52}$$

The $h(\boldsymbol{z}|\boldsymbol{x}_n)$ is called a *predictive distribution.*

In the following, we evaluate how well the predictive distribution approximates the distribution $g(\boldsymbol{z})$ that generates the data by using the expected log-likelihood

$$E_{G(\boldsymbol{z})}\left[\log h(\boldsymbol{Z}|\boldsymbol{x}_n)\right] = \int g(\boldsymbol{z}) \log h(\boldsymbol{z}|\boldsymbol{x}_n)d\boldsymbol{z}. \tag{9.53}$$

In actual modeling, the prior distribution $\pi(\boldsymbol{\theta})$ is rarely completely specified. In this section, we assume that the prior distribution of $\boldsymbol{\theta}$ is defined by a small number of parameters $\boldsymbol{\lambda} \in \Theta_\lambda \subset R^q$ called *hyperparameters* and that they are expressed as $\pi(\boldsymbol{\theta}|\boldsymbol{\lambda})$. In this situation, we denote the posterior distribution of $\boldsymbol{\theta}$, the predictive distribution of $\boldsymbol{z}$, and the marginal distribution of the data $\boldsymbol{x}_n$ by $\pi(\boldsymbol{\theta}|\boldsymbol{x}_n;\boldsymbol{\lambda})$, $h(\boldsymbol{z}|\boldsymbol{x}_n;\boldsymbol{\lambda})$, and $p(\boldsymbol{x}_n|\boldsymbol{\lambda})$, respectively.

For an ordinary parametric model $f(x|\boldsymbol{\theta})$, it is easy to see that

$$E_{G(\boldsymbol{x}_n)}\left[\log f(\boldsymbol{X}_n|\boldsymbol{\theta}) - E_{G(\boldsymbol{z})}\left[\log f(\boldsymbol{Z}|\boldsymbol{\theta})\right]\right] = 0, \tag{9.54}$$

as was shown in Chapter 3. Here, $E_{G(\boldsymbol{x}_n)}$ and $E_{G(\boldsymbol{z})}$ denote the expectations with respect to the data $\boldsymbol{x}_n$ and the future observations $\boldsymbol{z}$ obtained from the distribution G, respectively. Hence, in this case, the log-likelihood, $\log f(\boldsymbol{x}_n|\boldsymbol{\theta})$, is an unbiased estimator of the expected log-likelihood, and it provides a natural estimate of the expected log-likelihood. In the case of Bayesian models also, similar results can be derived with respect to the marginal distribution

$$p(\boldsymbol{z}) = \int f(\boldsymbol{z}|\boldsymbol{\theta})\pi(\boldsymbol{\theta})d\boldsymbol{\theta}. \tag{9.55}$$

This implies that the log-likelihood provides a natural criterion for estimation of parameters.

In contrast, the Bayesian predictive distribution $h(\boldsymbol{z}|\boldsymbol{x}_n;\boldsymbol{\lambda})$ constructed by a prior distribution with hyperparameters $\boldsymbol{\lambda}$ generally takes the form

$$b_P(G,\boldsymbol{\lambda}) \equiv E_{G(\boldsymbol{x}_n)}\left[\log h(\boldsymbol{X}_n|\boldsymbol{X}_n;\boldsymbol{\lambda}) - E_{G(\boldsymbol{z})}\left[\log h(\boldsymbol{Z}|\boldsymbol{X}_n;\boldsymbol{\lambda})\right]\right] \neq 0. \tag{9.56}$$

Consequently, the log-likelihood $\log h(\boldsymbol{x}_n|\boldsymbol{x}_n;\boldsymbol{\lambda})$ is not an unbiased estimator of the expected log-likelihood $E_{G(\boldsymbol{z})}[\log h(\boldsymbol{Z}|\boldsymbol{x}_n;\boldsymbol{\lambda})]$. Therefore, in the estimation of the hyperparameters $\boldsymbol{\lambda}$, maximizing the expression $\log h(\boldsymbol{x}_n|\boldsymbol{x}_n;\boldsymbol{\lambda})$ does not result in maximizing the expected log-likelihood, even approximately.

The reason for this difficulty lies in the fact that, as in the case of previous information criteria, the same data $\boldsymbol{x}_n$ are used twice in the expression $\log h(\boldsymbol{x}_n|\boldsymbol{x}_n;\boldsymbol{\lambda})$. Therefore, when evaluating the predictive distribution for the estimation of hyperparameters in a Bayesian model, it is more natural to use the bias-corrected log-likelihood

$$\log h(\boldsymbol{x}_n|\boldsymbol{x}_n,\boldsymbol{\lambda}) - b_P(G,\boldsymbol{\lambda}) \tag{9.57}$$

as an estimate of the expected log-likelihood [Akaike (1980a) and Kitagawa (1984)].

In this section, in a similar way as the information criteria that have been presented thus far, we define the *predictive information criterion* (PIC) for Bayesian models as

$$\text{PIC} = -2\log h(\boldsymbol{x}_n|\boldsymbol{x}_n;\boldsymbol{\lambda}) + 2b_P(G,\boldsymbol{\lambda}) \tag{9.58}$$

[Kitagawa (1997)]. If the hyperparameters $\boldsymbol{\lambda}$ are unknown, then the values of $\boldsymbol{\lambda}$ can be estimated by minimizing the PIC, in a manner similar to the maximum likelihood method described in Chapter 3. Given a predictive distribution of general Bayesian models, however, it is difficult to determine this bias analytically.

In the next section, we show that the bias correction term $b_P(G,\boldsymbol{\lambda})$ in (9.58) can be determined directly for a Bayesian normal linear model, and in Section 9.4, we describe how to use the Laplace integral approximation to determine it in the case of general Bayesian models.

9.3.2 Information Criterion for Bayesian Normal Linear Models

In this section, we consider a normal linear model in the Bayesian framework and determine the specific value of the bias term $b_P(G,\boldsymbol{\lambda})$.

Suppose that the n-dimensional observation vector $\boldsymbol{x}$ and the p-dimensional parameter vector $\boldsymbol{\theta}$ are both from multivariate normal distributions as follows:

$$\boldsymbol{X} \sim f(\boldsymbol{x}|\boldsymbol{\theta}) = N_n(A\boldsymbol{\theta},R), \qquad \boldsymbol{\theta} \sim \pi(\boldsymbol{\theta}|\boldsymbol{\lambda}) = N_p(\boldsymbol{\theta}_0,Q), \tag{9.59}$$

where A is an $n\times p$ matrix, and R and Q are $n\times n$ and $p\times p$ nonsingular matrices, respectively. It is further assumed that the matrices A and R and the hyperparameters $\boldsymbol{\lambda}=(\boldsymbol{\theta}_0,Q)$ are all known.

The bias term $b_P(G, \boldsymbol{\lambda})$ for the Bayesian model given by (9.56) varies depending on the nature of the true distribution. For simplicity in what follows, we assume that the true distribution may be expressed as $g(x) = f(x|\boldsymbol{\theta})$ and $G(x) = F(x|\boldsymbol{\theta})$. In addition, we consider the case in which we evaluate the goodness of fit of the parameters $\boldsymbol{\theta}$, but not that of the hyperparameters $\boldsymbol{\lambda}$. We also assume that the observed data $\boldsymbol{x}$ and the future data $\boldsymbol{z}$ follow distributions having the same parameter $\boldsymbol{\theta}$. Then the bias can be determined exactly by calculating

$$\begin{aligned} b_P(F, \boldsymbol{\lambda}) &= E_{\Pi(\boldsymbol{\theta}|\boldsymbol{\lambda})} E_{F(\boldsymbol{x}|\boldsymbol{\theta})} \left[\log h(\boldsymbol{X}|\boldsymbol{X}; \boldsymbol{\lambda}) - E_{F(\boldsymbol{z}|\boldsymbol{\theta})} [\log h(\boldsymbol{Z}|\boldsymbol{X}; \boldsymbol{\lambda})] \right] \\ &= \int \left[\int \left\{ \log h(\boldsymbol{x}|\boldsymbol{x}, \boldsymbol{\lambda}) - \int f(\boldsymbol{z}|\boldsymbol{\theta}) \log h(\boldsymbol{z}|\boldsymbol{x}; \boldsymbol{\lambda}) d\boldsymbol{z} \right\} \right. \\ &\quad \left. \times f(\boldsymbol{x}|\boldsymbol{\theta}) d\boldsymbol{x} \right] \pi(\boldsymbol{\theta}|\boldsymbol{\lambda}) d\boldsymbol{\theta}, \end{aligned} \tag{9.60}$$

where $\Pi(\boldsymbol{\theta}|\boldsymbol{\lambda})$ and $F(\boldsymbol{x}|\boldsymbol{\theta})$ are the distribution functions of $\pi(\boldsymbol{\theta}|\boldsymbol{\lambda})$ and $f(\boldsymbol{x}|\boldsymbol{\theta})$, respectively.

In the case of the Bayesian normal linear model, as will be shown in Subsection 9.3.3, we have the bias correction term

$$b_P(G, \boldsymbol{\lambda}) = \mathrm{tr}\left\{ (2W + R)^{-1} W \right\}, \tag{9.61}$$

where $W = AQA'$. Therefore, the PIC in this case is given by

$$\mathrm{PIC} = -2 \log f(\boldsymbol{x}|\boldsymbol{x}, \boldsymbol{\lambda}) + 2\mathrm{tr}\{(2W + R)^{-1} W\}. \tag{9.62}$$

Similarly, the bias correction term can also be determined when the parameters for the model $f(\boldsymbol{x}|\boldsymbol{\theta})$ depend on the MAP (maximum posterior estimate) defined by

$$\tilde{\boldsymbol{\theta}} = \arg\max_{\boldsymbol{\theta}} \pi(\boldsymbol{\theta}|\boldsymbol{x}), \tag{9.63}$$

and in this case we have

$$\tilde{b}_P(G, \boldsymbol{\lambda}) = \mathrm{tr}\left\{ (W + R)^{-1} W \right\}. \tag{9.64}$$

9.3.3 Derivation of the PIC

To derive the information criterion PIC for the Bayesian normal linear model in (9.59), we use the following lemma [Lindley and Smith (1972)]:

Lemma (Marginal and posterior distributions for normal models) Assume that the distribution $f(\boldsymbol{x}|\boldsymbol{\theta})$ of the n-dimensional vector $\boldsymbol{x}$ of random variables is an n-dimensional normal distribution $N_n(A\boldsymbol{\theta}, R)$ and that the distribution $\pi(\boldsymbol{\theta})$ of the p-dimensional parameter vector $\boldsymbol{\theta}$ is a p-dimensional normal distribution $N_p(\boldsymbol{\theta}_0, Q)$. Then we obtain the following results:

(i) The marginal distribution of $\boldsymbol{x}$ defined by

$$p(\boldsymbol{x}) = \int f(\boldsymbol{x}|\boldsymbol{\theta})\pi(\boldsymbol{\theta})d\boldsymbol{\theta} \tag{9.65}$$

is distributed normally as $N_n(A\boldsymbol{\theta}_0, W + R)$, where $W = AQA^T$.

(ii) The posterior distribution of $\boldsymbol{\theta}$ defined by

$$\pi(\boldsymbol{\theta}|\boldsymbol{x}) = \frac{f(\boldsymbol{x}|\boldsymbol{\theta})\pi(\boldsymbol{\theta})}{\displaystyle\int f(\boldsymbol{x}|\boldsymbol{\theta})\pi(\boldsymbol{\theta})d\boldsymbol{\theta}} \tag{9.66}$$

is distributed normally as $N_p(\boldsymbol{\xi}, V)$, where the mean vector $\boldsymbol{\xi}$ and the variance covariance matrix V are given by

$$\begin{aligned} \boldsymbol{\xi} &= \boldsymbol{\theta}_0 + QA^T(W + R)^{-1}(\boldsymbol{x} - A\boldsymbol{\theta}_0), \\ V &= Q - QA^T(W + R)^{-1}AQ \\ &= (A^TR^{-1}A + Q^{-1})^{-1}. \end{aligned} \tag{9.67}$$

For the prior distribution $\pi(\boldsymbol{\theta}|\boldsymbol{\lambda})$ in (9.59), we derive specific forms of the marginal and posterior distributions by using the above lemma. In this case, ξ, V, and W in (9.67) depend on the hyperparameters $\boldsymbol{\lambda}$ and should be written as $\xi(\boldsymbol{\lambda})$, $V(\boldsymbol{\lambda})$, and $W(\boldsymbol{\lambda})$. For the sake of simplicity, in the following we shall denote them simply as ξ, V, and W.

By applying the results (i) and (ii) in the lemma to the Bayesian normal linear model of (9.59), the marginal distribution $p(\boldsymbol{x}|\boldsymbol{\lambda})$ and the posterior distribution $\pi(\boldsymbol{\theta}|\boldsymbol{x};\boldsymbol{\lambda})$ are

$$p(\boldsymbol{x}|\boldsymbol{\lambda}) \sim N_n(A\boldsymbol{\theta}_0, W + R), \quad \pi(\boldsymbol{\theta}|\boldsymbol{x};\boldsymbol{\lambda}) \sim N_p(\boldsymbol{\xi}, V), \tag{9.68}$$

where $\boldsymbol{\xi}$ and V are respectively the mean vector and the variance-covariance matrix of the posterior distribution given in (9.67). Then the predictive distribution defined by (9.52) in terms of the posterior distribution $\pi(\boldsymbol{\theta}|\boldsymbol{x};\boldsymbol{\lambda})$ is an n-dimensional normal distribution, that is,

$$h(\boldsymbol{z}|\boldsymbol{x};\boldsymbol{\lambda}) = \int f(\boldsymbol{z}|\boldsymbol{\theta})\pi(\boldsymbol{\theta}|\boldsymbol{x};\boldsymbol{\lambda})d\boldsymbol{\theta} \sim N_n(\boldsymbol{\mu}, \Sigma), \tag{9.69}$$

where the mean vector $\boldsymbol{\mu}$ and the variance-covariance matrix Σ are given by

$$\begin{aligned} \boldsymbol{\mu} &= A\boldsymbol{\xi} \\ &= W(W + R)^{-1}\boldsymbol{x} + R(W + R)^{-1}A\boldsymbol{\theta}_0, \end{aligned} \tag{9.70}$$

$$\begin{aligned} \Sigma &= AVA^T + R \\ &= W(W + R)^{-1}R + R \\ &= (2W + R)(W + R)^{-1}R. \end{aligned} \tag{9.71}$$

Consequently, using the log-likelihood of the predictive distribution written as

$$\log h(\boldsymbol{z}|\boldsymbol{x};\boldsymbol{\lambda}) = -\frac{n}{2}\log(2\pi) - \frac{1}{2}\log|\Sigma| - \frac{1}{2}(\boldsymbol{z}-\boldsymbol{\mu})^T\Sigma^{-1}(\boldsymbol{z}-\boldsymbol{\mu}), \quad (9.72)$$

the expectation of the difference between the log-likelihood and the expected log-likelihood may be evaluated as follows:

$$\begin{aligned}
&E_{G(\boldsymbol{x})}\left[\log h(\boldsymbol{X}|\boldsymbol{X};\boldsymbol{\lambda}) - E_{G(\boldsymbol{z})}[\log h(\boldsymbol{Z}|\boldsymbol{X};\boldsymbol{\lambda})]\right] \quad (9.73)\\
&= -\frac{1}{2}E_{G(\boldsymbol{x})}\left[(\boldsymbol{X}-\boldsymbol{\mu})^T\Sigma^{-1}(\boldsymbol{X}-\boldsymbol{\mu}) - E_{G(\boldsymbol{z})}[(\boldsymbol{Z}-\boldsymbol{\mu})^T\Sigma^{-1}(\boldsymbol{Z}-\boldsymbol{\mu})]\right]\\
&= -\frac{1}{2}\mathrm{tr}\left\{\Sigma^{-1}E_{G(\boldsymbol{x})}\left[(\boldsymbol{X}-\boldsymbol{\mu})(\boldsymbol{X}-\boldsymbol{\mu})^T - E_{G(\boldsymbol{z})}[(\boldsymbol{Z}-\boldsymbol{\mu})(\boldsymbol{Z}-\boldsymbol{\mu})^T]\right]\right\}.
\end{aligned}$$

We note that $\boldsymbol{\mu}$ in (9.70) depends on $\boldsymbol{X}$.

In the particular situation that the true distribution $g(\boldsymbol{z})$ is given by $f(\boldsymbol{z}|\boldsymbol{\theta}_0) \sim N_n(A\boldsymbol{\theta}_0, R)$, we have

$$\begin{aligned}
&E_{F(\boldsymbol{z}|\boldsymbol{\theta})}\left[(\boldsymbol{Z}-\boldsymbol{\mu})(\boldsymbol{Z}-\boldsymbol{\mu})^T\right]\\
&\quad = E_{F(\boldsymbol{z}|\boldsymbol{\theta})}\left[(\boldsymbol{Z}-A\boldsymbol{\theta}_0)(\boldsymbol{Z}-A\boldsymbol{\theta}_0)^T\right] + (A\boldsymbol{\theta}_0-\boldsymbol{\mu})(A\boldsymbol{\theta}_0-\boldsymbol{\mu})^T\\
&\quad = R + (A\boldsymbol{\theta}_0-\boldsymbol{\mu})(A\boldsymbol{\theta}_0-\boldsymbol{\mu})^T. \quad (9.74)
\end{aligned}$$

Writing $\Delta\boldsymbol{\theta} \equiv \boldsymbol{\theta} - \boldsymbol{\theta}_0$, we can see that

$$\begin{aligned}
A\boldsymbol{\theta}_0 - \boldsymbol{\mu} &= W(W+R)^{-1}(A\boldsymbol{\theta}_0 - \boldsymbol{x}) + R(W+R)^{-1}A\Delta\boldsymbol{\theta},\\
\boldsymbol{x} - \boldsymbol{\mu} &= R(W+R)^{-1}\{(\boldsymbol{x} - A\boldsymbol{\theta}_0) + A\Delta\boldsymbol{\theta}\}. \quad (9.75)
\end{aligned}$$

Hence, by using $R = R(W+R)^{-1}W + R(W+R)^{-1}R$ and $\Sigma = R(W+R)^{-1}(2W+R)$, it follows from (9.74) and (9.75) that

$$\begin{aligned}
&E_{F(\boldsymbol{x}|\boldsymbol{\theta})}\left[E_{F(\boldsymbol{z}|\boldsymbol{\theta})}[(\boldsymbol{Z}-\boldsymbol{\mu})(\boldsymbol{Z}-\boldsymbol{\mu})^T] - (\boldsymbol{X}-\boldsymbol{\mu})(\boldsymbol{X}-\boldsymbol{\mu})^T\right]\\
&\quad = R + W(W+R)^{-1}R(W+R)^{-1}W - R(W+R)^{-1}R(W+R)^{-1}R\\
&\quad = W(W+R)^{-1}R + R(W+R)^{-1}W\\
&\quad = \Sigma - R(W+R)^{-1}R. \quad (9.76)
\end{aligned}$$

In this case, the bias correction term in (9.73) can be calculated exactly as

$$\begin{aligned}
b_P(F,\boldsymbol{\lambda}) &= E_{\Pi(\boldsymbol{\theta})}E_{F(\boldsymbol{x}|\boldsymbol{\theta})}\left[\log h(\boldsymbol{X}|\boldsymbol{X};\boldsymbol{\lambda}) - E_{F(\boldsymbol{z}|\boldsymbol{\theta})}[\log h(\boldsymbol{Z}|\boldsymbol{X};\boldsymbol{\lambda})]\right]\\
&= \frac{1}{2}\mathrm{tr}\left[\Sigma^{-1}\{\Sigma - R(W+R)^{-1}R\}\right]\\
&= \frac{1}{2}\mathrm{tr}\left\{I_n - (2W+R)^{-1}R\right\}\\
&= \mathrm{tr}\left\{(2W+R)^{-1}W\right\}. \quad (9.77)
\end{aligned}$$

Since the expectation with respect to $F(\boldsymbol{x}|\boldsymbol{\theta})$ is constant and does not depend on the value of $\boldsymbol{\theta}$, integration with respect to $\boldsymbol{\theta}$ is not required. In addition, the bias term does not depend on the individual observations $\boldsymbol{x}$ and is determined solely by the true variance covariance matrices R and Q.

By correcting the bias (9.77) for the log-likelihood of the predictive distribution in (9.72) and multiplying it by -2, we have the PIC for the Bayesian normal linear model in the form

$$\mathrm{PIC} = n \log(2\pi) + \log|\Sigma| + (\boldsymbol{x} - \boldsymbol{\mu})^T \Sigma^{-1} (\boldsymbol{x} - \boldsymbol{\mu}) + 2\mathrm{tr}\{(2W + R)^{-1} W\}, \tag{9.78}$$

where $\boldsymbol{\mu}$ and Σ are respectively given by (9.70) and (9.71).

9.3.4 Numerical Example

Suppose that we have n observations $\{x_\alpha; \alpha = 1, \ldots, n\}$ from a normal distribution model

$$x_\alpha = \mu_\alpha + w_\alpha, \qquad w_\alpha \sim N(0, \sigma^2), \tag{9.79}$$

where μ_α is the true mean and the variance σ^2 of the noise w_α is known. In order to estimate the mean-value function μ_α, we consider the trend model

$$x_\alpha = t_\alpha + w_\alpha, \qquad w_\alpha \sim N(0, \sigma^2). \tag{9.80}$$

For the trend component t_α, we assume a constraint model

$$t_\alpha = t_{\alpha-1} + v_\alpha, \qquad v_\alpha \sim N(0, \tau^2). \tag{9.81}$$

Then eqs. (9.80) and (9.81) can be formulated as the Bayesian model

$$\boldsymbol{x} = \boldsymbol{\theta} + \boldsymbol{w}, \qquad B\boldsymbol{\theta} = \boldsymbol{\theta}_* + \boldsymbol{v}, \tag{9.82}$$

where $\boldsymbol{x} = (x_1, \ldots, x_n)^T$, $\boldsymbol{\theta} = (t_1, \ldots, t_n)^T$, $\boldsymbol{w} = (w_1, \ldots, w_n)^T$, $\boldsymbol{v} = (v_1, \ldots, v_n)^T$, and B and $\boldsymbol{\theta}_*$ are, respectively, an $n \times n$ matrix and an n-dimensional vector given by

$$B = \begin{bmatrix} 1 & & & \\ -1 & 1 & & \\ & \ddots & \ddots & \\ & & -1 & 1 \end{bmatrix}, \qquad \boldsymbol{\theta}_* = \begin{bmatrix} t_0 \\ 0 \\ \vdots \\ 0 \end{bmatrix}. \tag{9.83}$$

In addition, for simplicity, we assume that $t_0 = \varepsilon_0$ ($\varepsilon_0 \sim N(0, 1)$) and that the random variables $\boldsymbol{\theta}$ and $\boldsymbol{w}$ and $\boldsymbol{\theta}_*$ and $\boldsymbol{v}$ are mutually independent.

Setting $Q_0 = \mathrm{diag}\{\tau^2 + 1, \tau^2, \ldots, \tau^2\}$ and $\boldsymbol{\theta}_0 = B^{-1}\boldsymbol{\theta}_*$, we have

$$\boldsymbol{\theta} \sim N_n(\boldsymbol{\theta}_0, B^{-1} Q_0 (B^{-1})^T). \tag{9.84}$$

Therefore, by taking $A = I_n$, $Q = B^{-1} Q_0 (B^{-1})^T$, and $R = \sigma^2 I_n$, where I_n is the n-dimensional identity matrix, this model turns out to be the Bayesian normal linear model of (9.59).

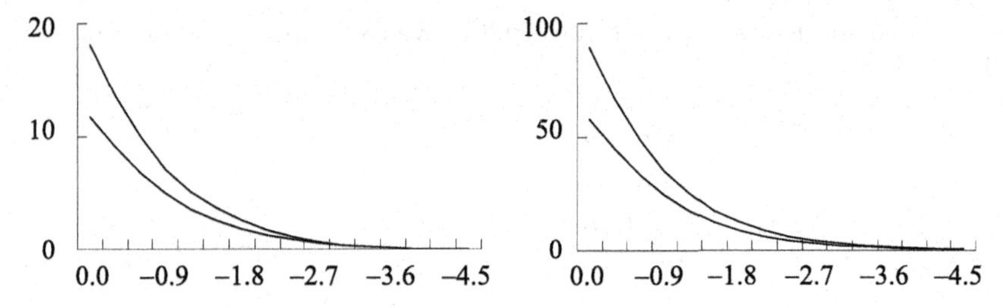

Fig. 9.3. Bias correction terms $2b_P(G,\lambda)$ and $2\tilde{b}_p(G,\lambda)$ for the Bayesian information criterion. The horizontal axis is λ, and the vertical axis shows the bias correction term. For the left graph, $n = 20$, and for the right graph, $n = 100$.

Figure 9.3 shows changes in the bias $2b_P(G,\lambda)$ and $2\tilde{b}_P(G,\lambda)$ as $n = 20$ and $n = 100$ for the values of $\lambda = \tau^2/\sigma^2 = 2^{-\ell}$ $(\ell = 0, 1, \ldots, 15)$, where $b_P(G,\lambda)$ and $\tilde{b}_P(G,\lambda)$ were obtained from (9.60) and (9.64), respectively. We note that, for a given value of n, the value of the bias depends solely on the variance ratio λ. As λ increases, the bias also increases significantly. In addition, the bias also increases as the number of observations increases, suggesting that the order is $O(n)$. From these results, we observe that the predictive likelihood without bias correction overestimates the goodness of fit when compared with the true predictive distribution, especially when the value of λ is large. Smoother estimates can be obtained by using a small λ that maximizes the predictive likelihood with a bias correction.

9.4 Bayesian Predictive Distributions by Laplace Approximation

This section considers a Bayesian model constructed from a parametric model $f(x|\boldsymbol{\theta})$ $(\boldsymbol{\theta} \in \Theta \subset R^p)$ and a prior distribution $\pi(\boldsymbol{\theta})$ for n observations $\boldsymbol{x}_n = \{x_1, \ldots, x_n\}$ that are generated from an unknown probability distribution $G(x)$ with density function $g(x)$.

For a future observation z that is randomly extracted independent of the data $\boldsymbol{x}_n$, we approximate the distribution $g(z)$ by the Bayesian predictive distribution

$$h(z|\boldsymbol{x}_n) = \int f(z|\boldsymbol{\theta})\pi(\boldsymbol{\theta}|\boldsymbol{x}_n)d\boldsymbol{\theta}, \tag{9.85}$$

where $\pi(\boldsymbol{\theta}|\boldsymbol{x}_n)$ is the posterior distribution of $\boldsymbol{\theta}$ given by

$$\pi(\boldsymbol{\theta}|\boldsymbol{x}_n) = \frac{f(\boldsymbol{x}_n|\boldsymbol{\theta})\pi(\boldsymbol{\theta})}{\displaystyle\int f(\boldsymbol{x}_n|\boldsymbol{\theta})\pi(\boldsymbol{\theta})d\boldsymbol{\theta}}. \tag{9.86}$$

By substituting this expression into (9.85), we can express the predictive distribution as

$$\begin{aligned} h(z|\boldsymbol{x}_n) &= \frac{\displaystyle\int f(z|\boldsymbol{\theta})f(\boldsymbol{x}_n|\boldsymbol{\theta})\pi(\boldsymbol{\theta})d\boldsymbol{\theta}}{\displaystyle\int f(\boldsymbol{x}_n|\boldsymbol{\theta})\pi(\boldsymbol{\theta})d\boldsymbol{\theta}} \\ &= \frac{\displaystyle\int \exp\left[n\left\{n^{-1}\log f(\boldsymbol{x}_n|\boldsymbol{\theta}) + n^{-1}\log \pi(\boldsymbol{\theta}) + n^{-1}\log f(z|\boldsymbol{\theta})\right\}\right] d\boldsymbol{\theta}}{\displaystyle\int \exp\left[n\left\{n^{-1}\log f(\boldsymbol{x}_n|\boldsymbol{\theta}) + n^{-1}\log \pi(\boldsymbol{\theta})\right\}\right] d\boldsymbol{\theta}} \\ &= \frac{\displaystyle\int \exp\left[n\left\{q(\boldsymbol{\theta}|\boldsymbol{x}_n) + n^{-1}\log f(z|\boldsymbol{\theta})\right\}\right] d\boldsymbol{\theta}}{\displaystyle\int \exp\left\{nq(\boldsymbol{\theta}|\boldsymbol{x}_n)\right\} d\boldsymbol{\theta}}, \end{aligned} \tag{9.87}$$

where

$$q(\boldsymbol{\theta}|\boldsymbol{x}_n) = \frac{1}{n}\log f(\boldsymbol{x}_n|\boldsymbol{\theta}) + \frac{1}{n}\log \pi(\boldsymbol{\theta}). \tag{9.88}$$

We will now show that we can apply the information criterion GIC_M in (5.114) to the evaluation of a Bayesian predictive distribution, using Laplace's method for integrals described in Subsection 9.1.2 to approximate the predictive distribution in (9.87).

Let $\hat{\boldsymbol{\theta}}_q$ be a mode of $q(\boldsymbol{\theta}|\boldsymbol{x}_n)$ in (9.88). By applying the Laplace approximation to the denominator of (9.87), we obtain

$$\begin{aligned} &\int \exp\left\{nq(\boldsymbol{\theta}|\boldsymbol{x}_n)\right\} d\boldsymbol{\theta} \\ &\quad = \frac{(2\pi)^{p/2}}{n^{p/2}\left|J_q(\hat{\boldsymbol{\theta}}_q)\right|^{1/2}} \exp\left\{nq(\hat{\boldsymbol{\theta}}_q|\boldsymbol{x}_n)\right\}\left\{1 + O_p(n^{-1})\right\}, \end{aligned} \tag{9.89}$$

where $J_q(\hat{\boldsymbol{\theta}}_q) = -\partial^2\{q(\hat{\boldsymbol{\theta}}_q|\boldsymbol{x}_n)\}/\partial\boldsymbol{\theta}\partial\boldsymbol{\theta}^T$. Similarly, by letting $\hat{\boldsymbol{\theta}}_q(z)$ be a mode of $q(\boldsymbol{\theta}|\boldsymbol{x}_n) + n^{-1}\log f(z|\boldsymbol{\theta})$, we obtain the following Laplace approximation to the integral in the numerator:

$$\begin{aligned} &\int \exp\left[n\left\{q(\boldsymbol{\theta}|\boldsymbol{x}_n) + \frac{1}{n}\log f(z|\boldsymbol{\theta})\right\}\right] d\boldsymbol{\theta} \\ &= \frac{(2\pi)^{p/2}}{n^{p/2}|J_{q(z)}(\hat{\boldsymbol{\theta}}_q(z))|^{1/2}} \exp\left[n\left\{q(\hat{\boldsymbol{\theta}}_q(z)|\boldsymbol{x}_n) + \frac{1}{n}\log f(z|\hat{\boldsymbol{\theta}}_q(z))\right\}\right] \\ &\quad \times \{1 + O_p(n^{-1})\}, \end{aligned} \tag{9.90}$$

where $J_{q(z)}(\hat{\boldsymbol{\theta}}_q(z)) = -\partial^2\{q(\hat{\boldsymbol{\theta}}_q(z)|\boldsymbol{x}_n) + n^{-1}\log f(z|\hat{\boldsymbol{\theta}}_q(z))\}/\partial\boldsymbol{\theta}\partial\boldsymbol{\theta}^T$.

It follows from (9.89) and (9.90) that the predictive distribution $h(z|\boldsymbol{x}_n)$ can be approximated as follows:

$$h(z|\boldsymbol{x}_n) = \left(\frac{|J_q(\hat{\boldsymbol{\theta}}_q)|}{|J_{q(z)}(\hat{\boldsymbol{\theta}}_q(z))|} \right)^{\frac{1}{2}} \exp \left[n \left\{ q(\hat{\boldsymbol{\theta}}_q(z)|\boldsymbol{x}_n) - q(\hat{\boldsymbol{\theta}}_q|\boldsymbol{x}_n) \right. \right.$$

$$\left. \left. + \frac{1}{n} \log f(z|\hat{\theta}_q(z)) \right\} \right] \times \{1 + O_p(n^{-2})\}. \tag{9.91}$$

Substituting functional Taylor series expansions for the modes $\hat{\boldsymbol{\theta}}_q$ and $\hat{\boldsymbol{\theta}}_q(z)$ into the resulting approximation and then simplifying the Laplace approximation (9.91) yield the Bayesian predictive distribution in the form

$$h(z|\boldsymbol{x}_n) = f(z|\hat{\boldsymbol{\theta}})\{1 + O_p(n^{-1})\}. \tag{9.92}$$

The form of the functional that defines the estimator $\hat{\boldsymbol{\theta}}$ is related to whether or not the prior distribution $\pi(\boldsymbol{\theta})$ depends upon the sample size n. Given a prior distribution, let us now consider two cases: (i) $\log \pi(\boldsymbol{\theta}) = O(1)$, (ii) $\log \pi(\boldsymbol{\theta}) = O(n)$. As can be seen from (9.88), in case (i), the estimator $\hat{\boldsymbol{\theta}}$ is the maximum likelihood estimator $\hat{\boldsymbol{\theta}}_{\mathrm{ML}}$, and in case (ii), it becomes the mode $\hat{\boldsymbol{\theta}}_B$ of a posterior distribution. Functionals that define these estimators are solutions of

$$\int \left. \frac{\partial \log f(\boldsymbol{x}|\boldsymbol{\theta})}{\partial \boldsymbol{\theta}} \right|_{\boldsymbol{\theta} = \boldsymbol{T}_{\mathrm{ML}}(G)} dG(x) = \mathbf{0},$$

$$\int \left. \frac{\partial \log \{f(\boldsymbol{x}|\boldsymbol{\theta})\pi(\boldsymbol{\theta})\}}{\partial \boldsymbol{\theta}} \right|_{\boldsymbol{\theta} = \boldsymbol{T}_B(G)} dG(x) = \mathbf{0}, \tag{9.93}$$

respectively.

In the information criterion GIC_M given by (5.114) in Subsection 5.2.3, by taking

$$\psi(x, \hat{\boldsymbol{\theta}}) = \left. \frac{\partial \log f(\boldsymbol{x}|\boldsymbol{\theta})}{\partial \boldsymbol{\theta}} \right|_{\boldsymbol{\theta} = \boldsymbol{T}_{ML}(\hat{G})}, \tag{9.94}$$

$$\psi(x, \hat{\boldsymbol{\theta}}) = \left. \frac{\partial \{\log f(\boldsymbol{x}|\boldsymbol{\theta}) + \log \pi(\boldsymbol{\theta})\}}{\partial \boldsymbol{\theta}} \right|_{\boldsymbol{\theta} = \boldsymbol{T}_B(\hat{G})}, \tag{9.95}$$

we obtain the information criterion for the Bayesian predictive distribution model $h(z|\boldsymbol{x}_n)$. It has the general form

$$\mathrm{GIC}_B = -2 \log h(\boldsymbol{x}_n|\boldsymbol{x}_n) + 2\mathrm{tr} \left\{ R(\psi, \hat{G})^{-1} Q(\psi, \hat{G}) \right\}. \tag{9.96}$$

In the case that $\log \pi(\boldsymbol{\theta}) = O(n)$, the asymptotic bias in (9.96) depends on the prior distribution through the partial derivatives of $\log \pi(\boldsymbol{\theta})$, while in the case that $\log \pi(\boldsymbol{\theta}) = O(1)$, the asymptotic bias does not depend on the prior distribution and has the same form as that of TIC in (3.99). In the latter case, a more refined result is required in the context of smooth functional estimators.

The strength of the influence exerted by the prior distribution $\pi(\boldsymbol{\theta})$ is principally captured by its first- and second-order derivatives, with the result that if the prior distribution is $\log \pi(\boldsymbol{\theta}) = O(1)$, it does not contribute its effect solely on the basis of the first-order bias correction term. In such a situation, by taking the higher-order bias correction terms into account, we obtain a more accurate result.

The second-order (asymptotic) bias correction term $b_{(2)}(\hat{G})$ is defined as an estimator of $b_{(2)}(G)$, which is generally given by

$$\begin{aligned} E_{G(\boldsymbol{x})} \left[\log h(\boldsymbol{X}_n|\boldsymbol{X}_n) - \mathrm{tr}\left\{ R(\boldsymbol{\psi}, \hat{G})^{-1} Q(\boldsymbol{\psi}, \hat{G}) \right\} - n E_{G(z)}[h(Z|\boldsymbol{X}_n)] \right] \\ = \frac{1}{n} b_{(2)}(G) + O(n^{-2}). \qquad (9.97) \end{aligned}$$

Then we have the second-order bias-corrected log-likelihood of the predictive distribution in the form

$$\mathrm{GIC_{BS}} = -2 \log h(\boldsymbol{x}_n|\boldsymbol{x}_n) + 2\mathrm{tr}\left\{ R(\boldsymbol{\psi}, \hat{G})^{-1} Q(\boldsymbol{\psi}, \hat{G}) \right\} + \frac{2}{n} b_{(2)}(\hat{G}). \qquad (9.98)$$

In fact, $b_{(2)}(G)$ is given by subtracting the asymptotic bias of the first-order correction term $\mathrm{tr}\{R(\boldsymbol{\psi}, \hat{G})^{-1} Q(\boldsymbol{\psi}, \hat{G})\}$ from the second-order asymptotic bias term of the log-likelihood of the model (see Subsection 7.2.2). Derivation of the second-order bias correction term includes log-likelihood, a high-order differentiation of the prior distribution, and a higher-order, compact differentiation of the estimator, and analytically it can be extremely complex. In such cases, bootstrap methods offer an alternative numerical approach to estimate the bias.

Example 5 (Bayesian predictive distribution) We use a normal distribution model

$$f(x|\mu, \tau^2) = \left(\frac{\tau^2}{2\pi} \right)^{\frac{1}{2}} \exp\left\{ -\frac{\tau^2}{2}(x-\mu)^2 \right\} \qquad (9.99)$$

that approximates the true distribution as a prior distribution of parameters μ and τ^2, we assume

$$\begin{aligned} \pi(\mu, \tau^2) &= N(\mu_0, \tau_0^{-2}\tau^{-2}) G_a(\tau^2|\lambda, \beta) \qquad (9.100) \\ &= \left(\frac{\tau_0^2 \tau^2}{2\pi} \right)^{\frac{1}{2}} \exp\left\{ -\frac{\tau_0^2 \tau^2}{2}(\mu - \mu_0)^2 \right\} \frac{\beta^\lambda}{\Gamma(\lambda)} \tau^{2(\lambda-1)} e^{-\beta\tau^2}. \end{aligned}$$

Then the predictive distribution is given by

$$h(z|\boldsymbol{x}) = \frac{\Gamma\left(\dfrac{b+1}{2}\right)}{\Gamma\left(\dfrac{b}{2}\right)} \left(\frac{a}{b\pi}\right)^{\frac{1}{2}} \left\{1 + \frac{a}{b}(z-c)^2\right\}^{-(a+1)/2}, \tag{9.101}$$

where $\overline{x} = \frac{1}{n}\sum_{\alpha=1}^{n} x_\alpha$C $s^2 = \frac{1}{n}\sum_{\alpha=1}^{n}(x_\alpha - \overline{x})^2$, and a, b, and c are defined as

$$a = \frac{(n+\tau_0^2)(\lambda + \frac{1}{2}n)}{(n+\tau_0+1)\left\{\beta + \dfrac{1}{2}ns^2 + \dfrac{\tau^2 n}{2(\tau_0^2+n)}(\mu_0 - \overline{x})^2\right\}},$$

$$b = 2\lambda + n, \quad c = \frac{\tau_0^2\mu_0 + n\overline{x}}{\tau_0^2 + n}, \tag{9.102}$$

respectively.

From (9.96), the information criterion for the evaluation of the predictive distribution is then given by

$$\mathrm{GIC}_B = -2\sum_{\alpha=1}^{n} \log h(x_\alpha|\boldsymbol{x}_n) + 2\left\{\frac{1}{2} + \frac{\hat{\mu}_4}{2(s^2)^2}\right\} \tag{9.103}$$

with

$$\hat{\mu}_4 = \frac{1}{n}\sum_{\alpha=1}^{n}(x_\alpha - \overline{x})^4. \tag{9.104}$$

It can be seen that GIC_B, which is an information criterion for the predictive distribution of a Bayesian model, takes a form similar to the TIC. In addition, the second-order bias correction term is given by

$$E_{G(\boldsymbol{x}_n)}\left[\log h(\boldsymbol{X}_n|\boldsymbol{X}_n) - \left\{\frac{1}{2} + \frac{\hat{\mu}_4}{2(s^2)^2}\right\} - n\int g(z)\log h(z|\boldsymbol{X}_n)dz\right]. \tag{9.105}$$

Example 6 (Numerical result) We compare the asymptotic bias estimate ($\mathrm{tr}\,\hat{I}\hat{J}^{-1}$) in (9.103), the bootstrap bias estimate (EIC), and the second-order corrected bias ($\mathrm{GIC_{BS}}$) with the bootstrap bias estimate in (9.105). In the simulation study, data $\{x_\alpha;\ \alpha = 1,\ldots,n)$ were generated from a mixture of normal distributions

$$g(x) = (1-\varepsilon)N(0,1) + \varepsilon N(0,d^2). \tag{9.106}$$

Table 9.2 shows changes in the values of the true bias $b(G)$, $\mathrm{tr}\{\hat{I}\hat{J}^{-1}\}$, and the biases for EIC and $\mathrm{GIC_{BS}}$ for various values of the mixture ratio ε. For

Table 9.2. Changes of true bias $b(G)$, $\mathrm{tr}\{\hat{I}\hat{J}^{-1}\}$, and the biases for EIC and $\mathrm{GIC_{BS}}$ for various values of the mixture ratio ε.

ε	$b(G)$	$\mathrm{tr}\{\hat{I}\hat{J}^{-1}\}$	EIC	$\mathrm{GIC_{BS}}$
0.00	2.07	1.89	1.97	2.01
0.04	2.96	2.41	2.52	2.76
0.08	3.50	2.73	2.89	3.24
0.12	3.79	2.90	3.13	3.52
0.16	3.95	2.99	3.28	3.68
0.20	4.02	3.01	3.35	3.73
0.24	3.96	2.99	3.39	3.73
0.28	3.92	2.95	3.38	3.69
0.32	3.77	2.89	3.40	3.69
0.36	3.72	2.82	3.31	3.56
0.40	3.60	2.74	3.29	3.51

model parameters, we set $d^2 = 10$, $\mu_0 = 1$, $\tau_0^2 = 1$, $\alpha = 4$, and $\beta = 1$ and ran Monte Carlo trials with 100,000 repetitions. In the bias estimation for EIC, we used $B = 10$ for the bootstrap replications.

It can be seen from the table that the bootstrap bias estimate of EIC is closer to the true bias than the bias correction term $\mathrm{tr}\{\hat{I}\hat{J}^{-1}\}$ for TIC or GIC_B. It can also be seen that the second-order correction term of $\mathrm{GIC_{BS}}$ is even more accurate than these other two correction terms.

9.5 Deviance Information Criterion (DIC)

Spiegelhalter et al. (2002) developed a deviance information criterion (DIC) from a Bayesian perspective, using an information-theoretic argument to motivate a complexity measure for the effective number of parameters in a model. Let $f(\boldsymbol{x}_n|\boldsymbol{\theta})$ ($\boldsymbol{\theta} \in \Theta \subset R^p$) and $\pi(\boldsymbol{\theta}|\boldsymbol{x}_n)$ be, respectively, a probability model and a posterior distribution for the observed data $\boldsymbol{x}_n$. Spiegelhalter et al. (2002) proposed the effective number of parameters with respect to a model in the form

$$p_D = -2E_{\pi(\boldsymbol{\theta}|\boldsymbol{x}_n)}\left[\log f(\boldsymbol{x}_n|\boldsymbol{\theta})\right] + 2\log f(\boldsymbol{x}_n|\hat{\boldsymbol{\theta}}), \tag{9.107}$$

where $\hat{\boldsymbol{\theta}}$ is an estimator of the parameter vector $\boldsymbol{\theta}$. Using the Bayesian deviance defined by

$$D(\boldsymbol{\theta}) = -2\log f(\boldsymbol{x}_n|\boldsymbol{\theta}) + 2\log h(\boldsymbol{x}_n), \tag{9.108}$$

where $h(\boldsymbol{x}_n)$ is some fully specified standardizing term that is a function of the data alone, eq. (9.107) can be written as

$$p_D = \overline{D(\boldsymbol{\theta})} - D(\overline{\boldsymbol{\theta}}), \tag{9.109}$$

where $\overline{\boldsymbol{\theta}}$ $(= \hat{\boldsymbol{\theta}})$ is the posterior mean defined by $\overline{\boldsymbol{\theta}} = E_{\pi(\boldsymbol{\theta}|\boldsymbol{x}_n)}[\boldsymbol{\theta}]$ and $\overline{D(\boldsymbol{\theta})}$ is the posterior mean of the deviance defined by $\overline{D(\boldsymbol{\theta})} = E_{\pi(\boldsymbol{\theta}|\boldsymbol{x}_n)}[D(\boldsymbol{\theta})]$.

This shows that a measure for the effective number of parameters in a model can be considered as the difference between the posterior mean of the deviance and the deviance at the posterior means of the parameters of interest. Note that when models are compared, the second term in the Bayesian deviance cancels out.

Spiegelhalter et al. (2002) defined DIC as

$$\begin{aligned} \text{DIC} &= \overline{D(\boldsymbol{\theta})} + p_D \\ &= -2E_{\pi(\boldsymbol{\theta}|\boldsymbol{x}_n)}\left[\log f(\boldsymbol{x}_n|\boldsymbol{\theta})\right] + p_D. \end{aligned} \tag{9.110}$$

It follows from (9.109) that the DIC can also be expressed as

$$\begin{aligned} \text{DIC} &= D(\overline{\boldsymbol{\theta}}) + 2p_D \\ &= -2\log f(\boldsymbol{x}_n|\overline{\boldsymbol{\theta}}) + 2p_D. \end{aligned} \tag{9.111}$$

The optimal model among a set of competing models is chosen by selecting one that minimizes the value of DIC. The DIC can be considered as a Bayesian measure of fit or adequacy, penalized by an additional complexity term p_D [Spiegelhalter et al. (2002)].

10

Various Model Evaluation Criteria

So far in this book, we have considered model selection and evaluation criteria from both an information-theoretic point of view and a Bayesian approach. The AIC-type criteria were constructed as estimators of the Kullback–Leibler information between a statistical model and the true distribution generating the data or equivalently the expected log-likelihood of a statistical model. In contrast, the Bayes approach for selecting a model was to choose the model with the largest posterior probability among a set of candidate models.

There are other model evaluation criteria based on various different points of view. This chapter describes cross-validation, generalized cross-validation, final predictive error (FPE), Mallows' C_p, the Hannan–Quinn criterion, and ICOMP. Cross-validation also provides an alternative approach to estimate the Kullback–Leibler information. We show that the cross-validation estimate is asymptotically equivalent to AIC-type criteria in a general setting.

10.1 Cross-Validation

10.1.1 Prediction and Cross-Validation

The objective of statistical modeling or data analysis is to obtain information about data that may arise in the future, rather than the observed data used in the model construction itself. Hence, in the model building process, model evaluation from a predictive point of view implies the evaluation of the goodness of fit of the model based on future data obtained independently of the observed data. In practice, however, it is difficult to consider situations in which future data can be obtained separately from the model construction data, and if, in fact, such data can be obtained, a better model would be constructed by combining such data with the observed data. As a way to circumvent this difficulty, *cross-validation* refers to a technique whereby evaluation from a predictive point of view is executed solely based on observed data while making modifications in order to preserve the accuracy of parameter estimation as much as possible.

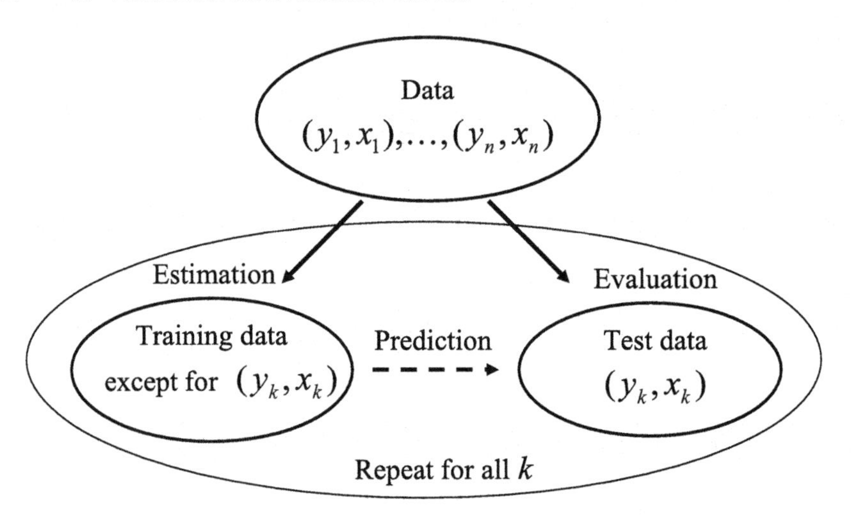

Fig. 10.1. Schematic of the cross-validation procedure.

Given a response variable y and p explanatory variables $\boldsymbol{x} = (x_1, x_2, \ldots, x_p)^T$, let us consider the regression model

$$y = u(\boldsymbol{x}) + \varepsilon, \tag{10.1}$$

where $E[\varepsilon] = 0$ and $E[\varepsilon^2] = \sigma^2$. Since $E[Y|\boldsymbol{x}] = u(\boldsymbol{x})$, the function $u(\boldsymbol{x})$ represents the mean structure. We estimate $u(\boldsymbol{x})$ based on n observations $\{(y_\alpha, \boldsymbol{x}_\alpha); \alpha = 1, 2, \ldots, n\}$ and write it as $\hat{u}(\boldsymbol{x})$. For example, when a linear regression model $y = \boldsymbol{\beta}^T \boldsymbol{x} + \varepsilon$ is assumed, an estimate of $u(\boldsymbol{x})$ is given by $\hat{u}(\boldsymbol{x}) = \hat{\boldsymbol{\beta}}^T \boldsymbol{x}$, using least squares estimates $\hat{\boldsymbol{\beta}} = (X^T X)^{-1} X^T \boldsymbol{y}$ of the regression coefficients $\boldsymbol{\beta}$, where $X^T = (\boldsymbol{x}_1, \ldots, \boldsymbol{x}_n)$.

The goodness of fit of the estimated regression function $\hat{u}(\boldsymbol{x})$ is measured using the (average) predictive mean square error (PSE)

$$\text{PSE} = \frac{1}{n} \sum_{\alpha=1}^{n} E\left[\{Y_\alpha - \hat{u}(\boldsymbol{x}_\alpha)\}^2 \right], \tag{10.2}$$

in terms of future observations Y_α that are randomly drawn at points $\boldsymbol{x}_\alpha$ according to (10.1) in a manner independent of the observed data. Here the residual sum of squares (RSS) is used to estimate the PSE by reusing the data y_α instead of the Y_α:

$$\text{RSS} = \frac{1}{n} \sum_{\alpha=1}^{n} \{y_\alpha - \hat{u}(\boldsymbol{x}_\alpha)\}^2. \tag{10.3}$$

If $\hat{u}(x)$ is a polynomial model, for example, the greater the order of the model, the smaller this value becomes, and the goodness of fit of the model seems to be improved. As a result, we end up selecting a polynomial of order $n-1$ that passes through all the observations, which defeats the purpose of an order selection criterion.

Cross-validation involves the estimation of a predictive mean square error by separating the data used for model estimation (training data) from the data used for model evaluation (test data). Cross-validation is executed in the following steps:

Cross-Validation

(1) From the n observed data values, remove the αth observation $(y_\alpha, \boldsymbol{x}_\alpha)$. Estimate the model based on the remaining $n-1$ observations, and denote this estimate by $\hat{u}^{(-\alpha)}(\boldsymbol{x})$.
(2) For the αth data value $(y_\alpha, \boldsymbol{x}_\alpha)$ removed in step 1, calculate the value of the predictive square error $\{y_\alpha - \hat{u}^{(-\alpha)}(\boldsymbol{x}_\alpha)\}^2$.
(3) Repeat steps 1 and 2 for all $\alpha \in \{1, \ldots, n\}$, and obtain

$$\mathrm{CV} = \frac{1}{n}\sum_{\alpha=1}^{n}\left\{y_\alpha - \hat{u}^{(-\alpha)}(\boldsymbol{x}_\alpha)\right\}^2 \tag{10.4}$$

as the estimated value of the predictive mean square error defined by (10.2). This process is known as *leave-one-out cross-validation.*

It can be shown that the cross-validation (CV) can be considered as an estimator of the predictive mean square error PSE as follows. First, the PSE in (10.2) can be rewritten as

$$\begin{aligned}
\mathrm{PSE} &= \frac{1}{n}\sum_{\alpha=1}^{n} E\left[\{Y_\alpha - \hat{u}(\boldsymbol{x}_\alpha)\}^2\right] \\
&= \frac{1}{n}\sum_{\alpha=1}^{n} E\left[\{Y_\alpha - u(\boldsymbol{x}_\alpha) + u(\boldsymbol{x}_\alpha) - \hat{u}(\boldsymbol{x}_\alpha)\}^2\right] \\
&= \frac{1}{n}\sum_{\alpha=1}^{n} E\Big[\{Y_\alpha - u(\boldsymbol{x}_\alpha)\}^2 + \{u(\boldsymbol{x}_\alpha) - \hat{u}(\boldsymbol{x}_\alpha)\}^2 \\
&\quad + 2\{Y_\alpha - u(\boldsymbol{x}_\alpha)\}\{u(\boldsymbol{x}_\alpha) - \hat{u}(\boldsymbol{x}_\alpha)\}\Big] \\
&= \frac{1}{n}\sum_{\alpha=1}^{n} E\left[\{Y_\alpha - u(\boldsymbol{x}_\alpha)\}^2\right] + \frac{1}{n}\sum_{\alpha=1}^{n} E\left[\{u(\boldsymbol{x}_\alpha) - \hat{u}(\boldsymbol{x}_\alpha)\}^2\right] \\
&= \sigma^2 + \frac{1}{n}\sum_{\alpha=1}^{n} E\left[\{u(\boldsymbol{x}_\alpha) - \hat{u}(\boldsymbol{x}_\alpha)\}^2\right].
\end{aligned} \tag{10.5}$$

On the other hand, the expectation of (10.4) is

$$
\begin{aligned}
E[\mathrm{CV}] &= E\left[\frac{1}{n}\sum_{\alpha=1}^{n}\left\{Y_\alpha - \hat{u}^{(-\alpha)}(\boldsymbol{x}_\alpha)\right\}^2\right] \\
&= \frac{1}{n}\sum_{\alpha=1}^{n} E\left[\left\{Y_\alpha - u(\boldsymbol{x}_\alpha) + u(\boldsymbol{x}_\alpha) - \hat{u}^{(-\alpha)}(\boldsymbol{x}_\alpha)\right\}^2\right] \\
&= \frac{1}{n}\sum_{\alpha=1}^{n} E\Big[\{Y_\alpha - u(\boldsymbol{x}_\alpha)\}^2 + 2\{Y_\alpha - u(\boldsymbol{x}_\alpha)\}\left\{u(\boldsymbol{x}_\alpha) - \hat{u}^{(-\alpha)}(\boldsymbol{x}_\alpha)\right\} \\
&\quad + \left\{u(\boldsymbol{x}_\alpha) - \hat{u}^{(-\alpha)}(\boldsymbol{x}_\alpha)\right\}^2\Big] \\
&= \sigma^2 + \frac{1}{n}\sum_{\alpha=1}^{n} E\left[\left\{u(\boldsymbol{x}_\alpha) - \hat{u}^{(-\alpha)}(\boldsymbol{x}_\alpha)\right\}^2\right]. \qquad (10.6)
\end{aligned}
$$

Hence, it follows from (10.5) and (10.6) that since $\hat{u}^{(-\alpha)}(\boldsymbol{x}_\alpha)$ and $\hat{u}(\boldsymbol{x}_\alpha)$ are asymptotically equal, the relationship $E\,[\mathrm{CV}] \approx \mathrm{PSE}$ holds. This implies that CV can be considered to be an estimator of predictive mean square error.

The leave-one-out cross-validation procedure can be generalized to the method called *K-fold cross-validation* as follows. The observed data are divided into K subsets. One of the K subsets is used as the test data for evaluating a model, and the union of the remaining $K-1$ subsets is taken as training data. The average prediction error across the K trials is then calculated.

For cross-validation, we refer to Stone (1974), Geisser (1975), and Efron (1982) among others.

10.1.2 Selecting a Smoothing Parameter by Cross-Validation

In order to estimate the mean structure $u(\boldsymbol{x})$ in (10.1), we consider the following regression model, which makes use of basis expansions (see Section 6.1):

$$
\begin{aligned}
y_\alpha &= \sum_{i=1}^{m} w_i b_i(\boldsymbol{x}_\alpha) + \varepsilon_\alpha \\
&= \boldsymbol{w}^T \boldsymbol{b}(\boldsymbol{x}_\alpha) + \varepsilon_\alpha, \qquad \alpha = 1, 2, \ldots, n, \qquad (10.7)
\end{aligned}
$$

where $\boldsymbol{w} = (w_1, w_2, \ldots, w_m)^T$, $\boldsymbol{b}(\boldsymbol{x}_\alpha) = (b_1(\boldsymbol{x}_\alpha), b_2(\boldsymbol{x}_\alpha), \ldots, b_m(\boldsymbol{x}_\alpha))^T$, and it is assumed that ε_α, $\alpha = 1, 2, \ldots, n$, are mutually independent and that $E[\varepsilon_\alpha] = 0$, $E[\varepsilon_\alpha^2] = \sigma^2$. We estimate the coefficient vector $\boldsymbol{w}$ of the basis functions by the regularized or penalized least squares method, that is, by minimizing the function of $\boldsymbol{w}$ given by

$$
\begin{aligned}
S_\lambda(\boldsymbol{w}) &= \sum_{\alpha=1}^{n}\left\{y_\alpha - \sum_{i=1}^{m} w_i b_i(\boldsymbol{x}_\alpha)\right\}^2 + \gamma \boldsymbol{w}^T K \boldsymbol{w} \\
&= (\boldsymbol{y} - B\boldsymbol{w})^T(\boldsymbol{y} - B\boldsymbol{w}) + \gamma \boldsymbol{w}^T K \boldsymbol{w}, \qquad (10.8)
\end{aligned}
$$

where $\boldsymbol{y} = (y_1, y_2, \ldots, y_n)^T$ and $B = (\boldsymbol{b}(\boldsymbol{x}_1), \boldsymbol{b}(\boldsymbol{x}_2), \ldots, \boldsymbol{b}(\boldsymbol{x}_n))^T$. The typical form of the matrix K was given in Subsection 5.2.4.

The regularized (penalized) least squares estimate is given by

$$\hat{\boldsymbol{w}} = (B^T B + \gamma K)^{-1} B^T \boldsymbol{y}, \tag{10.9}$$

which yields the estimate $\hat{u}(\boldsymbol{x}) = \hat{\boldsymbol{w}}^T \boldsymbol{b}(\boldsymbol{x})$ of the mean structure $u(\boldsymbol{x})$ in (10.1). Furthermore, for the predictive value $\hat{y}_\alpha = \hat{u}(\boldsymbol{x}_\alpha) = \hat{\boldsymbol{w}}^T \boldsymbol{b}(\boldsymbol{x}_\alpha)$ at each point $\boldsymbol{x}_\alpha$, we obtain the n-dimensional vector of predicted values

$$\hat{\boldsymbol{y}} = B\hat{\boldsymbol{w}} = B(B^T B + \gamma K)^{-1} B^T \boldsymbol{y}, \tag{10.10}$$

where $\hat{\boldsymbol{y}} = (\hat{y}_1, \hat{y}_2, \ldots, \hat{y}_n)^T$. The ridge type of estimator is given by taking $K = I_m$, where I_m is the m-dimensional identity matrix.

Since the estimated regression function $\hat{u}(\boldsymbol{x})$ depends on the smoothing parameter γ and also the number, m, of basis functions through the estimation of the coefficient vector $\boldsymbol{w}$, we need to select optimal values of these adjusted parameters. Applying cross-validation to this problem, we choose optimal values of the adjusted parameters as follows.

First, we specify the number m of basis functions and the value of a smoothing parameter γ. From n observations, remove the αth data point $(y_\alpha, \boldsymbol{x}_\alpha)$ and, based on the remaining $n-1$ observations, estimate $\boldsymbol{w}$ using the regularized least squares method and set it as $\hat{\boldsymbol{w}}^{(-\alpha)}$. The corresponding estimated regression function is given by $\hat{u}^{(-\alpha)}(\boldsymbol{x}) = \hat{\boldsymbol{w}}^{(-\alpha)^T} \boldsymbol{b}(\boldsymbol{x})$. Then the adjusted parameters $\{\gamma, m\}$ that minimize the equation

$$\mathrm{CV}(\gamma, m) = \frac{1}{n} \sum_{\alpha=1}^{n} \left\{ y_\alpha - \hat{u}^{(-\alpha)}(\boldsymbol{x}_\alpha) \right\}^2 \tag{10.11}$$

are selected as optimal values.

10.1.3 Generalized Cross-Validation

Selecting the optimal values of the number of basis functions and a smoothing parameter by applying cross-validation to a large data set can result in computational difficulties. If the predicted value $\hat{\boldsymbol{y}}$ is given in the form of $\hat{\boldsymbol{y}} = H\boldsymbol{y}$, where H is a matrix that does not depend on the data $\boldsymbol{y}$, then in cross-validation, the estimation process performed n times by removing observations one by one is not needed, and thus the amount of computation required can be reduced substantially.

Because the matrix H transforms observed data $\boldsymbol{y}$ to predicted values $\hat{\boldsymbol{y}}$, it is referred to as a *hat matrix*. In the case of fitting a curve or a surface, as in the case of a regression model constructed from basis expansions, it is called a *smoother matrix*. For example, since the predicted values for a linear regression model are given by $\hat{\boldsymbol{y}} = X(X^T X)^{-1} X^T \boldsymbol{y}$, the hat matrix is $H = X(X^TX)^{-1}X^T$. Similarly, since the predicted values for a nonlinear regression model based on basis expansions are given by (10.10), it follows that in this case the smoother matrix is

$$H(\gamma, m) = B(B^TB + \gamma K)^{-1}B^T, \tag{10.12}$$

which depends on the adjusted parameters.

Using either a hat matrix or a smoother matrix $H(\gamma, m)$, *generalized cross-validation* is given by

$$\mathrm{GCV}(\gamma, m) = \frac{1}{n} \frac{\displaystyle\sum_{\alpha=1}^{n} \{y_\alpha - \hat{u}(\boldsymbol{x}_\alpha)\}^2}{\left\{1 - \dfrac{1}{n}\mathrm{tr}H(\gamma, m)\right\}^2} \tag{10.13}$$

[Craven and Wahba (1979)]. As indicated by this formula, the need to execute repeated estimations n times by removing observations one by one is eliminated, thus permitting efficient computation.

The essential idea behind the generalized cross-validation may be described as follows [Green and Silverman (1994)]. First, based on the $n-1$ observations obtained by removing the αth data point $(y_\alpha, \boldsymbol{x}_\alpha)$ from n observed data points, estimate a regression function using the regularized least squares method, and thus define the regression function $\hat{u}^{(-\alpha)}(\boldsymbol{x}) = \hat{\boldsymbol{w}}^{(-\alpha)^T}\boldsymbol{b}(\boldsymbol{x})$. In the next step, we set $z_j = y_j$ and then replace the αth data point y_α with $z_\alpha = \hat{u}^{(-\alpha)}(\boldsymbol{x}_\alpha)$. In other words, define a new n-dimensional vector as

$$\boldsymbol{z} = (y_1, y_2, \ldots, \hat{u}^{(-\alpha)}(\boldsymbol{x}_\alpha), \ldots, y_n)^T. \tag{10.14}$$

Then the fact that the regression function $\hat{u}^{(-\alpha)}(\boldsymbol{x})$ estimated by removing the αth data point minimizes

$$\sum_{j=1}^{n} \{z_j - \boldsymbol{w}^T\boldsymbol{b}(\boldsymbol{x}_j)\}^2 + \gamma\boldsymbol{w}^TK\boldsymbol{w} \tag{10.15}$$

can be demonstrated based on the following inequality:

$$\begin{aligned}
&\sum_{j=1}^{n} \{z_j - \boldsymbol{w}^T\boldsymbol{b}(\boldsymbol{x}_j)\}^2 + \gamma\boldsymbol{w}^TK\boldsymbol{w} \\
&\qquad \geq \sum_{j\neq\alpha}^{n} \{z_j - \boldsymbol{w}^T\boldsymbol{b}(\boldsymbol{x}_j)\}^2 + \gamma\boldsymbol{w}^TK\boldsymbol{w} \\
&\qquad \geq \sum_{j\neq\alpha}^{n} \left\{z_j - \hat{u}^{(-\alpha)}(\boldsymbol{x}_j)\right\}^2 + \gamma\hat{\boldsymbol{w}}^{(-\alpha)^T}K\hat{\boldsymbol{w}}^{(-\alpha)} \\
&\qquad = \sum_{j=1}^{n} \left\{z_j - \hat{u}^{(-\alpha)}(\boldsymbol{x}_j)\right\}^2 + \gamma\hat{\boldsymbol{w}}^{(-\alpha)^T}K\hat{\boldsymbol{w}}^{(-\alpha)}.
\end{aligned} \tag{10.16}$$

Note here that $z_\alpha - \hat{u}^{(-\alpha)}(\boldsymbol{x}_\alpha) = 0$. Hence, it can be seen from the last expression that the term $\hat{u}^{(-\alpha)}(\boldsymbol{x})$ is a regression function that minimizes (10.15).

Let $h_{\alpha j}$ be the $(\alpha, j)^{th}$ component of the smoother matrix. Using this result leads to

$$\begin{aligned} \hat{u}^{(-\alpha)}(\boldsymbol{x}_\alpha) - y_\alpha &= \sum_{j=1}^{n} h_{\alpha j} z_j - y_\alpha \\ &= \sum_{j \neq \alpha}^{n} h_{\alpha j} y_j + h_{\alpha\alpha} \hat{u}^{(-\alpha)}(\boldsymbol{x}_\alpha) - y_\alpha \\ &= \sum_{j=1}^{n} h_{\alpha j} y_j - y_\alpha + h_{\alpha\alpha} \left\{ \hat{u}^{(-\alpha)}(\boldsymbol{x}_\alpha) - y_\alpha \right\} \\ &= \hat{u}(\boldsymbol{x}_\alpha) - y_\alpha + h_{\alpha\alpha} \left\{ \hat{u}^{(-\alpha)}(\boldsymbol{x}_\alpha) - y_\alpha \right\}, \end{aligned} \tag{10.17}$$

and hence we obtain

$$y_\alpha - \hat{u}^{(-\alpha)}(\boldsymbol{x}_\alpha) = \frac{y_\alpha - \hat{u}(\boldsymbol{x}_\alpha)}{1 - h_{\alpha\alpha}}. \tag{10.18}$$

By substituting this equation into (10.11), we obtain

$$\mathrm{CV}(\gamma, m) = \frac{1}{n} \sum_{\alpha=1}^{n} \left\{ \frac{y_\alpha - \hat{u}(\boldsymbol{x}_\alpha)}{1 - h_{\alpha\alpha}} \right\}^2. \tag{10.19}$$

The generalized cross-validation given by (10.13) is obtained by replacing the quantity $1 - h_{\alpha\alpha}$ contained in the denominator with its average $1 - n^{-1}\mathrm{tr}\, H(\gamma, m)$.

10.1.4 Asymptotic Equivalence Between AIC-Type Criteria and Cross-Validation

Cross-validation offers an alternative approach to estimate the Kullback–Leibler information from a predictive point of view. Suppose that n independent observations $\boldsymbol{y}_n = \{y_1, \ldots, y_n\}$ are generated from the true distribution $G(y)$. Consider a specified parametric model $f(y|\boldsymbol{\theta})$ $(\boldsymbol{\theta} \in \Theta \subset R^p)$. Let $f(y|\hat{\boldsymbol{\theta}})$ be a statistical model fitted to the observed data $\boldsymbol{y}$. The AIC-type criteria were constructed as estimators of the Kullback-Leibler information between the true distribution and the statistical model or equivalently the expected log-likelihood $E_{G(z)}[f(Z|\hat{\boldsymbol{\theta}})]$ for a future observation z that might be obtained on the same random structure. We know that the log-likelihood $\log f(\boldsymbol{y}_n|\hat{\boldsymbol{\theta}})(/n)$ yields an optimistic assessment (overestimation) of the expected log-likelihood, because the same data are used both to estimate the parameters of the model and to evaluate the expected log-likelihood.

Cross-validation can be used as a method for estimating the expected log-likelihood in terms of the predictive ability of the models. Let $f(y|\hat{\boldsymbol{\theta}}^{(-\alpha)})$ be a statistical model constructed by removing the αth observation y_α from n observed data and estimating the model based on the remaining $n-1$ observations. Then the cross-validation estimate of the expected log-likelihood $(nE_{G(z)}[f(Z|\hat{\boldsymbol{\theta}})])$ is

$$\text{IC}_{\text{CV}} = \sum_{\alpha=1}^{n} \log f(y_\alpha|\hat{\boldsymbol{\theta}}^{(-\alpha)}). \tag{10.20}$$

We now show that, in a general setting, cross-validation is asymptotically equivalent to AIC-type criteria.

Suppose that there exists a p-dimensional functional $\boldsymbol{T}(G)$ such that $\hat{\boldsymbol{\theta}} = \boldsymbol{T}(\hat{G})$, where $\hat{G}$ is the empirical distribution function based on n data points $\boldsymbol{y}_n$. Removing the αth data point y_α from $\boldsymbol{y}_n$ gives an empirical distribution function $\hat{G}^{(-\alpha)}$ and a corresponding estimator $\hat{\boldsymbol{\theta}}^{(-\alpha)} = \boldsymbol{T}(\hat{G}^{(-\alpha)})$. By expanding $\log f(y_\alpha|\hat{\boldsymbol{\theta}}^{(-\alpha)})$ in a Taylor series around $\hat{\boldsymbol{\theta}}$, we have

$$\begin{aligned}
&\sum_{\alpha=1}^{n} \log f(y_\alpha|\hat{\boldsymbol{\theta}}^{(-\alpha)}) \\
&= \sum_{\alpha=1}^{n} \log f(y_\alpha|\hat{\boldsymbol{\theta}}) + \sum_{\alpha=1}^{n} (\hat{\boldsymbol{\theta}}^{(-\alpha)} - \hat{\boldsymbol{\theta}})^T \left.\frac{\partial \log f(y_\alpha|\boldsymbol{\theta})}{\partial \boldsymbol{\theta}}\right|_{\boldsymbol{\theta}=\hat{\boldsymbol{\theta}}} \qquad (10.21) \\
&\quad + \frac{1}{2}\sum_{\alpha=1}^{n} (\hat{\boldsymbol{\theta}}^{(-\alpha)} - \hat{\boldsymbol{\theta}})^T \left.\frac{\partial^2 \log f(y_\alpha|\boldsymbol{\theta})}{\partial \boldsymbol{\theta}\partial \boldsymbol{\theta}^T}\right|_{\boldsymbol{\theta}=\hat{\boldsymbol{\theta}}} (\hat{\boldsymbol{\theta}}^{(-\alpha)} - \hat{\boldsymbol{\theta}}) + \cdots.
\end{aligned}$$

By taking $H = \hat{G}^{(-\alpha)}$ in (7.17), we have the functional Taylor series expansion of the estimator $\hat{\boldsymbol{\theta}}^{(-\alpha)} = \boldsymbol{T}(\hat{G}^{(-\alpha)})$ in the form

$$\begin{aligned}
\boldsymbol{T}(\hat{G}^{(-\alpha)}) &= \boldsymbol{T}(G) + \frac{1}{n-1}\sum_{i\neq\alpha}^{n} \boldsymbol{T}^{(1)}(y_i;G) \\
&\quad + \frac{1}{2(n-1)^2}\sum_{i\neq\alpha}^{n}\sum_{j\neq\alpha}^{n} \boldsymbol{T}^{(2)}(y_i,y_j;G) + o_p(n^{-1}) \\
&= \boldsymbol{T}(G) + \frac{1}{n}\sum_{i\neq\alpha}^{n} \boldsymbol{T}^{(1)}(y_i;G) \qquad (10.22) \\
&\quad + \frac{1}{n^2}\left\{\sum_{i\neq\alpha}^{n} \boldsymbol{T}^{(1)}(y_i;G) + \frac{1}{2}\sum_{i\neq\alpha}^{n}\sum_{j\neq\alpha}^{n} \boldsymbol{T}^{(2)}(y_i,y_j;G)\right\} + o_p(n^{-1}).
\end{aligned}$$

Using this stochastic expansion and the corresponding result for $\hat{\boldsymbol{\theta}}$ in (7.22) gives

$$\hat{\boldsymbol{\theta}}^{(-\alpha)} - \hat{\boldsymbol{\theta}} = -\frac{1}{n}\boldsymbol{T}^{(1)}(y_\alpha; G) + \frac{1}{n^2}\left\{\sum_{i \neq \alpha}^{n} \boldsymbol{T}^{(1)}(y_i; G)\right. \tag{10.23}$$

$$\left. + \frac{1}{2}\sum_{i \neq \alpha}^{n}\sum_{j \neq \alpha}^{n} \boldsymbol{T}^{(2)}(y_i, y_j; G) - \frac{1}{2}\sum_{i=1}^{n}\sum_{j=1}^{n} \boldsymbol{T}^{(2)}(y_i, y_j; G)\right\} + o_p(n^{-1}).$$

Substituting this stochastic expansion in (10.21) yields

$$\sum_{\alpha=1}^{n} \log f(y_\alpha | \hat{\boldsymbol{\theta}}^{(-\alpha)}) \tag{10.24}$$

$$= \sum_{\alpha=1}^{n} \log f(y_\alpha | \hat{\boldsymbol{\theta}}) - \frac{1}{n}\sum_{\alpha=1}^{n} \mathrm{tr}\left\{ \boldsymbol{T}^{(1)}(y_\alpha; G) \left.\frac{\partial \log f(y_\alpha | \boldsymbol{\theta})}{\partial \boldsymbol{\theta}^T}\right|_{\boldsymbol{\theta} = \hat{\boldsymbol{\theta}}} \right\} + o_p(1).$$

The second term on the right-hand side of (10.24) converges, as n goes to infinity, to

$$E_G\left[\mathrm{tr}\left\{ \boldsymbol{T}^{(1)}(y_\alpha; G) \left.\frac{\partial \log f(y_\alpha | \boldsymbol{\theta})}{\partial \boldsymbol{\theta}^T}\right|_{\boldsymbol{\theta} = \boldsymbol{T}(G)} \right\} \right]$$

$$= \mathrm{tr}\left\{ \int \boldsymbol{T}^{(1)}(z; G) \left.\frac{\partial \log f(z | \boldsymbol{\theta})}{\partial \boldsymbol{\theta}^T}\right|_{\boldsymbol{\theta} = \boldsymbol{T}(G)} dG(z) \right\}, \tag{10.25}$$

the bias correction term of GIC given by (5.62). Hence the cross-validation in (10.20) is asymptotically equivalent to GIC defined by (5.64). As described in Subsection 5.2.2, taking the p-dimensional influence function of the maximum likelihood estimator in (10.25) yields the TIC, and, further the AIC under the additional assumption that the specified parametric family of densities contains the true distribution.

Asymptotic equivalence between the cross-validation and AIC (TIC) was shown by Stone (1977) [see also Shibata (1989)]. We see that the cross-validation estimator of the expected log-likelihood has the same order of accuracy as the AIC-type criteria (see Subsection 7.2.1 for asymptotic accuracy). More refined results for criteria based on the cross-validation were given by Fujikoshi et al. (2003) for normal multivariate regression models and Yanagihara et al. (2006) in the general case.

10.2 Final Prediction Error (FPE)

10.2.1 FPE

In time series analysis, Akaike (1969, 1970) proposed a criterion called the *final prediction error* (FPE) for selection of the order of the AR model. This criterion was derived as an estimator of the expectation of the prediction

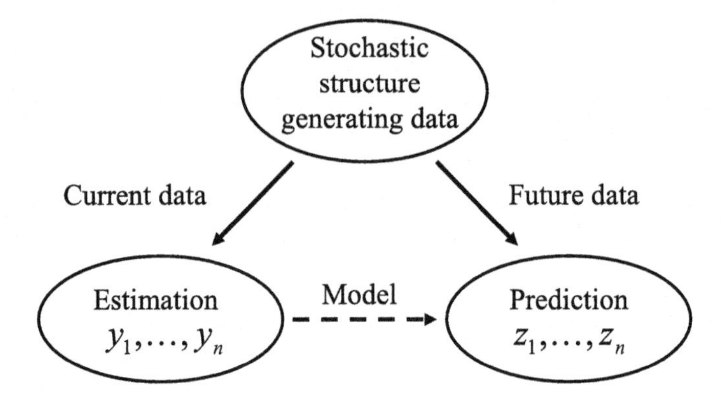

Fig. 10.2. Predictive evaluation scheme for the final prediction error, FPE.

error variance when the estimated model was used for the prediction of a future observation obtained independently from the same stochastic structure as the current time series data used for building the AR model. We explain the FPE in the more general framework of regression models.

Let us now fit a linear regression model to the n observations $\{(y_\alpha, \boldsymbol{x}_\alpha);\ \alpha = 1, \ldots, n\}$ drawn from the response variable Y and p explanatory variables $x_1, \ldots, x_p$. We have

$$\boldsymbol{y} = X\boldsymbol{\beta} + \boldsymbol{\varepsilon}, \qquad E[\boldsymbol{\varepsilon}] = \boldsymbol{0}, \quad V(\boldsymbol{\varepsilon}) = \sigma^2 I_n, \tag{10.26}$$

where $\boldsymbol{\beta} = (\beta_0, \beta_1, \ldots, \beta_p)^T$, $\boldsymbol{\varepsilon} = (\varepsilon_1, \ldots, \varepsilon_n)^T$, and X is an $n \times (p+1)$ design matrix given by

$$X^T = \begin{bmatrix} 1 & 1 & \cdots & 1 \\ \boldsymbol{x}_1 & \boldsymbol{x}_2 & \cdots & \boldsymbol{x}_n \end{bmatrix}_{(p+1)\times n}. \tag{10.27}$$

By estimating the unknown parameter vector $\boldsymbol{\beta}$ of the model by the least squares method, we obtain the predicted values $\hat{\boldsymbol{y}} = X\hat{\boldsymbol{\beta}}$, where $\hat{\boldsymbol{\beta}} = (X^TX)^{-1}X^T\boldsymbol{y}$. For these predicted values, let us consider the sum of squares of prediction errors

$$S_p^2 = (\boldsymbol{z}_0 - \hat{\boldsymbol{y}})^T(\boldsymbol{z}_0 - \hat{\boldsymbol{y}}), \tag{10.28}$$

where $\boldsymbol{z}_0$ is an n-dimensional future observation vector obtained independently of the current data $\boldsymbol{y}$ used for estimation of the model.

Put $H = X(X^TX)^{-1}X^T$. Then we have $\hat{\boldsymbol{y}} = H\boldsymbol{y}$ and $HX = X$. Hence, the expected value of S_p^2 can be calculated as follows:

$$\begin{aligned} E\left[S_p^2\right] &= E\left[(\boldsymbol{z}_0 - \hat{\boldsymbol{y}})^T(\boldsymbol{z}_0 - \hat{\boldsymbol{y}})\right] \\ &= E\left[\{\boldsymbol{z}_0 - X\boldsymbol{\beta} - (\hat{\boldsymbol{y}} - X\boldsymbol{\beta})\}^T \{\boldsymbol{z}_0 - X\boldsymbol{\beta} - (\hat{\boldsymbol{y}} - X\boldsymbol{\beta})\}\right] \end{aligned}$$

$$
\begin{aligned}
&= E\left[(\boldsymbol{z}_0 - X\boldsymbol{\beta})^T(\boldsymbol{z}_0 - X\boldsymbol{\beta})\right] + E\left[(\hat{\boldsymbol{y}} - X\boldsymbol{\beta})^T(\hat{\boldsymbol{y}} - X\boldsymbol{\beta})\right] \\
&= n\sigma^2 + E\left[(H\boldsymbol{y} - HX\boldsymbol{\beta})^T(H\boldsymbol{y} - HX\boldsymbol{\beta})\right] \\
&= n\sigma^2 + E\left[(\boldsymbol{y} - X\boldsymbol{\beta})^T H(\boldsymbol{y} - X\boldsymbol{\beta})\right] \\
&= n\sigma^2 + \mathrm{tr}\left\{HV(\boldsymbol{y})\right\} \\
&= n\sigma^2 + (p+1)\sigma^2. \qquad (10.29)
\end{aligned}
$$

Here we used the facts that H is an idempotent matrix ($H^2 = H$) and that $\alpha^T H\alpha = \mathrm{tr(H\alpha\alpha^T)}$.

By replacing the unknown error variance σ^2 with its unbiased estimate

$$
\frac{1}{n-p-1}S_e^2 = \frac{1}{n-p-1}(\boldsymbol{y} - \hat{\boldsymbol{y}})^T(\boldsymbol{y} - \hat{\boldsymbol{y}}), \qquad (10.30)
$$

we obtain

$$
\mathrm{FPE} = \frac{n+p+1}{n-p-1}S_e^2. \qquad (10.31)
$$

The model evaluation criterion based on the predicted error is called the *final prediction error* (FPE).

10.2.2 Relationship Between the AIC and FPE

The FPE, proposed prior to the information criterion AIC, is closely related to the AIC. In the case of an AR model of order p,

$$
y_n = \sum_{j=1}^{p} a_j y_{n-j} + \varepsilon_n, \qquad \varepsilon_n \sim N(0, \sigma_p^2), \qquad (10.32)
$$

the maximum log-likelihood is given by

$$
\ell(\hat{\theta}) = -\frac{n}{2}\log\hat{\sigma}_p^2 - \frac{n}{2}\log 2\pi - \frac{n}{2}. \qquad (10.33)
$$

Therefore, the AIC for an AR model of order p is given by

$$
\mathrm{AIC}_p = n\log\hat{\sigma}_p^2 + n(\log 2\pi + 1) + 2(p+1). \qquad (10.34)
$$

For comparing AR models with different orders, the constant terms in the equation are often omitted and a simplified version is used, namely,

$$
\mathrm{AIC}_p^* = n\log\hat{\sigma}_p^2 + 2p. \qquad (10.35)
$$

On the other hand, the FPE of an AR model with order p is

$$
\mathrm{FPE}_p = \frac{n+p}{n-p}\hat{\sigma}_p^2. \qquad (10.36)
$$

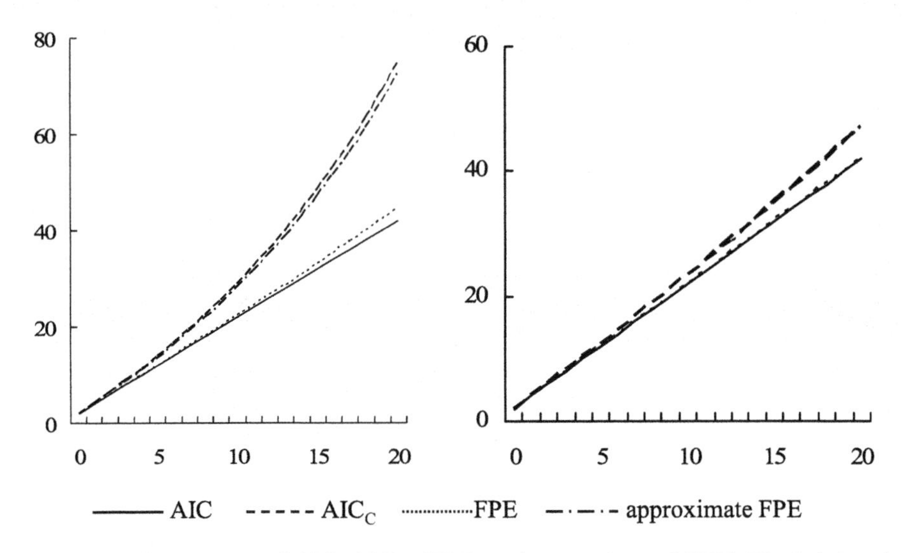

Fig. 10.3. Comparison of AIC, AIC_C, FPE, and approximated FPE. The left-hand plot shows the case $n = 50$, and the right-hand plot is for $n = 200$.

By multiplying by n after taking logarithms of both sides, we have

$$\begin{aligned} n \log \mathrm{FPE}_p &= n \log \left(\frac{n+p}{n-p} \right) + n \log \hat{\sigma}_p^2 \\ &= n \log \left(1 + \frac{2p}{n-p} \right) + n \log \hat{\sigma}_p^2 \\ &\approx n \frac{2p}{n-p} + n \log \hat{\sigma}_p^2 \\ &\approx 2p + n \log \hat{\sigma}_p^2 = \mathrm{AIC}_p^*. \end{aligned} \tag{10.37}$$

Therefore, we see that minimization of the AIC is approximately equivalent to minimization of the FPE and that, with regard to AR models, by minimizing the AIC, we obtain a model that approximately minimizes the final prediction error.

Figure 10.3 shows plots of bias correction terms for the AIC, AIC_C in (7.67), FPE, and approximated FPE for $n = 50$ and $n = 200$. For comparison with the AIC, the FPE is shown in terms of the logarithm of the correction term, $n \log\{1 + 2p/(n - p)\}$. The approximated FPE is shown in terms of the first term of its Taylor expansion, $2pn/(n - p)$. From the plots in Figure 10.3, it may be seen that the AIC, FPE, modified AIC, and approximated FPE each produce very similar correction terms.

10.3 Mallows' C_p

Suppose that we have n sets of data observations $\{(y_\alpha, \boldsymbol{x}_\alpha);\ \alpha = 1, \ldots, n\}$ drawn from a response variable Y and p explanatory variables $x_1, \ldots, x_p$. It is assumed that the expectation and the variance covariance matrix of the n-dimensional observation vector $\boldsymbol{y} = (y_1, \ldots, y_n)^T$ are

$$E[\boldsymbol{y}] = \boldsymbol{\mu}, \qquad V(\boldsymbol{y}) = E[(\boldsymbol{y}-\boldsymbol{\mu})(\boldsymbol{y}-\boldsymbol{\mu})^T] = \omega^2 I_n, \tag{10.38}$$

respectively.

We estimate the true expectation $\boldsymbol{\mu}$ by using the linear regression model

$$\boldsymbol{y} = X\boldsymbol{\beta} + \boldsymbol{\varepsilon}, \qquad E[\boldsymbol{\varepsilon}] = \boldsymbol{0}, \quad V(\boldsymbol{\varepsilon}) = \sigma^2 I_n, \tag{10.39}$$

where $\boldsymbol{\beta} = (\beta_0, \beta_1, \ldots, \beta_p)^T$, $\boldsymbol{\varepsilon} = (\varepsilon_1, \ldots, \varepsilon_n)^T$, and X is an $n \times (p+1)$ design matrix. Then, for the least squares estimator $\hat{\boldsymbol{\beta}} = (X^TX)^{-1}X^T\boldsymbol{y}$ of the regression coefficient vector $\boldsymbol{\beta}$, $\boldsymbol{\mu}$ is estimated by

$$\hat{\boldsymbol{\mu}} = X\hat{\boldsymbol{\beta}} = X(X^TX)^{-1}X^T\boldsymbol{y} \equiv H\boldsymbol{y}. \tag{10.40}$$

As a criterion to measure the effectiveness of the estimator, we consider the mean squared error defined by

$$\Delta_p = E[(\hat{\boldsymbol{\mu}}-\boldsymbol{\mu})^T(\hat{\boldsymbol{\mu}}-\boldsymbol{\mu})]. \tag{10.41}$$

Since the expectation of the estimator $\hat{\boldsymbol{\mu}}$ is

$$E[\hat{\boldsymbol{\mu}}] = X(X^TX)^{-1}X^TE[\boldsymbol{y}] \equiv H\boldsymbol{\mu}, \tag{10.42}$$

the mean squared error Δ_p can be expressed as

$$\begin{aligned}
\Delta_p &= E[(\hat{\boldsymbol{\mu}}-\boldsymbol{\mu})^T(\hat{\boldsymbol{\mu}}-\boldsymbol{\mu})] \\
&= E\left[\{H\boldsymbol{y} - H\boldsymbol{\mu} - (I_n - H)\boldsymbol{\mu}\}^T\{H\boldsymbol{y} - H\boldsymbol{\mu} - (I_n - H)\boldsymbol{\mu}\}\right] \\
&= E\left[(\boldsymbol{y}-\boldsymbol{\mu})^T H(\boldsymbol{y}-\boldsymbol{\mu})\right] + \boldsymbol{\mu}^T(I_n - H)\boldsymbol{\mu} \\
&= \mathrm{tr}\,\{HV(\boldsymbol{y})\} + \boldsymbol{\mu}^T(I_n - H)\boldsymbol{\mu} \\
&= (p+1)\omega^2 + \boldsymbol{\mu}^T(I_n - H)\boldsymbol{\mu}
\end{aligned} \tag{10.43}$$

(see Figure 10.4). Here, since H and $I_n - H$ are idempotent matrices, we have made use of the relationships $H^2 = H$, $(I_n - H)^2 = I_n - H$, $H(I_n - H) = 0$, and $\mathrm{tr}\, H = \mathrm{tr}\{X(X^TX)^{-1}X^T\} = \mathrm{tr}\, I_{p+1} = p+1$, $\mathrm{tr}(I_n - H) = n - p - 1$.

The first term of Δ_p, $(p+1)\omega^2$, increases as the number of parameters increases. The second term, $\boldsymbol{\mu}^T(I_n - H)\boldsymbol{\mu}$, is the sum of squared biases of the estimator $\hat{\boldsymbol{\mu}}$. This term decreases as the number of parameters increases. If Δ_p can be estimated, then it can be used as a criterion for model evaluation.

The expectation of the residual sum of squares can be calculated as

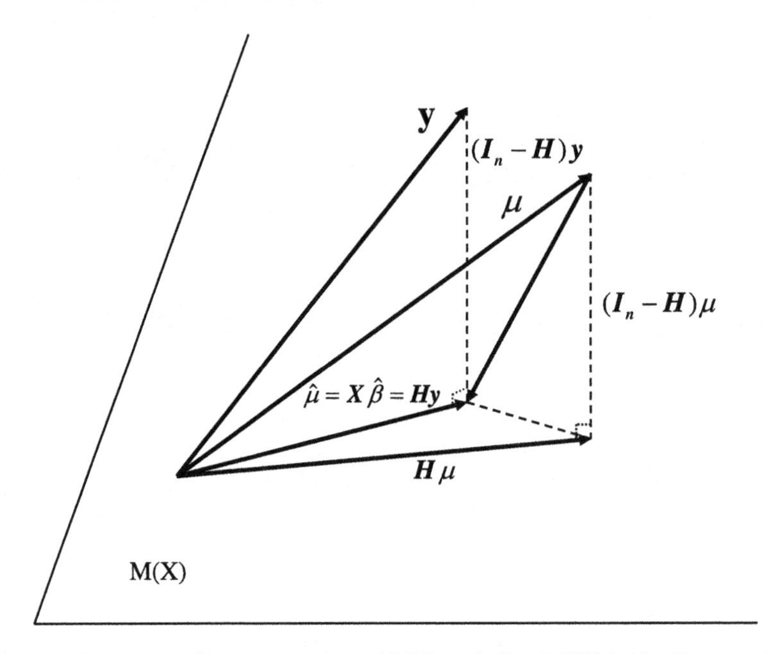

Fig. 10.4. Geometrical interpretation of Mallows' C_p; $\mathcal{M}(X)$ is the linear subspace spanned by the $p+1$ column vectors of the design matrix X.

$$\begin{aligned}
E[S_e^2] &= E[(\boldsymbol{y}-\hat{\boldsymbol{y}})^T(\boldsymbol{y}-\hat{\boldsymbol{y}})] \\
&= E[(\boldsymbol{y}-H\boldsymbol{y})^T(\boldsymbol{y}-H\boldsymbol{y})] \\
&= E[\{(I_n-H)(\boldsymbol{y}-\boldsymbol{\mu})+(I_n-H)\boldsymbol{\mu}\}^T\{(I_n-H)(\boldsymbol{y}-\boldsymbol{\mu})+(I_n-H)\boldsymbol{\mu}\}] \\
&= E[(\boldsymbol{y}-\boldsymbol{\mu})^T(I_n-H)(\boldsymbol{y}-\boldsymbol{\mu})]+\boldsymbol{\mu}^T(I_n-H)\boldsymbol{\mu} \\
&= \mathrm{tr}\{(I_n-H)V(\boldsymbol{y})\}+\boldsymbol{\mu}^T(I_n-H)\boldsymbol{\mu} \\
&= (n-p-1)\omega^2+\boldsymbol{\mu}^T(I_n-H)\boldsymbol{\mu}. \qquad (10.44)
\end{aligned}$$

Comparison between (10.43) and (10.44) reveals that, if ω^2 is assumed known, then the unbiased estimator of Δ_p is given by

$$\hat{\Delta}_p = S_e^2 + \{2(p+1)-n\}\omega^2. \qquad (10.45)$$

By dividing both sides of the above equation by the estimator $\hat{\omega}^2$ of ω^2, we obtain Mallows' C_p criterion, which is defined as

$$C_p = \frac{S_e^2}{\hat{\omega}^2} + \{2(p+1)-n\}. \qquad (10.46)$$

The smaller the value of the C_p criterion for a model, the better is the model. As an estimator $\hat{\omega}^2$, the unbiased estimator of the error variance of the most complex model is usually used.

10.4 Hannan–Quinn's Criterion

Addressing the autoregressive (AR) time series model of order p,

$$y_n = \sum_{j=1}^{p} a_j y_{n-j} + \varepsilon_n, \quad \varepsilon_n \sim N(0, \sigma_p^2), \tag{10.47}$$

Hannan–Quinn (1979) proposed an order selection criterion of the form

$$\log \hat{\sigma}_p^2 + n^{-1} 2pc \log\log n \tag{10.48}$$

that provides a consistent estimator of order p, where n is the number of observations and c is an arbitrary real number greater than 1. In what follows, for ease of comparison with other information criteria (IC), we multiply their criterion by n and consider

$$\mathrm{IC_{HQ}} = n \log \hat{\sigma}_p^2 + 2pc \log\log n. \tag{10.49}$$

Concerning the variance $\hat{\sigma}_p^2$ of the AR model, it follows from Levinson's formula [see for example, Kitagawa (1993)] that

$$\hat{\sigma}_p^2 = (1 - \hat{b}_p^2)\hat{\sigma}_{p-1}^2, \tag{10.50}$$

where $\hat{b}_p$, which is the pth coefficient of the AR model of order k, is referred to as a *partial autocorrelation coefficient*. By using this relation repeatedly, the $\mathrm{IC_{HQ}}$ can be expressed as

$$n \log \hat{\sigma}_0^2 + n \sum_{j=1}^{p} \log(1 - \hat{b}_j^2) + 2pc \log\log n, \tag{10.51}$$

where $\hat{\sigma}_0^2$ is the variance of the AR model of order 0, that is, the variance of the time series y_n. Consequently, when the order increases from $p-1$ to p, the value of $\mathrm{IC_{HQ}}$ changes by

$$\Delta \mathrm{IC} = \log(1 - \hat{b}_p^2) + 2c \log\log n. \tag{10.52}$$

We assume here that the model has actual order p_0, i.e., that $a_{p_0} \neq 0$ and $a_p = 0$ $(p > p_0)$. In this case, since $b_{p_0} = a_{p_0} \neq 0$, as $n \to \infty$, the following inequalities hold:

$$n \log(1 - \hat{b}_{p_0}^2) + 2c \log\log n < 0, \tag{10.53}$$

$$n \log(1 - \hat{b}_p^2) + 2c \log\log n \leq 0 \quad (p < p_0). \tag{10.54}$$

Hence, asymptotically $\mathrm{IC_{HQ}}$ never reaches its minimum value for $p < p_0$. On the other hand, for $p > p_0$, by virtue of the law of the iterated logarithm, for any $n > n_0$ there exists an n_0 such that the inequality

Table 10.1. Comparison of $\log n$ and $\log\log n$.

n	10	100	1,000	10,000
$\log n$	2.30	4.61	6.91	9.21
$\log\log n$	0.83	1.53	1.93	2.22

$$n \log(1 - \hat{b}_j^2) + 2c \log\log n > 0 \quad (p_0 < p \leq p) \tag{10.55}$$

holds. Therefore, for a sufficiently large n, $\mathrm{IC_{HQ}}$ always increases for $p > p_0$. This implies that $\mathrm{IC_{HQ}}$ provides a consistent estimator of order p.

Table 10.1 shows that the penalty term of $\mathrm{IC_{HQ}}$ yields a value smaller than $\log n$ of BIC. From the consistency argument above, it can be seen that a penalty term greater in order than $\log\log n$ gives a consistent estimator of order. Further, Hannan–Quinn has demonstrated that $\log n$ is not the smallest rate of increase necessary to ensure consistency and that it tends to underestimate the order if n is large. Although c (> 1) is assumed to be any real number, when dealing with finite data, the choice of c can have a significant effect on the result.

10.5 ICOMP

Bozdogan (1988, 1990) and Bozdogan and Haughton (1998) proposed an information-theoretic measure of complexity called ICOMP (I for informational and COMP for complexity) that takes into account lack of fit, lack of parsimony, and the profusion of complexity. It is defined by

$$\mathrm{ICOMP} = -2 \log L(\hat{\boldsymbol{\theta}}) + 2C(\hat{\Sigma}_{\mathrm{model}}), \tag{10.56}$$

where $L(\hat{\boldsymbol{\theta}})$ is the likelihood function of an estimated model, C represents a complexity measure, and $\hat{\Sigma}_{\mathrm{model}}$ represents the estimated variance-covariance matrix of the parameter vector. It can be seen that instead of the number of estimated parameters, ICOMP uses $C(\hat{\Sigma}_{\mathrm{model}})$ as a measure of complexity of a model. The optimal model among all candidate models is obtained by choosing one with a minimum value of ICOMP.

For multivariate normal 1inear and nonlinear structural models, the complexity measure is defined by

$$C(\hat{\Sigma}_{\mathrm{model}}) = \frac{p}{2} \log \frac{\mathrm{tr}(\hat{\Sigma}_p)}{p} - \frac{1}{2} \log |\hat{R}_p| + \frac{n}{2} \log \frac{\mathrm{tr}(\hat{R}_p)}{n} - \frac{1}{2} \log |\hat{\Sigma}_p|, \tag{10.57}$$

where $\hat{\Sigma}_p$ is the estimated variance covariance matrix and $\hat{R}_p$ is the model residual.

References

Akaike, H. (1969). Fitting autoregressive models for prediction. *Annals of the Institute of Statistical Mathematics* **21**, 243–247.

Akaike, H. (1970). Statistical predictor identification. *Annals of the Institute of Statistical Mathematics* **22**, 203–217.

Akaike, H. (1973). Information theory and an extension of the maximum likelihood principle. *2nd International Symposium on Information Theory* (Petrov, B. N. and Csaki, F., eds.), Akademiai Kiado, Budapest, 267–281. (Reproduced in *Breakthroughs in Statistics,* **1**, S. Kotz and N. L. Johnson, eds., Springer-Verlag, New York, 1992.)

Akaike, H. (1974). A new look at the statistical model identification. *IEEE Transactions on Automatic Control* **AC-19**, 716–723.

Akaike, H. (1977). On entropy maximization principle. *Applications of Statistics*, P. R. Krishnaiah, ed., North-Holland Publishing Company, 27–41.

Akaike, H. (1978). A new look at the Bayes procedure. *Biometrika* **65**, 53–59.

Akaike, H. (1980a). On the use of predictive likelihood of a Gaussian model. *Annals of the Institute of Statistical Mathematics* **32**, 311–324.

Akaike, H. (1980b). Likelihood and the Bayes procedure. In *Bayesian Statistics*, N. J. Bernardo, M. H. DeGroot, D. V. Lindley and A. F. M. Smith, eds., Valencia, Spain, University Press, 141–166.

Akaike, H. (1980c). Seasonal adjustment by a Bayesian modeling. *Journal of Time Series Analysis* **1**(1), 1–13.

Akaike, H. (1983a). Information measures and model selection. In *Proceedings 44th Session of the International Statistical Institute* **1**, 277–291.

Akaike, H. (1983b). Statistical inference and measurement of entropy. *Scientific Inference, Data Analysis, and Robustness*, Academic Press,

Cambridge, M.A., 165–189.

Akaike, H. (1985). Prediction and entropy. In *A Celebration of Statistics*, A. C. Atkinson and E. Fienberg. eds., Springer-Verlag, New York, 1–24.

Akaike, H. (1987). Factor analysis and AIC. *Psychometrika* **52**, 317–332.

Akaike, H. and Ishiguro, M. (1980a). A Bayesian approach to the trading-day adjustment of monthly data. In *Time Series Analysis*, O. D. Anderson and M. R. Perryman, eds., North-Holland, Amsterdam, 213–226.

Akaike, H. and Ishiguro, M. (1980b). Trend estimation with missing observation. *Annals of the Institute of Statistical Mathematics* **32**, 481–488.

Akaike, H. and Ishiguro, M. (1980c). BAYSEA, a Bayesian seasonal adjustment program. *Computer Science Monographs* **13**, The Institute of Statistical Mathematics, Tokyo.

Akaike, H. and Kitagawa, G. (eds.) (1998). *The Practice of Time Series Analysis.* Springer-Verlag, New York.

Anderson, T. W. (2003). *An Introduction to Multivariate Statistical Analysis* (3rd ed.). Wiley, New York.

Anderson, B. D. O. and Moore, J. B. (1979). *Optimal Filtering, Information and System Sciences Series.* Prentice-Hall, Englewood Cliffs.

Ando, T., Konishi, S. and Imoto, S. (2005). Nonlinear regression modeling via regularized radial basis function networks. To appear in *the special issue of Journal of Statistical Planning and Inference.*

Barndorff-Nielsen, O. E. and Cox, D. R. (1989). *Asymptotic Techniques for Use in Statistics.* Chapman and Hall, New York.

Berger, J. and Pericchi, L. (2001). Objective Bayesian methods for model selection: introduction and comparison (with discussion). In *Model Selection,* P. Lahiri, ed., Institute of Mathematical Statistics Lecture Notes – Monograph Series **38**, Beachwood, 135–207.

Bernardo, J. M. and Smith, A. F. M. (1994). *Bayesian Theory.* John Wiley & Sons, Chichester, UK.

Bhansali, R. J. (1986). A derivation of the information criteria for selecting autoregressive models. *Advances in Applied Probability* **18**, 360–387.

Bishop, C. M. (1995). *Neural Networks for Pattern Recognition.* Oxford University Press, Oxford.

de Boor, C. (1978). *A Practical Guide to Splines.* Springer-Verlag, Berlin.

Box, G. E. P. and Cox, D. R. (1964). An analysis of transformations. *Journal of the Royal Statistical Society* **B 26**, 211–252.

Bozdogan, H. (1987). Model selection and Akaike's information criterion (AIC): The general theory and its analytical extensions. *Psychometrika* **52**, 345–370.

Bozdogan, H. (1988). ICOMP: a new model-selection criterion. In *Classification and Related Methods Data Analysis.* H. H. Bock ed., Elsevier Science Publishers, Amsterdam, 599–608.

Bozdogan, H. (1990). On the information-based measure of covariance complexity and its application to the evaluation of multivariate linear models. *Communications in Statistics–Theory and Methods* **19**(1), 221–278.

Bozdogan, H. (ed.) (1994). *Proceedings of the first US/Japan conference on the frontiers of statistical modeling: an informational approach.* Kluwer Academic Publishers, the Netherlands.

Bozdogan, H. and Haughton, D. M. A. (1998). Informational complexity criteria for regression models. *Computational Statistics & Data Analysis* **28**, 51–76.

Brockwell, P. J. and Davis, R. A. (1991). *Time Series: Theory and Methods* (2nd ed.). Springer-Verlag, New York.

Broomhead, D. S. and Lowe, D. (1988). Multivariable functional interpolation and adaptive networks. *Complex Systems* **2**, 321–335.

Burnham, K. P. and Anderson D. R. (2002). *Model Selection and Multimodel Inference: A Practical Information-Theoretic Approach* (2nd ed.). Springer, New York.

Cavanaugh, J. E. and Shumway, R. H. (1997). A bootstrap variant of AIC for state-space model selection. *Statistica Sinica* **7**, 473–496.

Clarke, B. S. and Barron, A. R. (1994). Jeffreys' prior is asymptotically least favorable under entropy risk. *Journal of Statistical Planning and Inference* **41**, 37–40.

Craven, P. and Wahba, G. (1979). Optimal smoothing of noisy data with spline functions. *Numerische Mathematik* **31**, 377–403.

Cressie, N. (1991). *Statistics for Spatial Data.* Wiley, New York.

Davison, A. C. (1986). Approximate predictive likelihood. *Biometrika* **73**, 323–332.

Davison, A. C. and Hinkley, D. V. (1997). *Bootstrap Methods and Their Application.* Cambridge University Press, Cambridge, UK.

Diaconis, P. and Efron, B. (1983). Computer-intensive methods in statistics. *Scientific American* **248**, 116–130.

Durbin, J. and Koopman, S. J. (2001). *Time Series Analysis by State Space Methods.* Oxford University Press, Oxford.

Efron, B. (1979). Bootstrap methods: another look at the jackknife. *Annals of Statistics* **7**, 1–26.

Efron, B. (1982). *The jackknife, the bootstrap and other resampling plans.* Society for Industrial & Applied Mathematics, Philadelphia.

Efron, B. (1983). Estimating the error rate of a prediction rule: improvement on cross-validation. *Journal of the American Statistical Association* **78**(382), 316–331.

Efron, B. (1986). How biased is the apparent error rate of a prediction rule? *Journal of the American Statistical Association* **81**, 461–470.

Efron, B. and Gong, G. (1983). A leisurely look at the bootstrap, the jackknife, and cross-validation. *American Statistician* **37**, 36–48.

Efron, B. and Tibshirani, R. (1986). Bootstrap methods for standard errors, confidence intervals, and other measures of statistical accuracy. *Statistical Science* **1**, 54–77.

Efron, B. and Tibshirani, R. J. (1993). *An Introduction to the Bootstrap.* Chapman & Hall, New York.

Eilers, P. and Marx, B. (1996). Flexible smoothing with B-splines and penalties (*with discussion*). *Statistical Science* **11**, 89–121.

Fan, J. and Gijbels, I. (1996). *Local Polynomial Modeling and Its Applications.* Chapman & Hall, London.

Fernholz, L. T. (1983). *von Mises Calculus for Statistical Functionals.* Lecture Notes in Statistics 19, Springer-Verlag, New York.

Filippova, A. A. (1962). Mises' theorem of the asymptotic behavior of functionals of empirical distribution functions and its statistical applications. *Theory of Probability and Its Applications* **7**, 24–57.

Findley, D. F. (1985). On the unbiasedness property of AIC for exact or approximating linear stochastic time series models. *Journal of Time Series Analysis* **6**, 229–252.

Findley, D. F. and Wei, C. Z. (2002). AIC, overfitting principles, and the boundedness of moments of inverse matrices for vector autoregressions and related models. *Journal of Multivariate Analysis* **83**, 415–450.

Fujii, T. and Konishi, S. (2006). Nonlinear regression modeling via regularized wavelets and smoothing parameter selection. *Journal of Multivariate Analysis* **97**, 2023–2033.

Fujikoshi, Y. (1985). Selection of variables in two-group discriminant analysis by error rate and Akaike's information criteria. *Journal of Multivariate Analysis* **17**, 27–37.

Fijikoshi, Y., Noguchi, T., Ohtaki, M., and Yanagihara, H. (2003). Corrected versions of cross-varidation criteria for selecting multivariate regression and growth curve models, *Annals of the Institute of Statistical Mathematics*, **55**, 537–553.

Fujikoshi, Y. and Satoh, K. (1997). Modified AIC and C_p in multivariate linear regression. *Biometrika* **84**, 707–716.

Geisser, S. (1975). The predictive sample reuse method with applications. *Journal of the American Statistical Association* **70**, 320–328.

Golub, G. (1965). Numerical methods for solving linear least-squares problems. *Numerische Mathematik* **7**, 206–216.

Good, I. J. and Gaskins, R. A. (1971). Nonparametric roughness penalties for probability densities. *Biometrika* **58**, 255–277.

Good, I. J. and Gaskins, R. A. (1980). Density estimation and bump hunting by the penalized likelihood method exemplified by scattering and meteorite data. *Journal of the American Statistical Association* **75**, 42–56.

Green, P. J. and Silverman, B. W. (1994). *Nonparametric Regression and Generalized Linear Models.* Chapman and Hall, London.

Green, P. J. and Yandell, B. (1985). Semi-parametric generalized linear models. In *Generalized Linear Models*, R. Gilchrist, B. J. Francis, and J. Whittaker, eds., Lecture Notes in Statistics **32**, 44–55, Springer, Berlin.

Hall, P. (1992). *The Bootstrap and Edgeworth Expansion.* Springer-Verlag, New York.

Hampel, F. R., Ronchetti, E. M., Rousseeuw, P. J. and Stahel, W. A. (1986). *Robust Statistics, The Approach Based on Influence Functions.* John Wiley, New York.

Hannan, E. J. and Quinn, B. G. (1979). The determination of the order of an autoregression. *Journal of the Royal Statistical Society* **B-41**(2), 190–195.

Härdle (1990). *Applied Nonparametric Regression.* Cambridge University Press, Cambridge.

Harvey, A. C. (1989). *Forecasting, Structural Time Series Models and the Kalman Filter.* Cambridge University Press, Cambridge.

Hastie, T. J. and Tibshirani, R. J. (1990). *Generalized Additive Models.* Chapman and Hall, London.

Hastie, T., Tibshirani, R., and Friedman, J. (2001). *The Elements of Statistical Learning.* Springer, New York.

Huber, P. J. (1964). Robust estimation of a location parameter. *Annals of Mathematical Statistics* **35**, 73–101.

Huber, P. J. (1967). The behavior of maximum likelihood estimates under nonstandard conditions. In *Proceedings of the fifth Berkley Symposium on Statistics*, 221–233.

Huber, P. J. (1981). *Robust Statistics.* Wiley, New York.

Hurvich, C. M., Shumway, R., and Tsai, C. L. (1990). Improved estimators of Kullback–Leibler information for autoregressive model selection in small samples. *Biometrika* **77**(4), 709–719.

Hurvich, C. M., Simonoff, J. S., and Tsai, C.-L. (1998). Smoothing parameter selection in nonparametric regression using an improved Akaike information criterion. *Journal of the Royal Statistical Society* **B60**, 271–293.

Hurvich, C. M. and Tsai, C. L. (1989). Regression and time series model selection in small samples. *Biometrika* **76** 297–307.

Hurvich, C. M. and Tsai, C. L. (1991). Bias of the corrected AIC criterion for underfitted regression and time series models. *Biometrika* **78**, 499–509.

Hurvich, C. M. and Tsai, C. L. (1993). A corrected Akaike information criterion for vector autoregressive model selection. *Journal of Time Series Analysis* **14**, 271–279.

Ichikawa, M. and Konishi, S. (1999). Model evaluation and information criteria in covariance structure analysis. *British Journal of Mathematical and Statistical Psychology* **52**, 285–302.

Imoto, S. (2001). B-spline nonparametric regression models and information criteria. Ph.D. thesis, Kyushu University.

Imoto, S. and Konishi, S. (2003). Selection of smoothing parameters in B-spline nonparametric regression models using information criteria. *Annals of the Institute of Statistical Mathematics* **55**, 671–687.

Ishiguro, M. and Sakamoto, Y. (1984). A Bayesian approach to the probability density estimation. *Annals of the Institute of Statistical Mathematics* **B-36**, 523–538.

Ishiguro, M., Sakamoto, Y., and Kitagawa, G.(1997). Bootstrapping loglikelihood and EIC, an extension of AIC. *Annals of the Institute of Statistical Mathematics* **49**(3), 411–434.

Kallianpur, G. and Rao, C. R. (1955). On Fisher's lower bound to asymptotic variance of a consistent estimate, *Sankhya* **15**(3), 321–300.

Kass, R. E. and Raftery, A. E. (1995). Bayesian factors. *Journal of the American Statistical Association* **90**(430), 773–795.

Kass, R. E., Tierney, L., and Kadane, J. B. (1990). The validity of posterior expansions based on Laplace's method. In *Essays in Honor of George Barnard*, S. Geisser, J. S. Hodges, S. J. Press, and A. Zellner, eds., 473–488, North-Holland, Amsterdam.

Kass, R. E. and Wasserman, L. (1995). A reference Bayesian test for nested hypotheses and its relationship to the Schwarz criterion. *Journal of the American Statistical Association* **90**, 928–934.

Kawada, Y. (1987). Information and statistics (in Japanese with English abstract). In *Proceedings of the Institute of Statistical Mathematics* **35**(1), 1–57.

Kishino, H. and Hasegawa, M. (1989). Evaluation of the maximum likelihood estimate of the evolutionary tree topologies from DNA sequence data. *Journal of Molecular Evolution* **29**, 170–179.

Kitagawa, G. (1984). Bayesian analysis of outliers via Akaike's predictive likelihood of a model. *Communications in Statistics* Series B **13**(1), 107–126.

Kitagawa, G. (1987). Non-Gaussian state space modeling of nonstationary time series (with discussion). *Journal of the American Statistical Association* **82**, 1032–1063.

Kitagawa, G. (1993). A Monte-Carlo filtering and smoothing method for non-Gaussian nonlinear state space models. *Proceedings of the 2nd U. S.-Japan Joint Seminor on Statistical Time Series Analysis*, 110–131.

Kitagawa, G. (1997). Information criteria for the predictive evaluation of Bayesian models. *Communications in Statistics-Theory and Methods* **26**(9), 2223–2246.

Kitagawa, G. and Akaike, H. (1978). A procedure for the modeling of non-stationary time series, *Annals of the Institute of Statistical Mathematics* **30**, 351–363.

Kitagawa, G. and Gersch, W. (1996). *Smoothness Priors Analysis of Time Series.* Lecture Notes in Statistics **116**, Springer-Verlag, New York.

Kitagawa, G., Takanmai, T., and Matsumoto, N. (2001). Signal extraction problems in seismology. *International Statistical Review* **69**(1), 129–152.

Konishi, S. (1991). Normalizing transformations and bootstrap confidence intervals. *Annals of Statistics* **19**, 2209–2225.

Konishi, S. (1999). Statistical model evaluation and information criteria. In *Multivariate Analysis, Design of Experiments and Survey Sampling*, S. Ghosh, ed., 369–399, Marcel Dekker, New York.

Konishi, S. (2002). Theory for statistical modeling and information criteria –functional approach. *Sugaku Expositions* **15-1**, 89–106, American Mathematical Society.

Konishi, S., Ando, T., and Imoto, S. (2004). Bayesian information criterion and smoothing parameter selection in radial basis function network. *Biometrika* **91**, 27–43.

Konishi, S. and Kitagawa, G. (1996). Generalized information criteria in model selection. *Biometrika* **83**(4), 875–890.

Konishi, S. and Kitagawa, G. (2003). Asymptotic theory for information criteria in model selection-functional approach. *Journal of Statistical Planning and Inference* **114**, 45–61.

Kullback, S. and Leibler, R. A. (1951). On information and sufficiency. *Annals of Mathematical Statistics* **22**, 79–86.

Lanterman, A. D. (2001). Schwarz, Wallace, and Rissanen: intertwining themes in theories of model selection. *International Statistical Review* **69**, 185–212.

Lawley, D. N. and Maxwell, A. E. (1971). *Factor Analysis as a Statistical Method* (2nd ed.). Butterworths, London.

Linhart, H. (1988). A test whether two AICs differ significantly. *South African Statistical Journal* **22**, 153–161.

Linhart, H. and Zuccini, W. (1986). *Model Selection.* Wiley, New York.

Lindley, D. V. and Smith, A. F. M. (1972). Bayes estimates for the linear model (with discussion). *Journal of Royal Statistical Society* **B34**, 1–41.

Loader, C. R. (1999). *Local Regression and Likelihood.* Springer, New York.

MacKay, D. J. C. (1992). A practical Bayesian framework for backpropagation networks. *Neural Computation* **4**, 448–472.

Mallows, C.L. (1973). Some comments on C_p. *Technometrics* **15**, 661–675.

Martin, J. K. and McDonald, R. P. (1975). Bayesian estimation in unrestricted factor analysis: a treatment for Heywood cases. *Psychometrika* **40**, 505–517.

McCullagh, P. and Nelder, J. A. (1989). *Generalized Linear Models* (2nd ed.) Chapman and Hall, London.

McLachlan, G. J. (2004). *Discriminant Analysis and Statistical Pattern Recognition.* Wiley, New York.

McQuarrie, A. D. R. and Tsai, C.-L. (1998). *Regression and Time Series Model Selection.* World Scientific, Singapore.

Moody, J. (1992). The *effective* number of parameters: an analysis of generalization and regularization in nonlinear learning systems. In *Advances in Neural Information Processing System 4*, J. E. Moody, S. J. Hanson, and R. P. Lippmann, eds., 847–854, Morgan Kaufmann, San Mateo, CA.

Moody, J. and Darken, C. J. (1989). Fast learning in networks of locally-tuned processing units. *Neural Computation* **1**, 281–294.

Murata, N., Yoshizawa, S., and Amari, S. (1994). Network information criterion determining the number of hidden units for an artificial neural network model. *IEEE Transactions on Neural Networks* **5**, 865–872.

Nakamura, T. (1986). Bayesian cohort models for general cohort table analysis. *Annals of the Institute of Statistical Mathematics* **38**(2), 353–370.

Neath, A. A. and Cavanaugh, J. E. (1997). Regression and time series model selection using variants of the Schwarz information criterion. *Communications in Statistics* **A26**, 559–580.

Nelder, J. A. and Wedderburn, R. W. M. (1972). Generalized linear models. *Journal of the Royal Statistical Society* A**135**, 370–84.

Nishii, R. (1984). Asymptotic properties of criteria for selection of variables in multiple regression. *Annals of Statistics* **12**, 758–765.

Noda, K., Miyaoka, E., and Itoh,M. (1996). On bias correction of the Akaike information criterion in linear models. *Communications in Statistics* **25**, 1845–1857.

Nonaka, Y. and Konishi, S. (2005). Nonlinear regression modeling using regularized local likelihood method. *Annals of the Institute of Statistical Mathematics* **57**, 617–635.

O'Hagan, A. (1995). Fractional Bayes factors for model comparison (with discussion). *Journal of Royal Statistical Society* B**57**, 99–138.

O'Sullivan, F., Yandell, B. S., and Raynor, W. J. (1986). Automatic smoothing of regression functions in generalized linear models. *Journal of the American Statistical Association* **81**, 96–103.

Ozaki, T. and Tong, H. (1975). On the fitting of nonstationary autoregressive models in time series analysis. In *Proceedings 8th Hawaii International Conference on System Sciences*, 224–246.

Pauler, D. (1998). The Schwarz criterion and related methods for normal linear models. *Biometrika* **85**, 13–27.

Poggio, T. and Girosi, F. (1990). Networks for approximation and learning. *Proceedings of the IEEE* **78**, 1484–1487.

Rao, C. R. and Wu, Y. (2001). On model selection (with discussion). In *Model Selection*, P. Lahiri, ed., IMS Lecture Notes–Monograph Series **38**, 1–64.

Reeds, J. (1976). On the definition of von Mises functionals. Ph.D. dissertation, Harvard University.

Ripley, B. D. (1994). Neural networks and related methods for classification. *Journal of the Royal Statistical Society* B **50**(3), 409–456.

Ripley, B. D. (1996). *Pattern Recognition and Neural Networks.* Cambridge University Press, Cambridge.

Rissanen, J. (1978). Modeling by shortest data description. *Automatica* **14**, 465–471.

Rissanen, J. (1989). *Stochastic Complexity in Statistical Inquiry.* Series in Computer Science **15**, World Scientific, Singapore.

Roeder, K. (1990). Density estimation with confidence sets exemplified by superclusters and voids in the galaxies. *Journal of the American Statistical Association* **85**(411), 617–624.

Ronchetti, E. (1985). Robust model selection in regression. *Statistics and Probability Letters* **3**, 21–23.

Sakamoto, Y., Ishiguro, M., and Kitagawa, G.(1986). *Akaike Information Criterion Statistics.* D. Reidel Publishing Company, Dordrecht.

Satoh, K., Kobayashi, M., and Fujikoshi, Y. (1997). Variable selection for the growth curve model. *Journal of Multivariate Analysis* **60**, 277–292.

Sakamoto, Y. and Ishiguro, M. (1988). A Bayesian approach to nonparametric test problems. *Annals of the Institute of Statistical Mathematics* **40**(3), 587–602.

Schwarz, G. (1978). Estimating the dimension of a model. *Annals of Statistics* **6**, 461–464.

Serfling, R. J. (1980). *Approximation Theorems of Mathematical Statistics.* Wiley, New York.

Shao, J. (1996). Bootstrap model selection. *Journal of the American Statistical Association* **91**, 655–665.

Shao, J. and Tu, D. (1995). *The Jackknife and Bootstrap.* Springer Series in Statistics, Springer-Verlag, New York.

Shibata, R. (1976). Selection of the order of an autoregressive model by Akaike's information criterion. *Biometrika* **63**, 117–126.

Shibata, R. (1981). An optimal selection of regression variables. *Biometrika* **68**, 45–54.

Shibata, R. (1989). Statistical aspects of model selection. In *From Data to Model*, J. C. Willemsa ed., Springer-Verlag, New York, 215–240.

Shibata, R. (1997). Bootstrap estimate of Kullback–Leibler information for model selection. *Statistica Sinica* **7**, 375–394.

Shimodaira, H. (1997). Assessing the error probability of the model selection test. *Annals of the Institute of Statistical Mathematics* **49**, 395–410.

Shimodaira, H. (2004). Approximately unbiased tests of regions using multistep-multiscale bootstrap resampling. *Annals of Statistics* **32**, 2616–2641.

Shimodaira, H. and Hasegawa, M. (1999). Multiple comparisons of log-likelihoods with applications to phylogenetic inference. *Molecular Biology and Evolution* **16**, 1114–1116.

Silverman, B. W. (1985). Some aspects of the spline smoothing approach to nonparametric regression curve fitting (with discussion). *Journal of the Royal Statistical Society* **B36**, 1–52.

Simonoff, J. S. (1996). *Smoothing Methods in Statistics*. Springer-Verlag, New York.

Simonoff, J. S. (1998). Three sides of smoothing: categorical data smoothing, nonparametric regression, and density estimation. *International Statistical Review* **66**, 137–156.

Siotani, M., Hayakawa, T., and Fujikoshi, Y. (1985). *Modern Multivariate Statistical Analysis: A Graduate Course and Handbook.* American Sciences Press, Inc., Syracuse.

Spiegelhalter, D. J., Best, N. G., Carlin, B. P., and Linde, A. (2002). Bayesian measures of model complexity and fit (with discussion). *Journal of Royal Statistical Society* **B64**, 583–639.

Stone, C. J. (1974). Cross-validatory choice and assessment of statistical predictions (with discussion). *Journal of the Royal Statistical Society Series* **B36**, 111–147.

Stone, M. (1977). An asymptotic equivalence of choice of model by cross-validation and Akaike's criterion. *Journal of the Royal Statistical Society* **B39**, 44–47.

Stone, M. (1979). Comments on model selection criteria of Akaike and Schwarz. *Journal of the Royal Statistical Society* **B41**(2), 276–278.

Sugiura, N. (1978). Further analysis of the data by Akaike's information criterion and the finite corrections. *Communications in Statistics Series A* **7**(1), 13–26.

Takanami, T. (1991). ISM data 43-3-01: seismograms of foreshocks of 1982 Urakawa-Oki earthquake. *Annals of the Institute of Statistical Mathematics* **43**, 605.

Takanami, T. and Kitagawa, G. (1991). Estimation of the arrival times of seismic waves by multivariate time series model. *Annals of the Institute of Statistical Mathematics* **43**(3), 407–433.

Takeuchi, K. (1976). Distributions of information statistics and criteria for adequacy of models. *Mathematical Science* **153**, 12–18 (in Japanese).

Tierney, L. and Kadane, J. B. (1986). Accurate approximations for posterior moments and marginal densities. *Journal of the American Statistical Association* **81**, 82–86.

Tierney, L., Kass, R. E., and Kadane, J. B. (1989). Fully exponential Laplace approximations to expectations and variances of nonpositive functions. *Journal of the American Statistical Association* **84**, 710–716.

Uchida, M. and Yoshida, N. (2001). Information criteria in model selection for mixing processes. *Statistical Inference and Stochastic Processes* **4**, 73–98.

Uchida, M. and Yoshida, N. (2004). Information criteria for small diffusions via the theory of Malliavin–Watanabe. *Statistical Inference and Stochastic Processes* **7**, 35–67.

von Mises, R. (1947). On the asymptotic distribution of differentiable statistical functions. *Annals of Mathematical Statistics* **18**, 309–348.

Wahba, G. (1978). Improper priors, spline smoothing and the problem of guarding against model errors in regression. *Journal of the Royal Statistical Society* **B-40**, 364–372.

Wahba, G. (1990). *Spline Models for Observational Data.* Society for Industrial and Applied Mathematics, Philadelphia.

Wand, M. P. and Jones, M. C. (1995). *Kernel Smoothing.* Chapman & Hall, London.

Webb, A. (1999). *Statistical Pattern Recognition.* Arnold, London.

Whittaker, E. (1923). On a new method of graduation. *Proceedings of Edinburgh Mathematical Society* **41**, 63–75.

Withers, C. S. (1983). Expansions for the distribution and quantiles of a regular functional of the empirical distribution with applications tononparametricconfidenceintervals. *TheAnnalsofStatistics***11**(2),577–587.

Wong, W. (1983). A note on the modified likelihood for density estimation, *Journal of the American Statistical Association* **78**(382), 461–463.

Yanagihara, H., Tonda, T., and Matsumoto, C. (2006). Bias correction of cross-validation criterion based on Kullback–Leibler information under a general condition, *Journal of Multivariate Analysis* **97**, 1965–1975.

Ye, J. (1998). On measuring and correcting the effects of data mining and model selection. *Journal of the American Statistical Association* **93**, 120–131.

Yoshida, N. (1997). Malliavin calculus and asymptotic expansion for martingales. *Probability Theory and Related Fields* **109**, 301–342.

Index

ABIC, 222
accuracy of bias correction, 202
AIC, 51, 60, 68, 76, 80, 100, 115, 128
Akaike information criterion, 60
Akaike's Bayesian information criterion, 222, 223
AR model, 24, 43, 249, 253
ARMA model, 24, 26
arrival time of signal, 99
asymptotic accuracy of an information criterion, 176
asymptotic equivalence between AIC-type criteria and cross-validation, 245
asymptotic normality, 47, 48
asymptotic properties of information criteria, 176
asymptotic properties of the maximum likelihood estimator, 47
autoregressive moving average model, 26

B-spline, 143, 155
background noise model, 99
basis expansion, 139, 220, 242
basis function, 143
Bayes factor, 212
Bayes rule of allocation, 157
Bayesian information criterion, 211, 217
Bayesian modeling, 5
Bayesian predictive distribution, 224, 231
Bernoulli distribution, 13, 149
Bernoulli model, 39
bias correction of the log-likelihood, 52, 167
bias of the log-likelihood, 55, 120
BIC, 211, 217
bin size of a histogram, 77
binomial distribution, 12
Boltzmann's entropy, 33
bootstrap bias correction for robust estimation, 204
bootstrap estimate, 189
bootstrap estimation of bias, 192
bootstrap higher-order bias correction, 203, 204
bootstrap information criterion, 187, 192, 195
bootstrap method, 187
bootstrap sample, 188, 189, 195
bootstrap simulation, 191
Box–Cox transformation, 104
Box–plots of the bootstrap distributions, 200
Broyden–Fletcher–Goldfarb–Shanno (BFGS) algorithm, 41

calculating the bias correction term, 173
Canadian lynx data, 93
canonical link function, 90
Cardano's formula, 82
Cauchy distribution, 11, 42
change point, 97
change point model, 206
changing variance model, 20
comparison of shapes of distributions, 101

conditional distribution model, 17
continuous model, 29
continuous probability distribution, 10
cross-validation, 239, 241
cross-validation estimate of the expected log-likelihood, 246
cubic spline, 23

daily temperature data, 85
Davidon–Fletcher–Powell (DFP) algorithm, 41
definition of BIC, 211
definition of GIC, 119
degenerate normal distribution, 218
derivation of BIC, 215
derivation of GIC, 171
derivation of PIC, 227
derivation of the generalized information criterion, 167
derivative of the functional, 111, 170
detection of level shift, 96
detection of micro earthquake, 101
detection of structural change, 96
deviance information criterion, 236
DIC, 236
difference operator, 135, 136
discrete model, 29
discrete random variable, 10
distribution function, 10
distribution of order, 71

effective number of parameters, 162, 164
efficient bootstrap simulation, 196
efficient resampling method, 196
EIC, 195
empirical distribution function, 35, 110
empirical influence function, 120
equality of two discrete distributions, 75
equality of two multinomial distributions, 77
equality of two normal distributions, 79
estimation of a change point, 97
evaluation of statistical model, 4
exact maximum likelihood estimates of the AR model, 95
expected log-likelihood, 35, 51, 167
expected log-likelihood for normal model, 36
exponential family of distributions, 89
extended information criterion, 195
extension of BIC, 218
extraction of information, 3

factor analysis model, 67
family of probability model, 10
filter distribution, 27
final prediction error, 247
finite correction, 69, 181
first-order correct, 178
Fisher consistency, 117, 127
Fisher information matrix, 48, 128
Fisher's scoring method, 150
fluctuations of the maximum likelihood estimator, 44
Fourier series, 140
FPE, 247
functional, 168
functional for M-estimator, 110
functional for maximum likelihood estimator, 109
functional for sample mean, 108
functional for sample variance, 109
functional form of K-L information, 34
functional Taylor series expansion, 170
functional vector, 119

galaxy data, 79
Gaussian basis function, 146, 160
Gaussian linear regression model, 90, 180
GBIC, 219, 221, 222
generalized Bayesian information criterion, 219
generalized cross-validation, 243
generalized information criterion, 107, 118, 120
generalized linear model, 88
generalized state-space model, 27
Gibbs distribution, 28
GIC, 107, 116, 118–120, 167, 176
GIC for normal model, 121
GIC with a second-order bias correction, 180
gradient vector, 41

Hannan–Quinn's criterion, 253
hat matrix, 164, 243

Hessian matrix, 41
hierarchical Bayesian modeling, 6
higher-order bias correction, 176, 178
histogram, 79
histogram model, 14
hyperparameter, 222

ICOMP, 254
influence function, 111, 112, 119, 199
influence function for a maximum likelihood estimator, 126
influence function for the M-estimator, 113, 129
influence function for the maximum likelihood estimator, 114
influence function for the sample mean, 112
influence function for the sample variance, 113
information criterion, 4, 31, 51, 128
information criterion for a logistic model estimated by regularization, 152
information criterion for a model constructed by regularized basis expansion, 142
information criterion for a model estimated by M-estimation, 116
information criterion for a model estimated by regularization, 137
information criterion for a model estimated by robust procedure, 130
information criterion for a nonlinear logistic model by regularized basis expansion, 155
information criterion for Bayesian normal linear model, 226
information criterion for the Bayesian predictive distribution model, 233

K-fold cross-validation, 242
K-L information, 29
K-L information for normal and Laplace model, 32
K-L information for normal models, 32
K-L information for two discrete models, 33
k-means clustering algorithm, 147

Kalman filter, 43
knot, 23
Kullback–Leibler information, 4, 29

Laplace approximation, 213, 214
Laplace approximation for integrals, 213
Laplace distribution, 11, 183, 204
Laplace's method for integrals, 232
law of large numbers, 36
leave-one-out cross-validation, 241
likelihood equation, 38
linear logistic discrimination, 157
linear logistic regression model, 91, 149
linear predictor, 89
linear regression model, 19, 39, 90, 132, 180
link function, 89
log-likelihood, 36, 51
log-likelihood function, 37
log-likelihood of the time series model, 44
logistic discriminant analysis, 156
logistic discrimination, 157
logistic regression model, 91, 149

M-estimation, 132
M-estimator, 110, 114, 128
MAICE, 69
Mallows' C_p, 251
MAP, 227
marginal distribution, 211, 223
marginal likelihood, 211, 223
maximum likelihood estimator, 37, 109
maximum likelihood method, 37
maximum likelihood model, 37
maximum log-likelihood, 37
maximum penalized likelihood method, 134, 135
maximum posterior estimate, 227
MDL, 217
mean structure, 134
measure of the similarity between distributions, 31
median, 131, 204, 205
median absolute deviation, 131, 204
minimum description length, 217
mixture of normal distributions, 12, 15, 235

mixture of two normal distributions, 64
model, 10
model consistency, 71, 73, 253
model selection, 5
modeling, 10
motorcycle impact data, 20, 144
multinomial distribution, 17, 77
multivariate central limit theorem, 50
multivariate distribution, 16
multivariate normal distribution, 16

natural cubic spline, 24
Newton–Raphson method, 41
NIC, 138
nonlinear logistic discrimination, 159
nonlinear logistic regression model, 152, 221
nonlinear regression model, 19, 139, 220
normal distribution, 11, 203
normal distribution model, 11, 230
normal model, 38, 121, 182, 203
number of bootstrap samples, 192
numerical optimization, 40

order selection, 5, 19, 71, 92
order selection in linear regression model, 71

Pearson's family of distributions, 11, 102
penalized least squares method, 160, 162, 242
penalized log-likelihood function, 135, 218
penalty term, 135
PIC, 227, 230
Poisson distribution, 13
polynomial regression model, 19, 22, 65, 66
posterior distribution, 224, 225
posterior probability, 212
power spectrum estimate, 95
prediction error variance, 26
predictive distribution, 25, 224, 232
predictive information criterion, 226
predictive likelihood, 224
predictive mean square error, 240, 241
predictive point of view, 2
probability density function, 10
probability distribution model, 10
probability function, 10
probability model, 14
probability of occurrence of kyphosis, 155
properties of K-L information, 30
properties of MAICE, 69

quasi-Newton method, 41

radial basis function, 145
regression function, 134
regression model, 17, 18, 21, 134, 208
regularization, 5
regularization method, 135, 218
regularization parameter, 135
regularization term, 135
regularized least squares method, 162, 242
regularized log-likelihood function, 135
relation between bootstrap bias correction terms, 205
relationship among AIC, TIC and GIC, 124
relationship between AIC and FPE, 249
relationship between the matrices $I(\boldsymbol{\theta})$ and $J(\boldsymbol{\theta})$, 50
residual sum of squares, 240
RIC, 138
ridge regression estimate, 162
robust estimation, 128, 204
role of the smoothing parameter, 145

sample mean, 108
sample variance, 109
sampling with replacement of the observed data, 191
Schwarz's information criterion, 211
second-order accurate, 178
second-order bias correction term, 180
second-order correct, 178
second-order difference, 136
seismic signal model, 99
selection of order, 73
selection of order of AR model, 92
selection of parameter of Box–Cox transformation, 104
smoother matrix, 164, 243
smoothing parameter, 135

spatial model, 27
spline, 23
spline function, 23
state prediction distribution, 27
state-space model, 26, 43, 95
statistical functional, 107, 108
statistical model, 1, 9, 10, 21
stochastic expansion of an estimator, 170
subset regression model, 86
subset selection, 208
symbols O, O_p, o, and o_p, 169
synthetic data, 160

third-order accurate, 180
third-order correct, 178, 180
TIC, 60, 115, 127
TIC for normal model, 61
TIC for normal model versus t-distribution case, 65
time series model, 24, 42
trigonometric function model, 19
true distribution, 10, 29
true model, 10, 29

variable selection, 19, 84
variable selection for regression model, 84
variance reduction in bootstrap bias correction, 199
variance reduction method, 195, 199
vector of influence function, 120

Springer Series in Statistics (continued from page ii)

Küchler/Sørensen: Exponential Families of Stochastic Processes.
Kutoyants: Statistical Inference for Ergodic Diffusion Processes.
Lahiri: Resampling Methods for Dependent Data.
Lavallée: Indirect Sampling.
Le/Zidek: Statistical Analysis of Environmental Space-Time Processes.
Le Cam: Asymptotic Methods in Statistical Decision Theory.
Le Cam/Yang: Asymptotics in Statistics: Some Basic Concepts, 2nd edition.
Liese/Miescke: Statistical Decision Theory: Estimation, Testing, Selection.
Liu: Monte Carlo Strategies in Scientific Computing.
Manski: Partial Identification of Probability Distributions.
Mielke, Jr./Berry: Permutation Methods: A Distance Function Approach, 2nd edition.
Molenberghs/Verbeke: Models for Discrete Longitudinal Data.
Mukerjee/Wu: A Modern Theory of Factorial Designs.
Nelsen: An Introduction to Copulas, 2nd edition.
Pan/Fang: Growth Curve Models and Statistical Diagnostics.
Politis/Romano/Wolf: Subsampling.
Ramsay/Silverman: Applied Functional Data Analysis: Methods and Case Studies.
Ramsay/Silverman: Functional Data Analysis, 2nd edition.
Reinsel: Elements of Multivariate Time Series Analysis, 2nd edition.
Rosenbaum: Observational Studies, 2nd edition.
Rosenblatt: Gaussian and Non-Gaussian Linear Time Series and Random Fields.
Särndal/Swensson/Wretman: Model Assisted Survey Sampling.
Santner/Williams/Notz: The Design and Analysis of Computer Experiments.
Schervish: Theory of Statistics.
Shaked/Shanthikumar: Stochastic Orders.
Shao/Tu: The Jackknife and Bootstrap.
Simonoff: Smoothing Methods in Statistics.
Song: Correlated Data Analysis: Modeling, Analytics, and Applications.
Sprott: Statistical Inference in Science.
Stein: Interpolation of Spatial Data: Some Theory for Kriging.
Taniguchi/Kakizawa: Asymptotic Theory for Statistical Inference for Time Series.
Tanner: Tools for Statistical Inference: Methods for the Exploration of Posterior Distributions and Likelihood Functions, 3rd edition.
Tillé: Sampling Algorithms.
Tsaitis: Semiparametric Theory and Missing Data.
van der Laan/Robins: Unified Methods for Censored Longitudinal Data and Causality.
van der Vaart/Wellner: Weak Convergence and Empirical Processes: With Applications to Statistics.
Verbeke/Molenberghs: Linear Mixed Models for Longitudinal Data.
Weerahandi: Exact Statistical Methods for Data Analysis.